History, Philosophy and Theory of the Life Sciences

Volume 36

History, Philosophy and Theory of the Life Sciences is a space for dialogue between life scientists, philosophers and historians – welcoming both essays about the principles and domains of cutting-edge research in the life sciences, novel ways of tackling philosophical issues raised by the life sciences, as well as original research about the history of methods, ideas and tools, which constitute the genealogy of our current ways of understanding living phenomena.

The series is interested in receiving book proposals that • are aimed at academic audience of graduate level and up • combine historical and/or philosophical and/or theoretical studies with work from disciplines within the life sciences broadly conceived, including (but not limited to) the following areas: • Anatomy & Physiology • Behavioral Biology • Biochemistry • Bioscience and Society • Cell Biology • Conservation Biology • Developmental Biology • Ecology • Evolution & Diversity of Life • Genetics, Genomics & Disease • Genetics & Molecular Biology • Immunology & Medicine • Microbiology • Neuroscience • Plant Science • Psychiatry & Psychology • Structural Biology • Systems Biology • Systematic Biology, Phylogeny Reconstruction & Classification • Virology The series editors aim to make a first decision within 1 month of submission. In case of a positive first decision the work will be provisionally contracted: the final decision about publication will depend upon the result of the anonymous peer review of the complete manuscript. The series editors aim to have the work peer-reviewed within 3 months after submission of the complete manuscript. The series editors discourage the submission of manuscripts that contain reprints of previously published material and of manuscripts that are below 150 printed pages (75,000 words). For inquiries and submission of proposals prospective authors can contact one of the editors: Charles T. Wolfe: ctwolfe1@gmail.com Philippe Huneman: huneman@wanadoo.fr Thomas A.C. Reydon: reydon@ww.uni-hannover.de

Snait B. Gissis

Lamarckism and the Emergence of 'Scientific' Social Sciences in Nineteenth-Century Britain and France

Springer

Snait B. Gissis
Cohn Institute for the History and Philosophy
of Science and Ideas
Tel Aviv University
Ramat Aviv, Tel Aviv, Israel

ISSN 2211-1948 ISSN 2211-1956 (electronic)
History, Philosophy and Theory of the Life Sciences
ISBN 978-3-031-52755-5 ISBN 978-3-031-52756-2 (eBook)
https://doi.org/10.1007/978-3-031-52756-2

This Springer imprint is published by the registered company Springer Nature Switzerland AG
The registered company address is: Gewerbestrasse 11, 6330 Cham, Switzerland

In memory of Abba, Imma, Sam
To the next generation – Tamuz

Acknowledgments

Though research and scholarship in the humanities is often considered an individual endeavor, they are collectivity dependent. Thus I want to express my deep gratitude to the communities and individuals who sustained and encouraged me throughout the years of research and of the actual writing:

Those who attentively read relevant articles' drafts, chapters, or parts of them; checked translations; and offered good advice at various stages. Their questions, comments, and at times sharp critique helped me rethink, focus, sharpen, and elucidate. In alphabetical order: Gadi Algazi, Cornelius Borck, Daniel Brooks, Jose Brunner, Raz Chen-Morris, Pietro Corsi, Daniel Dor, Otniel Dror, Avishai Ehrlich, James Elwick, late Yaron Ezrahi, Menachem Fisch, Gad Freudenthal, Ofer Gal, James Griesemer, Anne Harrington, Dafna Hirsch, Eva Jablonka, late Marion Lamb, Ehud Lamm, Maurizio Meloni, Esti Micenmacher, Daniel Nicholson, Robert Nye, Lynn Nyhart, Ayelet Shavit, Amir Teicher, Alfred Tauber, Charles Wolfe.

The editors of the various journals who accepted my articles and the anonymous reviewers whose critical comments were often perspicacious and so very helpful.

The anonymous reviewers of the manuscript and their most useful comments and advice.

I learned so much from the students in my graduate seminars at the Cohn Institute throughout the years, whose questions helped me think in complex ways about the passages between the biological and the social, on transfer, and the roles of models, metaphors and analogies, and on the constitution of new disciplines. The questions of audiences in various talks inspired me to broaden the scope of my investigation.

I am so deeply indebted to the editor Dr. Ursula Wokoeck who strove tirelessly to turn the manuscript into a crystal-clear text.

The late Everett Mendelsohn and then Anne Harrington enabled me to reach important sources at Harvard libraries. M.D. Ines Adin had given me the English translation of Freud's oeuvre.

All the librarians at the diverse libraries and archives I had used in Berlin, Cambridge, Mass., London, Paris, and Tel Aviv were so helpfully patient with my queries and resourceful in unearthing my requests.

I am grateful to Lorraine Daston for a one month fellowship at the Max Planck Institute for the History of Science in Berlin, and to the French Government for two summer fellowships of work at Paris archives and libraries. Shaul Katzir, the head of the Cohn Institute, my home institute at Tel Aviv University, greatly helped with the language editing expenses.

I want to particularly thank friends who encouraged and spurred me on with unabated lively interest: Avishai Ehrlich, Dafna Hirsch, Eva Jablonka, Esti Micenmacher, Smadar Sharon.

Sam Schweber died before the book was written, but the long conversations and intense controversies throughout our years together echo in it. Last but certainly not least is my only son Tamuz, who consistently stood by me with affection and never-failing humor through the trials and tribulations of recent years.

Gratitude is due to all those who do not give up, thereby encouraging me to finish writing this book through difficult times. The outbreak of Covid-19 occurred shortly after I started the actual writing. While hitting the whole world, causing death, human and economic disasters, it exposed deeply unsettling inequalities—ethnic, national, local, and global. At present, October–December 2023, while I toil over last revisions of the manuscript, terrible events are taking place. In early October, the Palestinian organization Hamas executed a gruesome planned, systematic, indiscriminate carnage of civilians, including sexual-gendered violence as weapons of war, abduction of helpless civilians of all ages, burning down their homes. The State of Israel launched an appalling indiscriminate war on the Hamas organization in the Gaza Strip, which has brought about mass displacement and uprooting of the civilian population of Gaza and its surroundings, ruin and devastation, death of thousands of civilians of all ages, especially of children. For me as an Israeli the human and emotional toll of this sequence of events, the scope of traumas, destruction and loss caused to the civil populations on both sides, have been horrendous.

I believe that it is only seemingly and for very limited periods that human collectivities, their individuals, their culture, including sciences and humanities, can flourish under conditions of recurrent wars, erupting cruelties, negation of *basic human rights*, denial of *mutual recognition* of collective disasters and their traumatic memories, and of collective aspirations.

In hope against hope to bring about change, we should follow the Mishnaic saying: "it is not upon you to finish the work, but you are not free to give it up" (Avot 2:16).

Introduction

In his presidential address on science, the explanation of nature, and materialism, at the British Association for the Advancement of Science (BAAS) in Belfast in 1874, the prominent physicist John Tyndall said, in continuation of his defense of 'evolution', when relating to Herbert Spencer's *Principles of Psychology*, second edition: "There are two obvious factors to be taken into account—the creature and the medium in which it lives, or, as it is often expressed, *the organism and its environment. Mr Spencer's fundamental principle is, that between these two factors there is incessant interaction*" (1892, 195 [1874] –emphasis added).

Soon afterward, in 1875, John Hughlings Jackson, later considered as 'founder of British Neurology' and much else, wrote in one of his medical papers: "I have long thought that we shall be very much helped in our investigations of diseases of the nervous systems by considering them as reversals of *evolution*, that is as dissolutions. Dissolution is a term I take from Spencer" (1958, 1:45 – emphasis added), further specifying in 1888: "I accept *Herbert Spencer's doctrine of the inheritance of organized experiences*" (1958, 2:98 – emphasis added).

In 1888, young Émile Durkheim, later considered as 'the founder of French Sociology', wrote in the introductory lesson of his first course on social science: "*Biology* has become a veritable treasure of views and hypotheses for the sociologist, which no doubt he did not have the right to brutally grab, but he could at least wisely use" (1978, 56 [1888a]) – emphasis added).

In the third part of his controversial last work, *Moses and Monotheism* (1939), Freud wrote concerning his long time (since the 1890s) engagement with and adherence to neo-Lamarckian mechanisms and assumptions: "My position, no doubt, is made more difficult by the present attitude of biological science, which refuses to hear of the inheritance of acquired characters by succeeding generations. I must, however, in all modesty confess that nevertheless I cannot do without this factor in *biological evolution*" (1953–74, 23:100 – emphasis added).

These quotations seem to point toward a common feature—the significance of biology in the second half of the nineteenth century for new fields outside it: psychology, neurology, sociology, and even psychoanalysis. Furthermore, they indicate additional features of significance that characterize this biology: it is evolutionized;

it deals with the relations—interactions—between organism and environment; it deals with a particular mode of biological heredity, namely the inheritance of modifications acquired during the organism's lifetime and passed on trans-generationally.

How and why had biology, and that particular mode of evolutionized biology, become significant for constituting these novel fields? These are precisely the issues the book addresses. The book argues for the central role of diverse modes of engagement with evolutionized biology in the emergence of new fields dealing with 'the social', 'the mental', and even 'the neural', in nineteenth-century Britain and France. I present a framework, within which the late nineteenth-century emergence of social sciences aspiring to be constituted as autonomous, as 'scientific' disciplines, is described, analyzed, and explained, by pointing to conceptual and constructive commonalities conjoining significant founding figures, who were not grouped nor analyzed in this manner before. Thus the book offers a synthesis on the interactions of the social, the mental, and the evolutionary biological crystallizing into novel fields.

I argue for the central role of an entanglement with a particular brand of evolutionized biology, *Spencerian Lamarckism and/or Neo-Lamarckism* (SL/NL), discussing how and why Spencerian Lamarckism and then European, in particular French, neo-Lamarckism—their models, metaphors, mechanisms, and assumptions—were perceived as especially congenial resource of repertoires for emerging sociology and psychology. In what ways did transfer, i.e., engagement with and appropriation from this resource, become an enabling condition for their emergence?

Biology emerged as a cluster of autonomous fields by the beginning of the nineteenth century. By the last third of that century, it had become evolutionized and was considered a well-established discipline to serve as an *exemplum* and source of repertoires to other emerging fields, especially the Sciences of Man.[1]

In spite of the diversity of meanings of what Lamarckism was, both then and in present-day historiography, there was a certain 'logic', or 'stance', that distinguished the SL/NL approach, and its distinctive emphases had important implications for the kind of questions that could be asked. For present purposes, I refer to

[1] The book discusses European Lamarckism in Britain and France, thus omitting other significant sites in Europe and their historiography, such as the controversy around Haeckel as a Lamarckist or Darwinist, the impact of very late nineteenth and early twentieth centuries Lamarckism both in biology and in studies of mind-body relations in German-speaking countries. Another important omission is the neo-Lamarckists in the USA, who have been considered in the historiography as a more cohesive group, with a formulated agenda, a journal, and reliance on somewhat similar disciplinary sources. The scope of any book is limited, and so is mine. Recently a number of studies have appeared that deal with these groups and individuals within them either directly or indirectly. For example: W.D. Almon on paleontology and evolutionary thinking in USA and Britain (2020), M. Battran on Lamarckism in Germany (2023 – based on the earlier PhD. dissertation), D. Ceccarelli on Cope and Baldwin (2019), Hilton, Mitchell & Smith on Cope (2014), G.R. Mcghee on Cope (2023), T. Pearce on organism and environment in American Philosophy/Pragmatism (2020), M. Sommer on Osborn in her 2016 book, K. Taschwer on Kammerer (2016).

the SL/NL *problématique*.[2] Two notions became central for SL/NL theorizing and experimenting—plasticity and heredity. Contemporaneous engagement with the SL/NL *problématique* converged, with changing emphases, around three core tenets:

(a) The central role of the environment in adaptation, i.e., the relations between active organisms and the environment, and their looping effects, including behavioral ones
(b) Evolution understood as a development from less to more complex organization (and thus in some sense potentially involving 'internal guidance', notions of hierarchy, etc.)
(c) The inheritance of adaptively acquired traits

By focusing on the SL/NL *problématique* under the aspects of 'evolution', 'development', and 'heredity', the book reconfigures the social–biological interactions, the diverse intertwining, coexisting, and separating strands within the modes in which contemporaries experienced, conceptualized, and institutionalized these interactions. They are discussed in the various chapters below, in questions such as:

- Was the environment perceived as an active causal agent of/in adaptive change in the process of evolution, and respectively were organisms perceived as passive receptacles, or as active or interactive agents?
- What was the role of 'behavior' from the perspective of a one-way or a two-way flow of effects between organisms and their environments?
- How was the notion of evolutionary hierarchy used? Did events at one level affect events on another level, and in which direction?
- How was an increase in the complexity of organisms accounted for? With or without incorporating a progressivist evolutionary view?
- What was the function of changes acquired within the lifespan of a developing organism, through a use–inheritance mechanism? Were they transmitted trans-generationally?
- Was there any role for some kind of embryonic developmental specification mechanism, either as a 'divergence from archetype' or as 'recapitulation' mechanism, i.e., an ontogenetic sequence of processes in an individual that simulated the phylogeny of the lineage?
- Was there any role for chance in the structuring of the organic world?
- How were classificatory continuities—social, neural, psychological—deployed?

I propose that the SL/NL *problématique* was attractive for the emerging disciplines aspiring to establish themselves as proper sciences, for three main reasons: For one, it allowed for the use of specific SL/NL evolutionary mechanisms of change and

[2] I use '*problématique*' to delineate a space of clusters of questions rather than lists of topics; questions posited, debated, solved, and resolved, considered as legitimate or illegitimate, where the boundaries of such a space are changing and context dependent, while its core questions may be more stable; these questions are answered in diverse modes, e.g., metaphysically, methodologically, experimentally.

stabilization to make room for an otherwise implausible *plasticity* in the conception of human experience and cognition.

I deal solely with plasticity as discussed and elaborated upon within *evolutionized biology*, relating to living nature at large, and consequently also to humans, and transferred as such. I do not deal with the variety of non-evolutionary modes of permeability and impressionability of human bodies as posited and practiced in premodern as well as in contemporary modes, investigated in numerous recent studies. These studies present differing facets of the complex relations within human bodies between the social and the biological, aiming to intertwine, blur, annul the dichotomous conceptualizing of the relations between human bodies and environment.[3]

Every evolutionary framework positing adaptive relations between organisms and environments—be they organic or physical—that was transferred to the social/psychological could be interpreted in two alternative ways. Either it could be interpreted as having a central, deterministic, hard-core biological component (e.g., in criminology and anthropologies of the period considered as medicalized-biologized disciplines) or interpreted as positing the role of the environment as central in the formation and the transmission of social, cultural, and psychological functionally adaptive patterns. Consequently, a socially or psychologically acquired adaptive trait could be inherited biologically, and vice versa. But plasticity also implied the rising complexification of the nervous system, which made room for the emergence of new neural patterns. This perspective of enhanced plasticity allowed for innovating in the present and molding the present in light of a projected future. It emphasized the overall importance of the environment in adaptively shaping present and future generations through the inheritance of acquired characteristics, with 'use–repetition–habit' as its principal mechanism. 'Habit' of patterns formed as a result of repetition was an important explanatory tool. The resulting changes were regarded as developmental-evolutionary, passed on by hereditary transmission. This implied the possibility of shaping the future: present psychological (or social) changes could be bequeathed as prospective biological or cultural traits to be elaborated further and complexified in the future.

Second, in theorizing, the nervous system, particularly the brain, was regarded as a consequence of evolutionary processes forming inclusive and linear *hierarchical levels* of organization. These levels indicated levels of complexity. Disorders were conceived as a 'dissolution' of the evolutionary direction or 'reversion' (or fixation) of that order. This assumption was closely interwoven with, and dependent on, the

---

[3] E.g., the numerous works of W. Anderson on autoimmunity, on bodies in Southeast Asia; Samantha Frost on the mutual constitution of the body, on embodied subjectivity, on the uses of the 'biocultural'; G. Geltner on Galenic/humoral medicine and on medieval medicine at large; H. Landecker's works on the human body—in discussions of epigenetics, nutrition and metabolism; M. Lock's whole corpus of works on (anthropology of) biomedicine and the human body, all the more so in the last decade; M. Meloni from a sociological perspective—the biosocial as entangled with genetics/genomics/epigenetics, premodern non-evolutionary conceptions of permeability and plasticity; aspects of the body in G. Palsson 's works on the biosocial.

existence of such an order, and was thus not arbitrary. 'Dissolution/reversion' played an all-significant role, both methodologically, as a tool of investigation, and theoretically, as an object of studies of disorders. Description and analysis of pathological states of the nervous system, and the constitution of its diseases or of 'the mental', within these psychologies depended crucially on the existence of such a hierarchical evolutionary order.

Third, the predominance of an individualist ontology in this period impelled biologists, sociologists, and psychologists to cope methodologically and epistemologically with issues related to collectivity, collective frameworks, sociality. The SL/NL *problématique* allowed its users to construct, through the interplay between heredity and memory, conceptual bridges between *individuals and collectivities*, and use them as enabling conditions to posit and enhance their theorized assumed laws as universal and necessary, thereby constituting the various fields as scientific.

To my mind 'migration' among fields was crucial for the emergence of the new disciplines; and that 'migration' from the biological played a crucial role. I use 'transfer' as a generalized technical term for that migration, signifying the contemporary modes of appropriation of and engagement with the evolutionary-biological. Such modes enabled the passage and the matching of tools and a two-way exchange from one conceptual field/domain, the source, to another, the target. These tools—concepts, models, metaphors, analogies, and thus mechanisms, ordering of hierarchies, methodological strategies, etc.—were used to lay down foundational assumptions and methodologies for the emerging fields, to draw boundaries, to supply rhetorical repertoires, to legitimate claims to scientificity,[4] and to offer solutions to difficulties of dynamics and classification in the new fields. Transfer was often used, for somewhat different ends, also between already existing and well-established disciplines. Transfer processes can occur under culturally specific reasoning, justifying it in a particular historical context and its implicit assumptions, such as deep beliefs and preconceptions about the conceptual and cultural power and scientific status of both the source and the target domains, and about possible fits of *problématique*. They enable reinterpretations, change of emphasis, and transformations of both the source domain and its tools, and the target domain. Thus, certain features or properties of the source domain may disappear by becoming implicit, or be relegated to marginality, while others, perhaps implicit or peripheral in the source domain, may become central and explicit in the target one. These processes can bring about unforeseen restructuring and remodeling, i.e., emerging novelties.

The choice then to transfer and deploy biological mechanisms, models, and metaphors, whether Darwinian or Lamarckian, to any new field outside biology was consequential for that target-field. The outcome of transfers reflected not only internal scientific factors but also particular juxtapositions within emerging fields, their position vis-à-vis contemporaneous cultural and political institutions. Moreover, it

[4] E.g., the use of metaphors to authorize scientific claims has been notably discussed by feminist scholars. See, e.g. Keller, 1995; Martin, 1991.

reflected their attitude to both novel and existing disciplines and their practices, particularly those perceived as adjacent to and competing with the emerging ones in the specific national contexts and styles. Such processes could only work in a cultural context that allowed for the assumption of a fundamental similarity, or analogy, or correspondence, between organic nature and the social, between mechanisms of biological and social development, and between types of regularities or even *problématique* observed in these fields.[5] Ongoing debates on the meaning and impact of processes of industrialization and urbanization that were part and parcel of the concerns of emergent sociologies and psychologies were fashioned through modes of transfer from evolutionized biology that practitioners of these emerging fields deployed and appropriated.[6] A caveat: metaphors, analogies, etc. are to be read as their audiences understood, used, and negotiated their potential polysemy to their own purposes.

A brief comment on the significance of these transfers for my general argument may be in place here. Questions such as the following will accompany the discussion of transfer throughout: When does one find more metaphorization and use of analogy or less, and what were their justifications, if any? How and when did certain metaphors, analogies, models, become over-all emblems of the age, e.g., mechanism versus organism, leaving their mark on multiple disciplines, delimiting the field of legitimate questions and answers? Among the diverse elaborations on models and their use, the relevant aspect for my specific discussion of transfer is to view models as functioning as a dynamic multi-functional form of knowledge-seeking in the endeavor to reach generalizations within a relations framework, which is not that of deduction. The model is being tested on whether it works usefully as an investigative and/or intervening tool, e.g., so that in transfer, the 'naturalizing features' of the evolutionary biological source-domain model are carried over to the target domain.[7]

The accumulated literature analyzing metaphors from the early twentieth century until now, as well as the growing number of 'school controversies' on their status, features, usage, impact, etc. are almost immeasurable.[8] Yet, from the specific perspective of transfer from source to target relevant here, the principal questions would be: What were these metaphors to their nineteenth-century users? How and to what purpose were they identified and understood within diverse cultural and ideological contexts? I suggest that even when claimed to be a mere scaffolding,

[5] E.g., Barberis, 2003; Beck, 2014; Pearce, 2020.

[6] E.g., Campbell, 1975; Soskice and Harré, 1995, Harré, 1996.

[7] Avoiding the tangled controversies on models and modeling.

[8] E.g., I mention mostly collections rather than single articles from the very wide literature: Bartha, 2013; Black, 1962, 1979; Cameron and Graham, 1999; Fox-Keller, 1995, 2002; Gentner, 1988; Gentner and Markman, 1997; Gibbs, 2008; Hanks and Giora, 2012; Hesse, 1966; Kittay, 1987; Lakoff and Johnson, 1980; Maasen and Weingart, 2000; Maasen et al., 2002; Miall, 1982; Ortony, 1979; Reynolds, 2018. These approaches can roughly/simplistically be divided into two non-overlapping contrasting-duality views and then further subdivided, e.g., semantic/pragmatic, literal/constructivist, etc.

these metaphors were deployed as a cognitive tool, to constitute, to create, rather than merely expose, networks of relations of similarity and of difference among domains, objects, groups of phenomena, modes of behavior; to reorganize and to establish novel vocabularies through a repertoire of two-ways metaphorical transformative re-descriptions or at times analogical structural mapping. With further routine use, the results of these innovating actions could turn from innovative into 'technical' or the merely heuristic.

Following repertoires of transfer reveals how members of particular existing disciplines, e.g., contemporaneous French academic philosophy, looked upon themselves as being intellectually and even institutionally 'dispossessed' of their 'rightful *problématique*' by the novel conceptual frames, new cultural options, such as novel publication venues or instruction frameworks. The latter were posited by new groups claiming to deal with these *problématiques* 'scientifically', thereby laying the foundations for new discipline, implicitly subverting existing academic/administrative boundaries.

It seems to me that through transfer from biology, SL/NL was instrumental in devising and in announcing a new methodology; yet the role of methodology in an emerging field was quite different from its role in an existing one. In an emerging field, decisions, such as which phenomena were to be posited as primary, which repertoires of models and metaphors from which cultural and scientific resources would be deployed, seem to have had much greater urgency than methodological ones, whereas in an already existing one, they were closely related to prior ideological commitments or more general worldviews. Thus debates seemingly on transfer/methodology conducted between existing disciplines and the new ones at a stage of constituting would often serve to stake out legitimation claims.

In distinguishing between new fields in the process of emergence and constitution and established ones, I use the term 'discipline' in its later nineteenth-century common manifestation beyond national/local differences: self-and-other proclaimed collective identity, organized around common conceptual *problématique*, and transmitted agreed-upon canons, practices, methodologies, with internal differentiation. These were viewed as a professional/intellectual division of labor, structured modes of communication, controlling and regulating norms inculcated throughout the instructive bureaucratic-institutional hierarchical structures, primarily universities, institutes, laboratories, and hospitals. They were accompanied by organized forms of public activities (e.g., professional journals) that nurtured and sustained each other, offered the promise of public recognition, and provided employment for those educated within them.

The individuals I discuss strove to constitute novel conceptual *problématiques*, new objects, to organize autonomous communities, to invent canons, lay down research norms and practices, to redraw field boundaries. They strove to move their novel fields from professional marginality to the center, through gaining both scientific and cultural recognition, in order for these fields to finally become integrated into existing institutional frames. Thereby they hoped to change these institutional frameworks, their publics, and the cultural field at large in a meaningful manner, and to create novel 'scientific collectivities', and novel 'scientific personae'.

Rather than presenting this history in broad brush strokes, I propose a close examination of these processes by explicating the innovative founding endeavors, paying attention to the specificity of locality and period, as well as the features of bounded communities and their negotiated demarcations of scientific fields. Unfortunately, the contextualization cannot be fully elaborated here, due to the space limitations. Consequently, I remain within the specifically European context in explicating 'transfer' from the biological to the social, although most of the figures discussed lived during the age of global imperialism, and the possession of colonies and the metropoles' complex interrelations with them deeply impacted the materials dealt with in diverse disciplines. I do not place my discussion within that context: neither through the working-out of notions of plasticity in colonial contexts, especially in colonial medicine (e.g.the work of W. Anderson and of M. Osborne), nor through the working-out of notions of evolutionary hierarchy into the graded lists of races and racist practices (which has an ample historiography), thereby 'scientifically' rationalized. In a future study, I hope to examine 'race', nation and racism within such 'transfer' contexts.

I begin with Jean-Baptiste Lamarck's work, then discuss the works of Herbert Spencer, John Hughlings Jackson, Théodule Armand Ribot, David Émile Durkheim, and Sigmund Freud, as labor of 'scientific discipline' founders, spanning over half a century. I argue that a crucial facet of the endeavor of these founders should be regarded as stemming from their choice of field of transfer, namely to deploy biological concepts, models, analogies, and metaphors taken from the contemporaneous varieties of Lamarckism in novel fields outside biology. Their deployment of these repertoires reflected internal scientific considerations such as how to construct conceptual vocabulary, analytic, and classificatory tools and to constitute investigative practices. Moreover, it indicated an effort to supply their emerging fields with grounds for claims of scientific necessity and universality, as perceived then. The diverse outcomes of these endeavors indicate the differences in their particular positioning vis-à-vis their contemporary perceptions on how to investigate 'scientifically' within the politics of societal and institutional pressures, needs, and processes of change.

None of these individuals were born to families belonging to an elite group or the upper classes in their countries. They belonged to a lower (or descending) class (Lamarck, Ribot, Jackson, Durkheim, Freud), to a geographical periphery (Lamarck, Spencer, Jackson, Ribot, Durkheim), and/or to a religious minority (Spencer to some extent, but mostly Durkheim and Freud); and so were some members of their emerging new communities.[9]

I suggest to consider their founding endeavors concurrently as a carving-out of a new role, a type of person—a new persona—namely, the scientist of that new field. These new roles, collective templates, integrated components of repertoires of previous cultural categories, those of past scholarly and scientific behavior, of a novel formation of the field and profession-to-be, and of a calling, i.e., positioning of the

[9] For each, there is a concise biographical note in the Appendix.

scientist as a moral figure, together with the scientist as an institution builder—all converging into sets of rules and regulative ideals that both created authority and public recognition, and conferred them on individual figures. Thus, for example, Lamarck fulfilled the model of the biologist-philosopher, while Jackson actively fashioned a new role in contemporaneous British medicine, namely the clinical scientist, followed later on by Freud, and so on. Yet, such roles were fashioned within collective contexts, and more particularly in collectivities. These were made up of people who in complex ways participated—in a looser or a more specific and bounded sense—in sustained pursuit, in producing, transmitting, sharing 'situations', 'documents', 'facts', 'tools', 'routines', concerns and controversies, ending up distancing or cooperating—thus taking part in what could become 'a SL/NL transfer community of practice',[10] in which such new roles might be formed.

It is important to emphasize that SL/NL were not the cause of the emergence of the new fields; rather they can be viewed as a significant condition of the possibility of their emergence, which in its turn was dependent on the historical-cultural-scientific context. As a condition of possibility, SL/NL allowed for new *problématiques* and a new ordering of importance. 'Scientific' social sciences emerged, where individuals/collectivities innovated, while sharing assumptions and cognitive tools in a way that helped provide what were then perceived by some, and denied by others, as novel valid coherent systems of social sciences. Thus the general arguments of this book are part of an ongoing historiographic controversy on the making of social sciences in the late nineteenth century, while the individual chapters also engage with present-day and historiographical interpretations of these individuals and their enterprises.

Attending to these issues can perhaps draw historiographical attention to the long period of the controversially assumed purification of the biological from social sciences. Furthermore, it could call for deliberation on the bifurcated views on present-day features of social sciences as solely reflective, and on novel or renewed entanglements with the biological, not only through the 'neurologization' of several social sciences sub-disciplines, but also through possible deployment of resources from the genetic/genomic and epigenetic repertoire.

The book consists of a series of chapters, each of which introduces one central founding figure and also engages with present-day historiographical debates and interpretations of these individuals and their enterprises. After each couple of chapters there is a concise conceptual interlude, which sums up the meaning, use, and significance of clusters of concepts, central to the nineteenth-century intertwining of evolutionized biology and social thinking. Each of the three interludes is located after the chapters discussing principal contributors to the respective cluster. The book is girded by a theoretical introduction and concluding reflections. The ordering of the chapters though basically chronological takes into account intergenerational, conceptual, and cultural impacts. The appendix provides concise biographical portraits.

---

[10] E.g., Wenger, 1998; and also, Morgan, 2012, on travelling facts.

The first chapter discusses Jean-Baptiste Lamarck's innovative science of transforming living entities, '*la marche de la nature*', later called 'evolution'. His oeuvre has received various interpretations, mostly relating Lamarck to the natural history and physical sciences of his time. I put forward that the social discourses and practices of his time, a period of social and political turmoil, of novel institutions and shifting boundaries, were crucial for his innovations. Thus he resorted to a new repertoire of models about classification and change found in social-political discourses, and his model analyzed processes of development and of change, the impact of the environment on active living entities, the emergence of complexity and of complex organization.

The *second chapter* discusses Herbert Spencer's enterprise to lay the foundations for psychology and sociology as scientific disciplines through evolutionizing, focusing on both human individuality and sociality. His basic epistemological/methodological framework for living nature was a tripartite model. In *Principles of Biology* (1864–67), he posited a novel concept of biological individuality, and his evolutionary conception was based on a Lamarckian framework. In *Principles of Psychology* (1855, 1870/72), he refashioned the assumptions of contemporary human sciences through evolutionizing them. In *Principles of Sociology*, Spencer depicted individuals as organisms and society as 'superorganism'; I relate also to his use of 'race'.

Chapter 3 – the first Interlude, in which I explicate the cluster of concepts related to nineteenth-century use of plasticity, deployed as an inseparable 'conceptual bundle': 'environment/milieu' tropes.

Chapter 4 deals with John Hughlings Jackson as a clinical scientist, whose work formed a significant link in the transfer-network of SL/NL models and mechanisms that proved to be so generative for the emerging social sciences of the late nineteenth century, particularly psychology. I argue that Jackson created a new role, a clinical *scientist*. His search for epistemological foundations resulted in opening up a new investigative space—neurology. While applying Spencer's notions of evolutionary hierarchies to the nervous system, he constituted innovative 'medical-biological fact and classification' of epilepsy and of aphasia.

Chapter 5 discusses Théodule Armand Ribot's endeavor of Scientific Psychology. He was an important cultural-scientific agent of change in France, who strove to constitute and academize a psychology aspiring to be scientific, non-metaphysical, and experimental, positing 'facts' and 'laws' for it. His endeavor was based on SL/NL assumptions applied to the nervous system. He held a nuanced position on memory as biological, on the significance of the memory–heredity continuum, on conscious/unconscious states, and on voluntary/automatic behavior.

Chapter 6 – the second Interlude analyzes the role of the cluster of concepts related to hierarchy of levels within diverse nineteenth-century SL/NL frames, particularly relevant to the work of Spencer, Jackson, Ribot, and Freud.

Chapter 7 analyzes David Émile Durkheim's endeavor—a competitor for the position of French Sociology founder and the promoter of 'scientific sociology'. Two closely intertwined threads are discussed: his endeavor to constitute and academize a novel field, and his use of SL/French NL modes of evolutionizing as a principal, though not unique, source of repertoire for his construction of scientific tools, practice, and rhetoric. The relations between collectivity and individuals played an inestimable role in both threads.

Chapter 8 is a short coda, wherein Sigmund Freud is posited as a member of the group of nineteenth-century founders, whose endeavor was entwined with transfer from the SL/NL framework. The SL/NL heredity mechanism of back-and-forth movement between social/psychological and biological/neurological heredity and Haeckel's recapitulation mechanism were of particular significance to Freud in constructing his notions of archaic memory and his phylogenetic core-narrative sequence.

Chapter 9 – the third Interlude analyzes the ways, in which from Spencer on, the SL/NL perspective—models, concepts, and mechanisms—on adaptation, evolution, and heredity provided nineteenth-century modes of linking individuals and sociality, of collectivizing.

In the *Concluding reflections,* I summarize the principal threads connecting the various chapters and reflect on them.

References

Barberis, D. S. (2003). In search of an object: Organicist sociology and the reality of society in fin-de-Siècle France. *History of the Human Sciences, 16*(3), 51–72.

Bartha, P. F. A. (2013). *By parallel reasoning: The construction and evaluation of analogical arguments*. Oxford University Press.

Beck, N. (2014). *La Gauche évolutionniste: Spencer et ses lecteurs en France et en Italie*. Presses universitaires Franche Comté.

Black, M. (1962). *Models and metaphors: Studies in language and philosophy*. Cornell University Press.

Black, M. (1979). *On metaphor*. University of Chicago Press.

Cameron, L., & Low, G. (1999). *Researching and applying metaphor*. Cambridge University Press.

Campbell, D. T. (1975). On the conflicts between biological and social evolution and between psychology and moral tradition. *American Psychologist, 30*(12), 1103–1126.

Durkheim, E. ([1892] 1953). *Montesquieu et Rousseau, précurseurs de la sociologie* (S. A. Cuvillier, Trans.). Petite Bibliothèque sociologique international.

Durkheim, E. ([1885–1912] 1978). *Émile Durkheim on institutional analysis*. (Mark Traugott, Trans.). University of Chicago Press.

Gentner, D. (1988). Metaphor as structure mapping: The relational shift. *Child Development, 59*(1), 47–59.

Gentner, D., & Markman, A. B. (1997). Structure mapping in analogy and similarity. *The American Psychologist, 52*(1), 45–56.

Gibbs, R. W. (2008). *The Cambridge handbook of metaphor and thought*. Cambridge University Press.

Hanks, P., & Giora, R., Eds. (2012). *Metaphor and figurative language*. Vol. 1: *Theoretical issues*; Vol. 4: *Cognition and metaphor*; Vol. 6, part 28: *Applied metaphor studies*. Routledge.

Harré, R. (1996). Metaphor in the history of psychology. *The British Journal for the Philosophy of Science, 47*(1), 141–145.

Hesse, M. B. (1966). *Models and analogies in science*. University of Notre Dame Press.

Jackson, J. H. ([1931–1932] 1958). *Selected writings of John Hughlings-Jackson* (J. Taylor, G. Holmes, & F. M. R. Walshe, Eds.), 2 vols. Staples Press.

Keller, E. F. (1995). *Refiguring life: Metaphors of twentieth century biology*. Columbia University Press.
Keller, E. F. (2002). *Making sense of life: Explaining biological development with models, metaphors and machines*. Harvard University Press.
Kittay, E. F. (1987). *Metaphor: Its cognitive force and linguistic structure*. Clarendon.
Lakoff, G., & Johnson, M. (1980). *Metaphors we live by*. The University of Chicago Press.
Maasen, S., & Weingart, P. (2000). *Metaphors and the dynamics of knowledge*. Routledge.
Maasen, S., Mendelsohn, E., & Weingart, P. (2002). *Biology as society, society as biology: Metaphors*. Kluwer Academic Publishers.
Martin, E. (1991). The egg and the sperm: How science has constructed a romance based on stereotypical male-female roles. *Signs: Journal of Women in Culture and Society, 16*(3), 485–501.
Miall, D. (Ed.). (1982). *Metaphor*. The Harvester Press.
Morgan, M. S. (2012a). One observation or many? Justification or discovery? *Philosophy of Science, 79*(5), 667–677.
Morgan, M. S. (2012b). Travelling facts. In P. Howlett & M. S. Morgan (Eds.), *How well do facts travel? The dissemination of reliable knowledge*. Cambridge University Press.
Ortony, A. (Ed.). (1979). *Metaphor and thought*. Cambridge University Press.
Pearce, T. (2020). *Pragmatism's evolution: Organism and environment in American philosophy*. University of Chicago Press.
Reynolds, A. S. (2018). *The third lens: Metaphor and the creation of modern cell biology*. University of Chicago Press.
Soskice, J. M., & Harré, R. (1995). Chapter 11: Metaphor in science. In Z. Radman (Ed.), *From a metaphorical point of view* (pp. 289–307). De Gruyter.
Spencer, H. (1891). *Essays scientific, political and speculative* (Vol. I). Williams & Norgate.
Wenger, E. (1998). *Communities of practice: Learning, meaning and identity*. Cambridge University Press.

Contents

List of Figures

Chapter 1
Jean-Baptiste Lamarck: *La marche de la nature*

The chapter is an introduction to Lamarck, presenting facets relevant to the discussion of the interaction between the social and the biological from the perspective of Lamarckism. I start with a brief summary of the commonly agreed present-day interpretation of Lamarck's principal positions and then take a closer look at further topics relevant to the other chapters. Discussing Lamarck's transition period, I argue that Lamarck's *problématique* (clusters of core questions) was that of the natural history communities of his time, but that in his search for solutions to what he perceived as tensions, incompatibilities, dead-ends, and open questions and to new questions never asked before within that field, he *also* resorted to a repertoire of models about classification and change found in the social-political discourses of his time. An elucidation of some aspects of Lamarck's neuro-psychology follows, particularly the concepts of 'feeling' and '*sentiment intérieur*.' The chapter ends with a discussion of the Lamarck *historiography*, as related to the notion of *Biologie* (Fig. 1.1).[1]

1.1 Lamarck's Mature Positions

I outline those components of Lamarck's mature theorizing, i.e., from *Philosophie zoologique* (*Phil.Zo*, 1809) onward, the interpretation of which has gained general consensus among present-day historians of the period. I will not discuss here

[1] Besides the works mentioned in the chapter, I used a wide array of secondary sources, referred to in my much more detailed papers on Lamarck, some of more often consulted ones here are: Agulhon, 1979; Appel, 1987; Baker, 1990; Barthelmy-Madaule, 1982; Bret, 1991, 2002; Censer and Hunt, 2001; Duchesneau, 1982; Furet et al., 1988; Giglioni, 2013; Gohau, 2006; Gross, 1979; Hamy, 1893; Hodge and Jonathan, 1995; Jordanova, 1981; Kaitaro, 2008; Limoges, 1980; Moravia, 1972; Omodeo, 1997; Osborne, 1992; Pickstone, 2007; Reddy, 2001; Reil, 2005; Rey, 1994; Riskin, 2002; Sewell, 1987; Terrall, 2007; Tirard, 2006; Vidal, 2006; Vovelle, 1985; Williams, 2002 and Wokler, 1995.

S. B. Gissis, *Lamarckism and the Emergence of 'Scientific' Social Sciences in Nineteenth-Century Britain and France*, History, Philosophy and Theory of the Life Sciences 36, https://doi.org/10.1007/978-3-031-52756-2_1

Fig. 1.1 Jean-Baptiste de Lamarck portrait (1744–1828). Stipple engraving by A. Tardieu, 1821, after J. Boilly. Source: Wellcome collection, London

Lamarck's own internal theoretical difficulties and inconsistencies, nor present-day disagreements on interpreting him.

Lamarck was an eighteenth-century materialist. He assumed the overall presence of physiochemical laws, but he acknowledged the difference between living/organized and non-living entities. He attributed a great impact on living entities to a variety of *circumstances*—environmental and behavioral (*milieux*). This impact could be direct or mediated, depending on the level of complexity of the living being. Living entities adapted their *behavior* under the pressure of *changing* circumstances, i.e., by developing novel habits and neglecting others. (Cheung, 2006; Wolfe, 2019). The new behavior patterns were a response to fundamental *needs*, differentially felt or spurring to act, to which organisms responded, e.g., food, sexual reproduction, avoidance of pain, and seeking of well-being/pleasure (1809, 352).

There was the possibility, under particular and specific physiological conditions, of passing on or transmitting these modified habitual and often functional patterns between generations so that living entities gradually and incrementally transformed—first (use and disuse) and second (transmission) laws (1809, 235). The scope of change was unlimited and involved the organization and structures of the living entities, with a contingent transforming process. Lamarck assumed that all individuals of a species under similar circumstances would be similarly modified, thus transforming the species.

Lamarck also assumed that 'a power of life' pushed living entities to greater organizational complexity—he did not believe that power to be an innate teleological force, but rather the result of physicochemical action in an endless variety of changing circumstances. Combining these two factors, Lamarck claimed that there had been a continuous, successive, gradual production and development of all life-forms, starting from the simplest, thus most elementary ones, to *progressively* more complex ones. This combination of factors accounted for diversity, complexity, and specialization. Lamarck named this process '*La marche de la nature*' (later 'transformism,' 'evolution').[2] Thus, nature had a *history* of immense time-span. The model describing the *causal* mechanism assumed for these processes was hydraulic—the action of various kinds of fluids. His living entities could be looked upon, as Wolfe (2019) suggests, as hybrid of the mechanist and organicist models, as having agency.

Methodologically, Lamarck stressed that one ought to start one's analysis and description from those living entities, which were the most elementary, developmentally, in order to explicate laws and mechanisms, to reconstruct the network of relationships and thus of classifications.

1.2 Lamarck's Transition Period

1.2.1 Contexts—Conceptual and Institutional

I argue that Lamarck's views on change within the organic world, in particular his conception of *la marche de la nature*, should be viewed as resulting in part from interactions with, and transfers from, the *social thought*-modes of thinking—ways of conceptualizing, models, metaphors, and analogies—of the decades just before the French revolution and of the revolutionary decade itself. Lamarck's involvement with the new institutional frameworks initiated during the revolutionary period brought him into direct contact with the prevalent modes of discourse on things social, as argued in detail Gissis (2009).

From around the mid-eighteenth century, there were *commonalities of discontent* among those investigating the phenomena related to 'living entities' and those looking into 'social' phenomena, during a period of gradual conceptual separation of the 'social,' first as an adjective (in Rousseau's work) and later as a noun, 'society,' denoting something which was separate from the state (see Baker, 1973, 1990; Carrithers, 1995; Staum, 1996). At the roots of this common disaffection were the limited abilities of explanatory mechanist models to deal with time-dependent processes and the historical dimensions of 'living nature,' with the dynamics of

---

[2] I changed some terms into present-day researcher's terms, where deemed useful, e.g., 'physiological' rather than 'physical,' even though in Lamarck's time 'physiology' was deemed to be more closely related to medicine. 'Evolution'—processes related to the dynamics of Lamarck's *transformism* or his 'the march of nature.'

whole–part relations, with modes of cohesiveness, with the notion of functions and their changing over time, with the distinctness of living entities.

Investigating 'living nature' was considered a combined empirical and theoretical endeavor. Later in the nineteenth century, this gave the field a privileged position and helped shape a context, both scientific and social-political-cultural, that allowed the assumptions that there was a fundamental correspondence between organic nature, social life, and mental life, and between their mechanisms and modes of development. Later, this correspondence became embedded in evolutionary features, foundational entities, and types of lawfulness involving modes of heredity that enabled such relations with models and metaphors *seemingly* streaming one way, from the biological field to the social one. Yet, during the period that I deal with the interaction between various discourses was, in my view, one of 'multidirectional traffic,' i.e., transfer. The extensive interactions and consequent transfers between social thought and biological thought occurred within specific social, political, cultural, and institutional environments and served as enabling conditions for the rise of novel conceptual frameworks. This applies to Lamarck's views on change within the organic world, and in particular, to Lamarck's conception of *la marche de la nature*.

Consider the period between the 1750s and the late 1780s in France with the following question in mind: Where did an amalgamation and assimilation of the discourses of the different sciences and the non-scientific forms of organizing knowledge (e.g., literary, philosophical) take place? What sort of sites? Literary salons, learned societies, learning and corresponding circles, coffee houses, and even free masons' lodges. The common feature of the sites of amalgamation was the mixing of people and the exchanges of concepts that facilitated theoretical mediations and the transfer of strategies of particular modes of discourse. This has been argued by many authors, who have dealt with specific circles and learned societies in the revolutionary period, with such frameworks in the provinces, and with the new societies constituted during the revolutionary period, etc. The debate on the public sphere then is largely tangential to this (Broman, 1998). In this mixing and assimilation, the concepts of A could become the metaphors of B. Looking at the texts produced during this period from the perspective of 'transfer' (that is, not only in terms of contents but also in terms of the 'actions' performed by achieving certain forms of contents), a number of features stand out, such as models, practices, and epistemological considerations.

The application of models to various fields and discourses—In addition to the physical–mechanist–mathematical Cartesian model of knowledge/explanation, the generalized Newtonian 'attraction' model, and the beginnings of the formation of an inductive statistical model, there emerged one, which for lack of a better name, could be called the 'historical' model. It is exemplified by the many books on the history of the human mind, the history of sensibilities, the history of a single nation, of the earth, and, later, of an individual.

The practice of applying theories outside their specific field—Most conspicuously, the practice of stretching the sensibilist or sensibilist-*cum*-sentiment assumption to fields beyond their original application, as a polysemic one used throughout the latter part of the eighteenth century in numerous diverging fields. The primary signification of 'sensibility' related to the senses and the significant status assigned to the ability to use the senses, to have sensations, as a venue in the world of phenomena and of meaning; this over and above that of reflection, on the one hand, and the 'response' of sentiment to the sensation, on the other. Toward the end of the eighteenth century, sensibility was conceived as the locus of interactions between the mind and body. Thus, sensory experience and sentiment constituted the foundations of the self, of both subjectivity and sociability. 'Sensibility' was translated into the moral-political idiom during the revolutionary decade.

Epistemological considerations—Fewer attempts at (logical or rhetorical) 'proofs' were made. Instead, there was more reliance on experience, observation and experiment, accumulated knowledge, language analysis, and the presentation of causal chains or filiations.

The period from the onset of the revolution till the very early years of the nineteenth century was likewise characterized by intense interactions among various sciences and forms of knowledge, though the sites of these interactions had radically changed. The intense politicization of numerous aspects of intellectual and cultural activities and institutions brought men of science into direct contact with questions of social and political theorizing and practice. Thus, 'transfers' and resonances among the various fields became even less intentional, less self-conscious than they had been previously. The rise of the *Idéologues* as a more or less cohesive and influential group should similarly be viewed within this context. There have been claims to the contrary, pointing to that same period as one of growing professionalization and of the emergence of clear disciplinary boundaries (Heilbron, 1995). The picture needs to be complexified somewhat by distinguishing developments in the first revolutionary decade and those in later periods; distinguishing between *parallel* developments—such as the mingling of discourses and the professionalization of sciences and systems—which took place side by side for a while, and by analyzing the political-economic-cultural context that made these possible (Corsi, 2018).

The restructuring of frameworks created a number of state-funded scientific institutions and a host of state-funded educational ones in which scientists lectured/taught. These, and the double role—political and scientific—that numerous scientists played, in turn created new hierarchies, networks, clusters of patronage, etc. (D'Hombres, 1989). Two such major institutions founded in quick succession were the Muséum national d'Histoire naturelle (MuséumNHN, 1793), where Lamarck worked and taught, and the Institut national des Sciences et des Arts (1795), which replaced the Academie royale des sciences, and to whose first class—'for mathematical and physical sciences' Lamarck was soon nominated.

The MuséumNHN became an amalgam of a teaching institution, an exposition that offered guided tours for the general public, and a research center. All the major sub-fields that constituted natural history were embodied in the various interests of its members. Yet, there was no overall agreement on natural–historical–literary practices, nor on classificatory methods and schemes; in fact, not even on such issues as what constitutes a species or other taxonomic groups, nor on such fundamental matters as how to demarcate the living from the non-living. My point is that the MuséumNHN became a site where diverse discourses within natural history could coexist. Its staff members did not necessarily agree on the appropriate classificatory scheme of nature, nor did they agree on a general philosophical world view. Yet, they tacitly agreed on something that was close to a general world view, namely, on what constituted natural history, and were all committed to a very generalized notion of the utility of science for the betterment of human existence (Osborne, 1992). This in turn had important implications for the relationship between the natural historian and the state. The constant need to be attentive to rapidly shifting governmental practices, ideologies, and rhetoric concerning scientific practices and institutions in order to collectively react to them, attuned the members of that community to things both social and political.

The MuséumNHN provided Lamarck a place within a structure that grew during the revolutionary decade to become the richest institution in collections. It was an institution in which he became intimately familiar with several models of doing natural history and one in which he could, in spite of significant constraints, develop his own research topics and objects and adopt or develop new classificatory procedures and explanatory schemes that were in conceptual confrontation with rival ones. In addition, teaching and having to confront the teaching practices of his colleagues, both directly and through his students, were valuable aspects of his association with the MuséumNHN.

Lamarck's special position within the MuséumNHN should also be considered. He was not appointed to make use of his long-established skills and knowledge of botany. Instead, he was appointed to the zoology department, where he was expected to address a rather vaguely defined, less coveted sub-field, one considered less productive and less 'utile,' namely 'insects, worms, and microscopic animals.' But Lamarck looked upon the distinction between invertebrates and vertebrates as one of his important original contributions. It effectively changed both the boundaries of, and the perspective on, the realm of animals he had been nominated to investigate at the MuséumNHN. He tended to posit the invertebrates as 'model organisms'; thus, they served as an index to the group of instances and to the set of *problématiques*. He applied fine distinctions and divisions among them in a mode analogous to distinctions and divisions in living nature at large. And the same was true for the relationships between functions and structures. For example, in one of the rare semi-autobiographical paragraphs in an opening course lecture, Lamarck stated (1817) that only gradually did he realize *what a rare glimpse into the working of nature the investigation of invertebrates had offered him*—"*il existoit réellement un ordre à reconnoitre, relativement à la composition de chaque sorte d'organisation animale*" (Lamarck, 1972, 29).

Furthermore, because of his age, his position on the 'new scientific attitudes' in chemistry and physics, and his rather unstable economic situation, Lamarck gradually became more marginalized within MuséumNHN. This marginalization increased as Cuvier established himself as a major figure in the MuséumNHN, in French natural history, and in the scientific community at large. As the Napoleonic empire grew and extended, so did those of Cuvier and Laplace. As Corsi (1997a, b,

2012) has shown, however, for a number of years, Lamarck had a wide audience of students from all over Europe for his courses and a large and growing international readership for some of his works, mainly outside the MuséumNHN and its coterie. Lamarck's marginalization within the Muséum became more marked toward the end of the first decade of the nineteenth century. The number and volume of his articles appearing in the annals of the Muséum decreased. His works tended to be published privately. His meteorological surveys were harshly criticized, or worse still, ignored or hardly mentioned, e.g., in the summary reports on the field that appeared regularly in *La Décade philosophique*.

The same was true for other general culture and sciences journals (Hesse, 1991). Corsi claims that the number of his students also decreased. Furthermore, the change in the cultural field at large was marked by political instability of one kind before 1795, and of another later. Within this context, the methods and 'the language of sciences' changed rather dramatically. As Napoleon's power became all-pervasive, the influence of the reinstated Catholic church once again became extensive, and these factors had their effect on Lamarck. He became ever more isolated by virtue of his materialist position, which persisted in the deep structure of his writings even when 'God' was referred to. His refusal to accept the new physics and chemistry and his views of 'nature' were incompatible with both the Catholic dogma and with the contemporary trend toward a traditionalist nostalgia for the past and an emphasis on 'matters spiritual' such as in Chateaubriand. These positions stood in opposition to the 'irreligion' of the revolutionary years.

Lamarck has been re-instituted as an important contributor to 'evolutionism' in the early nineteenth century, and in particular, as an important influence on Geoffroy Saint Hilaire, Frédéric Cuvier, Grant, Chambers, Spencer, Darwin, Baden Powell, and others, who concerned themselves with the history of life on earth during the first half of the nineteenth century in France and in Great Britain. Corsi (1998, 2005, 2012) has corroborated this by making clear the role Lamarck and his students played in making the history of life on earth a central issue. Corsi and Barsanti (1995, 1997) have also pointed to the extensive considerations of the subject matter instigated or influenced by Lamarck or his auditors, *outside* of France and Great Britain. By contrast, the latter two countries were the locations usually addressed by scholars in the field until the last decades of the twentieth century.

Going back from the institutional to the conceptual context that was reflected in the work done at the MuséumNHN, certain clusters of concepts are identifiable that recurred in the discourses on 'the social' during those years, some of which originated in much earlier controversies, while others were of more recent coinage. Their substantive and rhetorical resonances within the political and scientific discourses of the time can be located. These clusters and the discourses, in which they were embedded, the models for explaining social interaction and organization, social growth and transformation, were available to Lamarck as a result of his position, his involvement with the various activities of the MuséumNHN and other institutions during the early revolutionary period. I believe that the impact of these clusters played a meaningful role in changing his views on living nature. For example, the uses of

- 'Laws of nature,' 'regularities of nature,' or, less frequently, 'natural law' and the occasional, rather rare use at that stage of 'state of nature,' Rousseauian strains of environment and interiority
- Sensibility and sentiment; perfectibility of human nature, of reason, and of society; sociability as based on sentiment, on obedience to laws, on conventions; looking upon either individuals or collectivities as the basic units of society
- 'Organization' and 'self-organization'; balance, equilibrium, harmony; 'simple and complex'; the relations between structure and function of social institutions; reciprocal interaction and interdependence
- Utility and usability; '*les progrès*'—change, fields or capacities of progressive change, laws of change, inevitability of change, epochs or stages of change, varying paces of change, continuity of change; ruptures in change, regression.

Most thinkers during the second half of the eighteenth century, particularly in France, looked upon the individual as the basic unit of description and analysis, though some, like Ferguson, argued that the individual had to be considered already as socialized. The dynamic character of being human became emphasized as the foremost feature at both the individual and the collective levels. The dynamic character of the individual was embodied in the notion of perfectibility (or in that of *Bildung* in its German version), and in the notion of progress for the collectivity. Two marked changes occurred in the latter part of the eighteenth century: first, the notion of processes was brought in as part and parcel of explanatory models; and second, diversity and complexity were accepted as inevitable components of the object to be explained. This meant that the difficulty of producing certain and fully predictable results was recognized and discussed. For example, Condorcet's discussion of the relationship between moral and physical sciences and their progress in his acceptance speech at the Academy *Discours* February 1782 (Condorcet vol I, 1847–49). Lamarck was already a member by that time. Also relevant was Condorcet's later analogy between the new science of society and the new science of meteorology in relation to the calculus of probabilities.

Diversity and complexity also meant that causal explanations had to take into account filiations of events in historical time and that greater emphasis was placed on observation and comparison. The social patterns or orders resulting from processes of change became viewed either as unintended (and unpredictable in this sense), planned (intended yet not fully predictable), or emergent (the whole could not be explained or predicted from knowledge of the component parts). The malleability of either individuals or societies was, however, bounded by 'structural' constraints on human nature, on the possible forms of human sociability. Individual and social behavior were conceived as revealing regularities, even when these were gradually developing, emerging ones. Thus, even though both perfectibility and progress were presented as being open-ended, having unbounded horizons, they were actually bounded by the assumptions on their possible horizons (even in the Jacobin vision). The most prevalent models during the 'revolutionary decade' incorporated components of these diverse conceptual patterns (Fig. 1.2).

The Revolution turned the discourse of societal progress that encompassed the progress of individuals into an undisputed and essential component of the revolutionary worldview. Questions that had been asked about society before the revolution gained great urgency, for the answers were to be translated into daily political and cultural practices and could apply to all the sections and factions in society. The discourse about societal progress found expression not only in the purple passages of various leaders but also in quotidian rhetoric (Gissis, 2009). The totalizing experience of the revolutionary decade for its contemporaries also had the effect of relativizing the notion of 'stability.'

Until 1795, Lamarck was actively participating in a wide range of activities, all of which were connected to innovations of the new revolutionary regime or to the results of the nationalization of church and private property dealing with 'natural history.' For example, he was preparing memoranda, taking part in revolutionary learned societies and committees, organizing, classifying, and even guarding confiscated natural history treasures, writing and teaching within revolutionary frameworks, co-editing what was perhaps the first natural history journal, which he chose to publish with a Girondin-inclined publishing house, *le cercle social*.

Fig. 1.2 View of the public works at Field of Mars, by the Parisians 1st year of Liberty July 12, 1790: "patriotism overcoming all obstacles." Engraving Pierre Gentot 1790. Source: gallica.bnf.fr

Making a broad generalization, I would say that even before the turn of the century, doing science, including dealing with living entities, already meant both gathering phenomena under laws that had universal application, as Laplace, whose scientific work and institutional politics Lamarck had shunned, was considered the outstanding contemporaneous representative of a 'scientist.' Science meant decomposing the whole into parts, seeking solutions at lower and lower levels in the hierarchy, quantifying and reaching unto ever smaller entities. A 'biological temperament' could be expressed either in an ontology of dynamic processes or in an ontology of stable, non-changing entities. Lamarck chose dynamic processes over 'dumb,' static entities. He deployed the tool of decomposition with his living entities, but he refused to quantify and to measure the stages of the living processes he studied in the way promoted by the 'new sciences' and their 'new language,' which was epitomized for him in the figures of Lavoisier and Laplace. When looking at his use of both '*mésurer*' and '*quantité*,' the latter was rare, mostly referring to an undistinguished large number of the units and/or the matter discussed, e.g., acts, rain, and electricity, while the former was rarer still, mostly found in earlier works relating to meteorology, or later on to irritability.

Lamarck also does not fit neatly into the more conventional narrative of the tension between natural history and laboratory physiology, which posited the latter as seeking to emulate physical sciences and establish universalistic knowledge. This dichotomous division fits better the somewhat later nineteenth-century positions and practices rather than the checkered landscape of the late eighteenth and very early nineteenth centuries. All the more so in recalling the positions then taken by the young Geoffroy, Lamarck, and Cuvier concerning field, cabinet, and comparative laboratory practices, respectively. Lamarck did not experiment, though there are indications that he followed and was acquainted with at least some of the experimental work of his immediate predecessors and contemporaries. From early on, even before the crucial turn to transformism, Lamarck looked upon living entities as distinctively different from non-living ones, but not because of the makeup of their molecules; and this was in line with his physical–chemical assumptions of the makeup of matter. His later constructions of the makeup, the causalities involved, the evolvement and history of the plurality, and the diversity of living entities did not change that position.

From the turn of the century until his last work, the issues that Lamarck worried about and tried to address in his work on living entities bounded and filled *his* field of *biologie*. They included the *chemical and physical understanding of life processes* (recall that his own chemistry and physics were opposed to the new science of Laplace and Lavoisier), the *hydraulic mechanisms of active fluids* and their transferring agents/body structures, the role of the *environment*, the notion of '*organization*' (and self-organization), and his *taxonomic enterprise*.

1.2.2 Texts

At the beginning of the period, around the mid-1780s, Lamarck was a well-known botanist, who held firm views on the fixity of species and considered species as the most real entities in ontological terms. To repeat succinctly, he defined 'life' as a state of matter infused with a vital element expressed in a special kind of motion (1794, 2: 186). He did not look upon change as being directional, progressing, in time, and believed, along with Buffon, that change within the living world was limited by inner-boundaries within the species and the individual. With regard to issues contested in physics and chemistry, Lamarck was against the new 'pneumatic' theory propounded by Lavoisier and others, yet he adhered to a mechanist model of explanation. Like most members of the intellectual communities of France at the time, he held sensationalist views both on epistemological issues and on questions regarding the proper method of doing science.

By 1803, Lamarck concluded that the earth in general had a history of continual change and suggested a specific developmental-historical narrative on the emergence of various forms of life that wove together their ordering, their filiation, and their organizational features. He believed that species changed over time into new ones, although they had relative stability that enabled their ordering and classifying into categories. Ontological 'reality' was attributed in an absolute sense only to individuals. Whether considered from the perspective of the order of becoming or that of knowing, the individual became the unit of the analysis, with 'masses' becoming the principal category of ordering. 'Masses' were non-classified collections of individuals that fell under very generalized divisions of living entities, viewed from the perspective of combined timelines, of diversified environmental conditions and complexifying organizational form lines (e.g., already in the *Discours préliminaire* an x [1802], published 1907).

Lamarck's theoretical work on living nature during 1800–1802 included his *Système des animaux sans vertèbres* (1801b), *Hydrogéologie* (1802), *Recherches sur l'organisation des corps vivans* (1802), and his unpublished *Biologie* (1800/02). After these writings, Lamarck's evolving position can be found in the introductions to the courses he taught at the Muséum, in hints in his articles on fossils, published in the MuseeumNHN *Annales* between 1802 and 1806, and then his mature position in the *Philosophie zoologique* of 1809. Later works relevant for this discussion include the items for the *Nouveau dictionnaire d'histoire naturelle* (1817–18) edited by Virey, where his non-classificatory articles were followed by Virey's articles with identical titles, his *Histoire naturelle des animaux sans vertèbres* (1815–1822), and his *Système analytique* (1820) (see the items in http://www.lamarck.cnrs.fr/index.php?lang=fr and Lamarck 1812, 1907, 1933, 1972, 1990).

Let us look at the notion of temporally ordered complexifying processes as an example of the positions elaborated in these works. One of the most noticeable changes after the turn of the century in Lamarck's presentation in his courses, especially since 1803, was his insistence that naturalists should formulate a philosophy and his assertion that he was a *naturalist-philosophe* who produced, rather than reproduced, knowledge of nature. In almost every introductory lecture from 1803 onward, Lamarck called upon his listeners to exercise their imagination in order to encompass the vast variety and diversity of living nature. In 1806 (1972[1806]), he suggested to his audience the following thought experiment (repeated almost verbatim in later texts): Suppose that the whole of nature was laid out for a spectator to contemplate as if it were a collection, could one then classify nature in a way that "would cut it really at its joints"? This is the same terminology of diversity, bountifulness, and variety, which he had formulated, much earlier, in the *Discours preliminaire* to his *Flore* in 1779, but there, only in relation to plants. In the *Flore*, the human mind was supposed to introduce methodical classifications that were useful, but not necessarily 'cutting nature at its joints.' For Lamarck then, most of the components of the botanical hierarchy of classification, excepting that of species, were to be considered '*fictional and arbitrary divisions*,' or '*artificial means*,' or at best useful 'rules of convention,' which consistently were considered antithetical to '*the real/truthful relations*,' that 'really' describe *la marche de la nature*.[3] In numerous introductory lectures, Lamarck started from a classificatory scheme, asking what could count as a common feature, thus calling upon his students to delve more deeply into the foundations upon which one could reconstruct the natural *rapports*. In this particular work and in those that were published in the *Encyclopédie méthodique*, the progeny of the *Grande Encyclopédie*, Lamarck described the 'true *rapports*' as graduated and nuanced differences, impossible to 'cut,' yet their knowledge, once arrived at, was solid. In 1794, when a work of his dating from 1776 was first published (*Recherches sur les causes principaux des faits physiques*, 2 vols), and in subsequent works during the second half of the 1790s, Lamarck introduced the reader to his alternative physical and chemical theories.

Lamarck posited a distinction between living and non-living entities, in which the concept of organization played a subtle role: at first, in distinguishing between living and non-living, and somewhat later in drawing the line between different sorts of living entities. Instead of just 'graduated nuanced differences' of the early botanical period, one was offered differences in accordance with the degree of complexity of the organization. The latter was to be decided according to faculties, functions, and modes of activity, and later according to the existence of specific organs shaped by these, thus the other way around from Cuvier. This coincided in meaningful ways with the 'organizational' discourse and rhetoric of both the revolutionary period and the early post-Jacobin regime of the Thermidorian rule. Until 1799/1800, Lamarck talked about a graduated continuous series of living entities, distinguished by demarcations that had been suggested by contemporaries' new modes of dividing natural entities in general and the living variety and diversity in particular (e.g., division into vertebrates and invertebrates, and the further divisions in each of these classes). This was still a view of a static chain going from the complex to the simple.

[3] '*divisions factices et arbitraires*,' or '*moyen artificial*,' '*règles de convention*, '*les vrais rapports*.'

A striking distinguishing feature of Lamarck's version was his insistence on a clearly delineated continuum with the end points incorporated within it and considered as components, i.e., part of the materialist assumptions. It meant that both the simplest entity and the most complex one could be looked upon as starting points. Both had to be included in the mechanisms that were to account for the whole series.

The change in his position is evident in two writings dating from around 1799/1800 (the manuscript *Prodorme de l'histoire naturelle des animaux invertèbres, discours d'ouverture an VIII*, 1800). It consisted in reversing the order from one of degradation (from complex to simple ones) to one of ascent-progress and by adding time, genealogy, and the beginnings of a causal filiation. The lines dividing species got blurred and then crumbled, and in the three following years, mechanisms of change of individuals, and thereby of species, were introduced, and spontaneous generation was brought in to account for the beginnings of life. The latter made possible separate causal series for the most general 'realms' of nature but retained the clear-cut boundary ('hiatus' in Lamarck's words), dividing the living and the non-living. The stability and reality of the species became a relative one: it was relative to the particular environmental conditions, at a particular time, in which the individuals of a species found themselves. Another issue closely related to this was the changing status and role of the individual in Lamarck's writings after 1800. Thus, searching through Corsi's site reveals that most of the occurrences (994 in all) of 'individual' are in works from 1802 onward, particularly in the *Rechèrches sur l'organization des corps vivants…*; *Philosophie zoologique*; and in the introductory course from 1806. To discuss his novel conceptualization of an individual, one had to include both its temporal locus and time-directionality and the role of the environment in the shaping of a living entity and its life span.

A comment is in order here on a significant departure by Lamarck from a known practice of many naturalists of his time, including his teacher, Buffon, who used the human individual, usually termed 'man,' as the starting point of their discussion of living nature. Unlike them, Lamarck started with the panorama of living nature. He was a thorough-going materialist, who, when he finally discussed man, deemed it necessary to emphasize that all a naturalist—i.e., he Lamarck could do was *to apply uniformly the laws of physics*, i.e., his Lamarckian chemistry, geology, and physics, *to Man*. For Lamarck, the faculty of reason was looked upon as causally dependent on matter, and matter was the sole entity the naturalist could investigate and explicate. In both *Flore* and the *Encyclopédie méthodique*, species were fixed by virtue of certain features that appeared in all individuals, and were perpetuated in the process of reproduction, a view that falls neatly within the Buffonian definition.

In the various physicochemical articles that followed, Lamarck's position concerning living bodies that were to be considered as individuals was gradually modified. This falls within the search for forces that would explain the organization and generation of systems, particularly of living systems, whose very description implied history (their genealogy), which was seen as time-directed irreversible processes, and as a developmental 'progressive' aspect. Individuality became invariably related to life, with life being the condition of becoming individualized.

Lamarck called this process 'animalization.' Only living bodies produced themselves. Living bodies also produced other bodies either directly or indirectly. When, within two years, the transformist notions were much more clearly formulated, the concept of the individual crystallized as well. The lowest boundary of the living entities was also the almost inconceivable limit of animalization and thereby of living structure. It would be wrong, however, to say that there was 'more or less individuation.' It was in relation to this end point of the continuum of the living—of the animalized, the individualized—that the role of time became manifest. Time was that through which the gradual emergences and changes occurred and through which the ordering took place in an environmentally causal manner: via particular mechanisms of change, specifically those of use and habit (*Discours preliminaire* an viii, 1800).

A proper individual was a living entity that bore resemblances to other individuals, and this collection of individuals constituted a species. This was sharply formulated in his *Biologie* of 1800: "A living body is a natural body, circumscribed in its duration, in its parts, at least in the principal ones, possessing what is called life… this body *constitutes an individual*" (1944, 7—emphasis added).[4] In the period from 1800 to 1803, the attribution of reality was gradually transferred from the species to individuals. It evolved as the entities that changed were perceived to be individuals. Qua individuals, these entities had progressively complexified structures; they could be considered as belonging to a species only by virtue of secondary consequences.

Given that in Lamarck's conception both the environment and the living entities within it were in constant change, the question arises as to what could account for the stable component in that system. I would argue that in Lamarck's conception, stability resided in a pattern rather than in its contents, i.e., in the pattern of a time-directed process of change from the simple toward further complexification. Looking at social thinking at that time, the movement to view society either as some kind of aggregate or as an emergent collectivity was related to choosing the social model of progress, which in almost all its then-current variations in France was based on the ontological primacy of the individual. It would have been extremely difficult to conceptualize time-directional causal progressive change based on ontologically stable groups, the species. Since 1800, Lamarck used fossils, particularly those of marine organisms, as evidence for the gradual, constant, continuous, and uniform change of the earth's surface, i.e., as evidence for both the continuous impact of the environment on organisms and of their transformation during an immeasurably long span of time. This view had to do with the attempt to constitute a coherent conceptual framework that would harbor *progressive change* and thus complexification and emergence, together with *order*. In contrast to Cuvier, Lamarck initially rejected extinction, but between 1800 and 1802 he came to accept the extinction of the simplest life-forms, which were continuously spontaneously generated.

---

[4] "*Un corps vivant est un corps naturel borné dans sa durée, dans ses parties au moins dans les principales, possédant ce qu'on nomme la vie... Ce corps,* ***constitue un individu***."

1.2.3 Concluding Remarks

When Lamarck had reached the limits of then prevalent modes and models of explanation of living nature, he transferred the model of progress, which could be found not only in the actual writing of prominent contemporary figures such as Turgot and Condorcet, but also in contemporaneous public festivals, political speeches and orations, pamphlets, drafts for regulations, laws, and constitutions. Progress was part and parcel of the administrative discourse as well as of the political-cultural programs and projects for reforming and reshaping social, cultural, and scientific institutions, society at large, and every individual in it. This was a model combining time, genealogy, and causality that could provide a better description and analysis of processes of emergence of complexity and of complex organizations, which Lamarck came to see as the principal feature of living entities. The *problématique* that Lamarck dealt with was certainly that of the natural history communities of the time, but his innovative solutions should also be understood as resulting from transfers of notions, concepts, models, and metaphors from the social discourses of his time. His was not a straightforward and simplistic transposition, but rather a deployment of the elements, transferred and indirectly translated, *within* his natural history context, thus changing their referents and their role in the overall narrative thereby constructed.

During the destabilizing, innovative, revolutionary times, Lamarck was both at the center and in some significant respects, personally semi-peripheral. Others, as well, were going in the same general direction as Lamarck in solving natural-history *problematiques*. There was dissatisfaction within natural-history communities with the prevalent accounts of what was understood as time-bound change, process, the emergence of unrecognized forms, the distinct features of living entities, etc., but no one else did it as thoroughly, in a consequential and systematic fashion, as Lamarck. As he formulated an innovative theory on a grand scale, encompassing the whole of living nature, the scientific field in particular and the cultural field at large were beginning to undergo a transformation that soon enough eliminated either the sites or the possibilities of the mingling of discourses, especially of social and natural history ones, and gradually hardened disciplinary boundaries. Moreover, Catholicism returned to the cultural field.

Lamarck was unique in some respects: his career and biography attest to a non-conventional path. He moved from one field, botany, where he was an established naturalist, to another, zoology, and to that part of it which was less well-investigated and more disorderly, namely 'white blooded animals,' at a late stage in his professional and personal life. In some limited way, he can be regarded as an example of what the sociologist of science Ben David (1991) called role hybridization. He tended to locate the issues he was dealing with within overall frameworks ('*systemes*'), and yet to discuss them in a very focused and detailed manner. In that sense he can be seen as both inheriting the questions posed within the Buffonian tradition and as going outside its methodological and conceptual boundaries in reformulating and answering them. He belonged to the most coveted and stable natural history community of the time—the *Muséum national d'historie naturelle*—and was a member of the most prestigious scientific institution, the *Institut national*. But he was, or soon became in many ways, marginal, both politically and scientifically, in both institutions, and found his audience mainly in scientific and non-scientific communities outside them. Yet, the very fact that he belonged to these institutions allowed him to disseminate his ideas as a teacher as well as an author, and to some extent forced his colleagues, who belonged to the canonical center, to relate to his work, even if derisively, as, for example, Cuvier (1810, 1815/1989, 1860).

Lamarck's model, one facet of which I have analyzed, was radically novel within the context of the natural history of his time; novel because it was fashioned from a new amalgam of components, some of which, I have argued, were taken from the social discourses of the time.

1.3 Lamarck's Neuro-physiological Psychology

1.3.1 'Feelings'

I argue that Lamarck's innovative stance in this field stemmed from his evolutionary perspective on 'mental events' and that his conception of 'self' was derived from it (Gissis, 2010; on the emergence of new scientific objects, see Daston, 2000). The focus of my discussion is on Lamarck's writings from 1806 onward that primarily deal with '*Sentimens*'—'Feelings' (explicated through an evolutionary sieve), and '*Sentiment intérieur*' (inner feeling)—an emergent evolving 'self' (standing for some of the more traditional functions of the 'soul') accompanied by a short note on 'Attention.' Higher mental functions will not be discussed, nor Lamarck's general view of individual and social human nature, except in so far as it is relevant to support my argument on the emergent evolving concept of 'self.' His discussion of mind was anchored in the discourse of 'faculties' that had been firmly established in the planning of the *Encyclopédie*, but became somewhat attenuated in its entries. Thus, though presented evolutionarily, Lamarck's discussion of the faculties was devoid of historical context-dependence. In fact, they were treated as 'natural kinds.'

Lamarck assumed that the 'mental events' (in the empiricist signification) subsumed under the experiencing of 'feelings' were direct consequences of physiological processes. He argued that the capacity of experiencing was dependent on bodily structures that were evolutionarily emergent. Consequently, the resulting experiencing and 'feelings' were also an emergent process. Organisms, as well as their component parts, were assumed to be in constant interaction with their specific environment. The feelings-events in the body were integral parts of it and bore its evolutionary specificities. Given that Lamarck belonged to the dwindling community of naturalists, for whom the elementary entities grounding physics and chemistry were still part of earlier systems, e.g., caloric and electric fluids, he did not reduce his mechanisms and processes of life and experiencing to the elementary units of the new physics and chemistry of his time. The presence or absence of a bodily system that enabled the activity of the various body fluids, such as blood, was considered a distinguishing feature, marking the invertebrates, and used later on for internal classification within that group (Cheung, 2006). Yet, Lamarck's endeavor was in a deep sense analogous to that of the 'new science,' but with a marked emphasis on the complexity and the time-directedness characterizing the phenomena of life. As mentioned earlier, numerous contemporaries, particularly the *Idéologues*, and especially Cabanis and physicians in general, used the human individual, usually termed 'man,' as the starting point of their discussion of living nature. Even though they looked upon 'man' as part of nature, they devoted long and protracted discussions to man's special, at times even separate, status within nature and to the particular conceptual tools needed to analyze his unique faculties. In this sense, some of the members of the *Société des observateurs de l'homme* were within the Buffonian tradition.

By contrast, Lamarck felt it necessary to provide explanations for experiencing, feeling, and intellection across the branching scale of animals only. When he finally discussed man, the faculty of reason itself was looked upon as depending causally—physically, functionally—on matter (for an analogous case, Kaitaro, 2008). To give an account of how the world was apprehended by living organisms/organized bodies within such a framework meant relating it to their specific bodily functioning and bodily structures. That is to say, 'sensations' and 'feelings' had to be embodied in order to be; they could not be related to semi-external sense-organs whose outputs would then become located and worked out in a disembodied mind, psyche, or an analogous mediating organ. Though 'man' had been mentioned earlier, it was only in the *Philosophie zoologique* of 1809 that Lamarck started detailed discussions of humans.

Lamarck's discussion of feelings wove together the environment, living body (i.e., an organism), nervous system, and behavior as components and attempted to draw a systematic evolutionary account of the relations and interactions (the '*rapports*') among them. Loosely speaking, it was conducted within the framework of sensationalism, later also called associationism, and the events and the processes at its focus were considered, I argue, neither solely physical nor solely mental. They were explicated in physiological (in Lamarck, 'physical') terms and presented as events and processes of the material organism, yet were understood as the discursive equivalents of the traditionally mental, all the more so for taking place solely inside the body. These events and processes occurred inside the organism and were to be detected through the reactions and actions of its various subsystems and parts in both space and time. Having adopted an empirical stance, Lamarck was not unique in trying to discuss the phenomena of the traditional category of mind within the empiricist range of available models. In Lamarck's case, being empirical meant using the materials in the rich natural-history cabinets, making use of the diverse findings (experimental or observational) of other contemporary or earlier eighteenth-century naturalists and zoologists, as well as those of anatomists and physicians, though he carefully distinguished his positions from all these practitioners. In

Lamarck's later writings, the scientific status of the work and the conclusions of others were dealt with in two ways: in the discussion of the multilayered use of imagination (not only in his later works but also in the dictionary item by that name); and in the distinction made between the immutable laws of nature and their representation in the laws of science which bear the mark of humans' limited means (e.g., *Apperçu des connaissances humaines*, 1972, 137 [1810–181]).[5]

In Lamarck's earlier 'transformist' writings (1800–1806), the influence of the (constantly changing) environment on living organisms was considered to be direct. Holding transformist position meant establishing the connection between the order of appearance in evolutionary time and the number of functions of the organism on that level, as in *Prodorme* (1800, 4). Lamarck thus suggested that taxonomic distinctions between vertebrates and invertebrates, and between plants and animals, in general, should be based on the most general feature found in the *lowest* forms of organization, rather than in the most sophisticated ones. The presence (or absence) of irritability, a particular 'nervous' activity (considered elementary, which by 1806 became totally separated from sensibility, contrary to the Hallerian tradition, directly criticized only in *Histoire*), was chosen as the predominant function serving as a tool for the distinction between plants and animals. 'Sensibility,' indicated by the presence or absence of any form of *a specialized system of nerves*, even a rudimentary one, e.g., *Phil.Zo* (1809, 274), served as the locus of the active agent, the '*le fluide nerveux*.' Thus, its presence pointed to a distinction within the animal taxa.

Lamarck used a method of analysis characteristic of the empiricist-sensational-associationist schools, namely, choosing basic units that were assumed to be the most elementary, primitive ones, with all other features being regarded as combinations and compounds of such units. The evolutionary mechanism he adopted operated on these basic units. This allowed him to point to a near zero-point, to suggest a gradual compounding across the ascending scale of organisms, and to exemplify an increasing *complexification*. Based on the kinds of experiencing and movement available to groups of animals, which in turn depended on the degree of sophistication of their nervous system and its parts, the animal realm was divided into four groups in Lamarck's writings up to 1810 (e.g., 1809, 48); and into three groups from 1812 onward (e.g., 1815–1822, 267–69). Burkhardt (1995) argues that although Lamarck's theoretical position was that structures were dependent on functions, in actual practice he derived function from structure, and that his work was less like that of a contemporary ethologist and more like that of a comparative anatomist.

Lamarck himself, though, viewed his practice differently. In his later writings, he emphasized the insights culled by any zoologist, who could observe the functioning of organs, and who presumably would also make inferences (within the 'faculties discourse') that were not open to an anatomist (1816/1972, 41–2). The principal components of Lamarck's explanatory model of the 'basic units' and their compounding were the nervous system at large, certain specific spaces within it (*centres*, '*foyers*'), particularly the brain, and the (hydraulic model of) nervous fluids.

[5] Daston (2005, 2007); Daston and Sibum (2003).

These fluids also molded and gave form to the body structures and traveled in pipes, tubes, canals, and other equivalent spaces. These fluids functioned as connections between various parts of the inside as well as linkages between the inside and the outside, i.e., they enabled the '*rapports*' between the environment and the body. Given that the nervous fluids were the active agent in the explanatory model, in the 1809 writings, there were also two degrees of their mode of activity which mapped unto two modes of growing complexity and sophistication of the physio-mental apparatus (*Philosophie zoologique* and the introductory lecture for the 1809 course, 1972, 201–14, or *l'avancement de la zoologie*, 1808–10, 1972, 215–26).

Lamarck's similarity to and distinction from some of his contemporaries, most conspicuously Cabanis, turned on the explanations proffered for sensations and feelings, instinct and inner-feeling or *sentiment interieur*. There was a persistent effort throughout the *Philosophie zoologique* to distinguish his positions from those of other naturalists, even though they were rarely mentioned by name. Thus, the book can be read as a series of exercises in '*dialogues manqués*.' I will not take this path here because my focus lies elsewhere, but note that only three of his interlocutors—Cabanis, his disciple Richerand, and the phrenologist Gall—were repeatedly mentioned by name and that their positions, particularly that of Cabanis, were read by contemporaries as close to Lamarck's (Corsi, 1988, esp. 22 and chap. 6; Staum 1980, esp. 182–89; indirectly Blanckaert, 2002). Thus, Lamarck's effort to differentiate himself and disconnect from their positions were both more difficult and more nuanced than from those positions deemed totally different (see also "Gall's system" in "Harvard MS," 1933).

In the *Philosophie zoologique*, Lamarck delineated a multi-leveled system for conveying, transferring, and maintaining external and internal data, transformed into various embodied informational modes that had specific ways of activation attached to them. The basic contemporary distinction between within/outside the body was somewhat blunted by the very fact that everything could only be processed and become an event inside the body. Thus, excitation, perception, sensation, and feeling (and some muscular movement) would be one 'organic' mode, termed 'sensibility'; excitation and contraction would be another 'organic' mode termed 'irritability.' These two modes were considered separate, differing, in their extension over the animal world and independent of one another. Only the first was seen as depending on a specialized organic sub-system, the nervous system. But in both modes, it was the movement of various fluids within the body that made it a living organism. This narrative could be characterized as being one of accumulation and at the same time of complexification. Within this narrative, humans were distinguished by highly specialized activities. Though these could theoretically be found in organisms of lower complexity, they stood out in humans: attention, imagination, understanding, thinking, and recalling (memory).

'Feeling' was defined from within the discourse on organic functions. Lamarck was clearly juggling to constitute 'feelings' as a basic physiological-mental unit, defined from the point of view of that discourse, and not the one of the faculties. "Feeling, or the capacity for experiencing sensations" (1809, 266); "Feeling is not a faculty … feeling is only an effect; that is to say the result of an organic act and not a faculty inherent in any of the substances, which enter into the composition of a body that can experience it . . . . none of our humors and none of our organs, not even our nerves, have the faculty of feeling. It is only by an illusion that we attribute the singular effect which we call sensation or feeling to a definite part of our body ... But the very remarkable effect called sensation, or, when more intense, pain, is the product of the function of a very special system of organs, the activity of which is dependent on the circumstances which provoke it" (1809, 2:ix, 273). In those parts of the *Philosophie zoologique*, where feelings were discussed, Lamarck elaborated: "*it is our entire being that feels* or rather undergoes a general effect ... it is a commotion throughout the entire sensitive system" (1809, 3:321—emphasis added).

In his 1815 *Histoire* (primarily a work of classification) and subsequent publications, 'first principles' were posited in which Lamarck tried to tie together method, model, and a rhetoric adequate to the task, so as to distinguish his position clearly from those of vitalists, *Idéologues*, phrenologists, proponents of the 'new science' and of the new comparative anatomists, e.g., Destutt de Tracy 1817–1818 and numerous articles in *La Décade* (Goetz, 1986, 1993; Moravia, 1980; Young, 1970). The fundamental relationship proposed was that of attributing functions to specialized organs. The novelty was the emphasis on the fact that this framework worked adequately, fittingly, and differentially, *all along the evolutionary path* (e.g., 1815–1822, 23–4, 252).

1.3.2 *Sentiment intérieur*

I will now attempt to analyze a special form of feelings in which, once again, the evolutionary perspective was posited as new, even when somewhat similar notions appeared elsewhere. Lamarck characterized the notion of *sentiment intérieur* thus:

> every sensing being, i.e., a being that is endowed with the faculty of sensing, and nowhere else but in the animal-realm that beings of that sort exist, possesses a *sentiment intérieur*, which it enjoys without discerning it and gives it (the being) a very obscure notion of its existence, or put differently, constitutes in it the feeling of its being, and in this way gives room [*donne lieu*] to that 'I' so familiar to us, because we have the capacity to pay attention to it. This intimate feeling of existence, in a word, this particular 'I' [*moi*] is well known, as I have said; but to my mind it seems that the *sentiment intérieur* that allows it [the *moi*], constitutes a power, which on the one hand is susceptible to being changed by [i.e., responsive to] any felt need and, on the other hand, is capable of immediately causing action, has not been recognized by anyone before me. (1817, "Instinct," 332)[6]

[6] "*Tout être sensible, c'est-à-dire, doué de la faculté de sentir, et ce n'est que dans le règne animal qu'il en existe de cette sorte, possède un sentiment intérieur, dont il jouit sans le discerner, qui lui donne une notion très-obscure de son existence, ou autrement, qui constitue en lui le sentiment de son être, et qui y donne lieu à ce moi si connu de nous, parce que nous avons le pouvoir d'y donner de l'attention. Ce sentiment intime d'existence, en un mot, ce moi en question nous étoit bien connu, comme je viens de le dire; mais le sentiment intérieur qui y donne lieu, constituant une puissance, d'une part, susceptible d'être émue par tout besoin senti, et de l'autre, capable de faire agir immédiatement, ne me paroît avoir été reconnu par personne avant moi*" (*Dictionnaire*, "Instinct," 332).

This short paragraph encapsulates what I consider one of Lamarck's most original ways of applying evolutionary thinking to contemporary biological, philosophical, and cultural issues. In it, Lamarck produced the first version of an evolutionary 'self' and a new conception of 'internality.'

The inner feeling was, on the one hand, a feeling, i.e., it fell under the explanatory mechanisms of feelings and was emergent just as other feeling within Lamarck's theory were emergent. Yet, on the other hand, it differed in remarkable ways from other feelings, and this tended to put into question it being 'a feeling.' A plethora of programs and disciplinary frameworks were developed in the eighteenth century to deal with the enduring self-identity of individuals and with the explanatory mechanisms for perception and intellection. Some historians view the period as that of the emergence of the 'self' both in high and low culture and as a scientific object (Goldstein, 2002, 2005; Wahrman, 2004; also Staum, 1996; Stedman Jones, 2002). The 'self' was problematized in both the 'social' and the 'natural' discourses of the time. When Lamarck called this 'self' a feeling, he was using an option already present in the repertoire of models of enduring cohesive identity. In common with the physician-vitalist, the more severe monist materialism, such as d'Holbach's, and the mixed options delineated by some members of the *Idéologues* group and by Diderot, Lamarck shunned introspection as a privileged venue to the 'self' and attempted to construct physical–physiological explanatory mechanisms, preferably non-reductive ones. Once again, his innovativeness lay in the evolutionary grid applied to this notion.

Lamarck used both terms—*sentiment intérieur* and *sentiment d'existence*, discussed in detail in *Phil.Z.* From 1800 to 1806, this *sentiment*, in the form of *sentiment d'existence*, is mentioned only once, as a component of the events that 'happen' in that part of the nervous system related to feeling as in *Recherches sur l'organisations des corps vivants* ([1802] 1986, 108). The *sentiment intérieur* was much more frequent in later years, and when using the *sentiment d'existence* it was often accompanied by further adjectives such as, 'the intimate sense of,' 'the obscure sense of,' 'the obscure notion of,' and from 1817 onward also '*conscience de*,' accompanied by the distinction between conscious and non-conscious animals, and between humans and animals in the *Système*.

Sentiment intérieur was defined as a feeling 'aroused by needs' and was seen as a causal agent of both movements and other actions related to needs. The assumption of special organs for specialized functions meant that there was a particular space within Lamarck's construction of the nervous system, where the processes of a need were felt through 'emotion' as well as the processes resulting in the formation of an action and/or reaction—since 'emotion' had to do with having a power to affect movement. Being activated by a need meant, within the Lamarckian explanation, that it played an important role in the formation of habits, which were passed on trans-generationally and thus were acquired traits. This 'feeling' of one's existence was there continually, pervading the whole organism as it gathered feelings produced by the activities of living and was thus based on the interconnectedness of the nervous system. This interconnectedness served to bind together processes, replacing what earlier in the eighteenth century had been considered two substances, a bodily substance and some non-bodily substance ruling and activating it. For Lamarck, the organism functioned reactively and efficaciously. It meant that it was a stable system that was constantly referring to the ever-present needs of the living organism arising in its interaction with the environment. This system was both self-referring and a referral system, a cohesive whole in relation to its inner working and in its distributed functioning, in relation to needs, particularly survival and adaptational needs. The assumption of an emergent, gradual *sentiment intérieur* was intended to resolve the empiricist cul-de-sac of identity-and-continuity of the self on many levels—which J. Goldstein (2005) called 'horizontal fragmentation of the self.'

Lamarck looked upon the nervous system as having distinct and separate parts for sensing and for movement. The uniqueness of the *sentiment intérieur* was, for him, its ability to be both a transmitter, as indicated above, and an initiator of movement, a necessary condition for higher functioning, such as a variety of modes of thought of which volitional acts were a derivation. Viewing the *sentiment intérieur* through the evolutionary grid meant that it was emergent, absent in very low invertebrates, and appearing in crude form in invertebrates from insects up. The source and cause of the movement of lower organisms were external pressures. They had tropism-like movements, and the reason for it, mentioned in the *Philosophie zoologique* (elaborated upon in much more detail in *Système*) was the inability to have feelings and thus to enjoy the concerted bodily movements related to *sentiment intérieur*: "Now, the very imperfect animals under consideration, since they do not in any way possess the *sentiment intérieur* in question, would not know having or have means of eliciting in them the instigating cause of their movements. Obviously, it comes to them from the outside, and certainly it is not at their disposal; also, none of their needs necessitate it being there; that is what I have already shown. All that they need is found within their reach; they are animals only by virtue of being irritable" (1820, 251).[7] Since it was 'a feeling,' the inner feeling could not exist without a nervous system and the fluids and transporting mechanisms that support it, and both its degree of sophistication and efficacy in bodily movement (i.e., constraint, compulsion, control, and choice) depended on the complexity of this supportive system. It meant that vertebrates in general, mammals and humans in particular, which had a highly complex supportive system, also had a more complex inner feeling. Given that for Lamarck there was an evolutionary fit between possible needs, the specific faculties that would satisfy them, and their supporting organs, the absence of *sentiment intérieur* was indicative of the narrow range of the organism's needs, and vice versa, and of the place of its series (a term often used, and much less loaded for him than species).

---

[7] "*Or, les animaux très-imparfaits dont il s'agit, ne possédant nullement le sentiment interieur en question, ne sauraient avoir ou faire naître en eux la cause excitatrice de leurs mouvemens. Elle leur vient donc évidemment du dehors, et dès lors elle n'est assurément pas à leur disposition; aussi aucun de leurs besoins n'exige qu'elle le soit; ce que j'ai déjà fait voir. Tout ce qu'il leur faut se trouve à leur portée; ce ne sont des animaux que parce qu'ils sont irritables*" (*Système* 251).

Lamarck discussed some of the enabling conditions for the higher mental functions in the *Philosophie zoologique*, the *Histoire*, the dictionary items, the *Système analytique* based on them, and two articles from the second decade of the nineteenth century (e.g., *Discours d'ouverture* de, 1816, *Apperçu analytique des connaissances humaines*). One enabling condition was '*attention*,' posited as a narrow and specialized function closely and directly related to the *sentiment interiéur.* 'Attention' acquired, in addition to its more generalized use during that period, also a specialized signification related to the practice of naturalists during the eighteenth century, namely, voluntary focusing and rapt concentration and absorption. It had a corresponding physiological mechanism to enable 'the soul' to do that (Daston, 2004, 117; 2005, 240–41).

Lamarck's use of 'attention' was based on, but also differed from, its more accepted use, which was employed by the *Idéologues* (particularly Cabanis), in order to deploy it for a different purpose. Cabanis discussed the possibility of partial, plural, '*moi*' related to different nervous centers, but he declared that one could say very little about them because "all our sensations of the self (*moi*) relate/are reported exclusively to the general center and because our means of acquiring exact notions/ideas" depend on grasping the circumstances "in their linked sequence" (1805, 503).[8] Contrary to Cabanis, Lamarck assumed a unitary organic individual and denied the possibility of a scattered or divided '*moi*.' The rudimentary sense of existence did not entail thinking nor judging capacities ordinarily characterizing human singularity at that time. 'Attention' was seen as one of the necessary preconditions for enabling the physiological arrangements that catered for the formation of ideas out of impressions, and thereby also for memory. This is where Cabanis' and Lamarck's positions were similar: Cabanis only dealt with humans and wrote profusely on 'pain' and its important role in medicine. Thus, 'attention' was posited as a necessary condition not only for the reception and transmission of an impression but also for the production of pain. These matters are discussed in the first three memoirs of Cabanis and in the tenth as well in *Rapports* (Cheung, 2006; Staum, 1980). One could view *attention* as a mechanism for fixing the object of sensation, prolonging its duration in the relevant receptor parts of the nervous system, thereby allowing a pause for focusing. That activity of making the pause possible could be seen as a time component in the spatial-physiological description of thinking of the formation of ideas. It allowed for gathering-and-enhancing activity in order to enable a move from one order of mental functioning to the next (higher, i.e., more complex one). This could only occur in specific regions within the nervous system that were present only in the 'highest' among the four groups of organisms endowed with *sentiment intérieur.*

Already in the *Philosophie zoologique*, but much more so in his *Histoire* and in his *Système*, as well as in the articles written for the Deterville *Dictionnaire*, Lamarck tried to stretch the epistemological and cultural ('moral') work that the

[8] "*puisque toutes nos sensations de moi rapportent exclusivement au centre général et que nos moyens d'acquérir des notions exactes*" depend on grasping the circumstances "*dans leur enchaînement*" (Cabanis & Georges, 1805, 503).

sentiment interieur could do within his system. He turned it into a full-blown social-cultural self, including morals, which pointed to a wide and diverse range of needs characteristic of humans that were amplified by civilization. In addition, Lamarck also tried to relate it to memory, imagination, and intelligence and saw its mode of functioning as initiating and bringing about human tendencies, first and foremost, the tendency to preserve oneself. In fact, in higher mammals and humans, a wide range of social behaviors could be derived from that tendency. Thus, as far as humans were concerned, Lamarck posited a feeling of an active 'self' within the sensualist framework, which, being a necessary result of organic structures, was in that sense 'innate.' However, this feeling of an active 'self' was meaningless within a human culture without 'experiencing' within a specific environment—anachronistically put, a Piaget-like structure, as, for example: "No doubt, man is born without ideas, without enlightenment, in possession of nothing but a *sentiment intérieur* and general proclivities which tend to exert themselves mechanically. It is only with time and through education, experience, and the circumstances which he encounters that he acquires ideas and knowledge" (1815–1822, 279).[9]

1.3.3 Concluding Remarks

In conceiving the *sentiment intérieur*, Lamarck was responding to questions at the focus of eighteenth-century debates (particularly late eighteenth-century debates) and using eighteenth-century means, i.e., materialization of the mental. The *Idéologues* as well as Lamarck certainly worked under the constraint of the evolving conception of science as produced in the First Class of the Institut. Without the theological self, the social understanding of man's place in nature also became a matter of scientific knowledge. Yet, Lamarck's evolutionary perspective, which assumed both growing complexification and openness to the environment, compelled him to look for a unifying cohering foundation for the individualized entities that populated his theory. This foundation had to cater to the demands of 'science' as well as those of 'experience,' and thus he produced the first version of an evolutionary 'self' and a new conception of 'internality.' To those who are familiar with it, Damasio's table of self, which delineates an evolutionary path from proto-self to a core-self (with consciousness), can help in grasping what Lamarck was attempting (Damasio, 1999, 199; also, Damasio & Damasio, 2006; Edelman, 2006).

Lamarck's model was a systematic and thoroughgoing application of an evolutionary perspective on the living world that aimed to encompass all the known functioning of the human mind and psyche, situating humanity as part of a natural

[9] "*Sans doute, l'homme naît sans idées, sans lumières, ne possédant alors qu'un sentiment intérieur et des penchans généraux qui tendent machinalement à s'exercer. Ce n'est qu'avec le temps et par l'éducation, l'expérience, et les circonstances dans lesquelles il se rencontre, qu'il acquiert des idées et des connaissances*" (*Histoire*, 279).

continuity. The basic analytical units, the mechanisms, the generalizations, and the laws, would all have to apply to living organisms at large in order to apply to humans. It 'naturalized' human behavior by putting it on a par with the behavior of any other living organism. The patterns and mechanisms that were applied to living nature were applied to human perception and consciousness as well as to unconscious, purposeful, planned, and spontaneous human activities; to thoughts, feelings, desires, and in fact to both individual and social behavior. With that Lamarck's theorizing laid possible foundations for a novel, generative framework.

1.4 Notes on Lamarck's Historiography

1.4.1 Periodization

I present a schematized periodization of the historiography and then comment briefly on how to analyze the later twentieth-century historiography. I start with Lamarck's historiography in the West in the last third of the nineteenth century because this is the period when Lamarck's writings were gradually reprinted. The principal assumptions of neo-Lamarckists (though they got that name much later) became a topic of discussion, both in the public debate within the field of life sciences and in its popularizations. Moreover, these assumptions were applied, interpreted, and elaborated upon by authors outside the life sciences, mainly in the emergent fields of sociology, anthropology, and psychology and in the general cultural and political field. In France, where Lamarck was becoming something of a national figure, with celebrations, the erection of a statute (1909) etc., but also in the USA and to some extent in Germany, the neo-Lamarckian following produced a number of biographical studies of his life and work, and a newly reawakened interest in putting together his various writings and other documentary materials related to him.

The meanings of 'evolution,' the relationship between its range of significations and usages, and that of 'progress' became contested, as Conry (1981) argues. As mechanisms of heredity became one of the principal foci of investigation in biology, analogies, metaphors, and modes of discourse derived from them played increasingly complex roles in the political and cultural discourses of that period. They were used to justify or to undermine the existing order, its hierarchies, the prevalent economic and social preferences, and cooperation or conflict among classes, races, and states.

A stream of writing on Lamarck, collections of articles related to his oeuvre, and edited selections from his more well-known texts started to appear sometime in the late 1870s and continued well until the beginning of the 1900s. Though one can find books published a bit later, which were written in a similar vein, these were authors whose upbringing and ideological roots were grounded in the above period.

Thus, I offer a rough periodization of the Lamarck historiography: 1. *Last third of the nineteenth century until the outbreak of World War I.* The most outstanding biographers of that period would be Alpheus Packard (1901), Marcel Landrieu (1909), and E. T. Hamy, a collector of biographical documents (1907). These biographies can be seen as a crystallization of a continuous stream of writing on the issue of Darwinian and Lamarckian conceptions of evolution and their biological and cultural significance, starting in the late 1860s and going on until the turn of the century outside France, and somewhat later in France. The following writings are relevant: those of scientists such as Haeckel (1868, with repeated and then enlarged editions until 1911, his lectures in 1882 and in 1909), Lang (1889), Hutton (1899), Perrier (1884, 1888, 1909), Le Dantec (1899, 1903), and of those philosophers, who tried to locate Lamarck among other 'transformism,' 'descent,' or 'evolution' thinkers—such as the essay by C. Royer (1870), Haeckel (in the 1868 edition), S. Butler (1879), and A. Wagner (1908). It was during this period that an investigation of Lamarck's '*herbier*' was carried out (1902), a number of selections from his writing appeared, and an exhibition of his work and an accompanying catalog was organized in Geneve (1910).

2. *During the period from World War 1 until the 1950s*, there were fewer studies: a major watershed in that period was the publication of H. Daudin's dissertation in 1926, which, rather than treating the former well-trodden topic of Lamarck versus Darwin, attempted a philosophical and methodological analysis of naturalists' thought from Linnaeus to Lamarck's death. In many ways, Daudin's study set the future stage for contextualized writing, raising the conceptual issues of classification, class, species, and series within the context of the naturalist thought of the period and within the Muséum. In another sense, his choice of topic is a continuation of the debate before World War I, as is the biography by Perrier (which was not published until 1925). Some writings that appeared during this period used the occasion of the centenary of Lamarck's death (around 1930), and most, but not all, dealt with the non-biological, i.e., ideological and moral aspects of his theories.

3. *The period soon after World War II, starting in the late 1950s.* Work was done from a historical perspective of French science, within, before, and during the Revolution, especially by Gillespie (1956, 1959), Goodfield (1960), or of early modern science in general such as Hooykaas (1959), as well as work related to the contemporary developments in evolutionary theories. A number of particular issues were addressed: a cluster of questions concerning the development, acceptance, and diffusion of Darwin's theories and the cultural and conceptual preconditions for these (J Greene, 1959; Glass et al., 1959). The narrative presented by E. Mayr's essay in 1972 was significant in promoting the partial rehabilitation given to Lamarck in the English-speaking world somewhat later.

From the early 1960s onward, there was an interesting relationship between the growing body of research on Darwin and the much smaller body of Lamarck scholarship, both in French (in which there was a certain continuity in writing since the nineteenth century, though in fluctuating quantity) and in English, and later in Italian too, which was more focused on the natural-history context than solely on individual biography. Moreover, the period from the late 1950s to the mid-1970s was also notable for the recovery/discovery and publication of new materials of Lamarck (e.g., the Lamarck conference of 1971). One consequence of this was the gradual construction of a continuous narrative of 'evolution,' rather than that of precursors and anticipators of that notion. Questions such as how contemporaries' understanding of varieties of 'evolution' developed and changed throughout the nineteenth century, how they interpreted 'nature,' its place in culture, and man's place in nature (within whatever evolutionary framework they had assumed) became the implicit framework underlining even those works that were seemingly devoted to a particular person's work and biography. Consequently, studies began to be geared toward these questions, rather than toward one particular figure within that constructed history. There emerged a host of new questions on the varied readings of the main corpus of evolutionary thought, on their interrelations, their impact on non-biological thought throughout the nineteenth century (e.g., R.M. Young, 1970, 1973; Bowler, 1983, 1988).

Lamarck was incorporated into that narrative, rather than posited as an odd lone figure, a narrative which was later pushed back into the eighteenth century. What was meant by 'context' underwent significant modifications, which brought in the inclusion of interactions among practices, power and learning institutions, as well as interactions among the various fields that were then considered as sciences. Though in one sense, the whole period starting from the 1960s can be subsumed under one heading, from another perspective the period beginning at the end of the 1970s/early 1980s marks a somewhat different historical approach with novel questions, methods, and materials being brought in.

In the nineteenth and early twentieth centuries, some of the leading questions were: Was Lamarck the first evolutionist? Was he the only one for a long while? How could one proceed to describe his intellectual development as well as his biography?—with slightly more emphasis on precursors and contemporaries toward the end of that period. Another leading question was how to evaluate the significance of his theory, the implications drawn from it—biological and non-biological—in comparison with that of Darwin (as then understood). The extra-historical interest of the various authors played an important, albeit implicit, role.

During the period between the world wars, the emphasis markedly shifted to a search for causal explanations for the change in Lamarck's view of nature. There was an attempt to understand the intricate mechanisms underlying this change in order to describe, analyze, and explain the seeming incongruence between his earlier career as a botanist and his later one as a zoologist. Various explications and explanations have been introduced since then, which can be classified, to some extent at least, according to the origin of the sources utilized: personality and personal history, scientific institutions and contemporary power-structures in them, schools in contemporary natural history, and history of other fields of the sciences of the period.

1.4.2 Lamarck's Biologie and Recent Historiography

I suggest looking into the principal interpretations of Lamarck that have been proffered in an attempt to crystallize commonalities and delineate disagreements on the object, the method, and the nature of Lamarck's study of living entities, within a field which was at times called the science of '*biologie*' (Gissis, 2017).

Almost all present-day writings on Lamarck's endeavor regarding *biologie* agree that understanding Lamarck's deployment of *his* physics and *his* chemistry is crucial. They also agree that his 'biology'-views were closely intertwined with his views on the *geological history* and *structure* of the 'crust of the earth,' thus employing the differential time-scales involved with each of the scientific fields (Gohau, 1997a, b, 2006; Laurent, 1987; Rudwick, 2005, 2009). Some commentators emphasize the rejection of the Cartesian mechanistic model on the one hand and on the other the adoption of the (specifically late eighteenth-century) Newtonian model, in a manner implicitly or explicitly described as analogous (Conry, 1981; Corsi, 1982; Roger, 1979a, b; Wolfe, 2014).

Assumed, implied, or explicitly stated is the following: in order to understand Lamarck's thought, his innovative stance, as well as the unresolved inner tensions within his theoretical framework, one has to relate his writings to the overall gradually evolving conceptual scheme he devised, as well as to the *changing* contemporaneous *natural-history problématique*. It includes questions concerning the conditions for the existence of living entities; the attributes of the material that constituted living entities; whether there were common functions and features of all living entities; and the properties of the environments of living entities that made their existence and reproduction possible. These, it is argued, will help clarify the presuppositions and theoretical constraints with which he struggled. One also has to relate Lamarck's framework to the prevalent contemporaneous conceptualizations of physics and chemistry. It is argued that the Newtonian tradition in 'natural philosophy,' i.e., in the physical sciences, served a double role, namely as a basic conceptual scheme and as a paradigmatic model of how to do science and what a scientific field should look like (Biener & Schliesser, 2014). And that one of the features of the Newtonian tradition is relevant, namely, that the scientific search was for laws of nature that had universal extension and universal applicability and were to account most generally for the totality of phenomena as regulated under them. These dynamical laws were assumed to hold at all levels. Yet, the 'individuals/entities' dealt with in Lamarck's *biologie*, contrary to those in the Newtonian model, had history, carried history within their different organizations, were transforming, and were nature-produced (and not god-produced), and their interactions with their specific environments were neither determined nor unchanging.

There is also common agreement regarding the importance of understanding Lamarck's use of fluids—subtle, imponderable fluids. Within his conception (and more generally, within those investigating and making use of them during the second half of the eighteenth century), these fluids, whether 'heat,' 'caloric,' or 'electricity,' were assumed to be able to expand, transfer physical material attributes.

Thus, they were assumed to act as causes and be a locus of causality, although these properties could not be translated quantitatively into properties of microscopic, foundational, entities. In the general eighteenth-century debate on the assignment of bodily structures and of 'agents' and mechanisms of living entities, there was a broad choice of models. In line with his geological and physical views, Lamarck chose a hydraulic model for the dynamics of various active fluids as transferring agents. All interpreters agree that it is within specific parts of the structures (which carried fluids) of living entities that the transformation into signals of inner and outer cues was being carried out. However, as is sharply brought out by Corsi (1988, chap. 5, and 2006), within a few years of these publications, Lamarck's emphasis had moved from elaborating the generalized dynamics of atmospheric and terrestrial fluids and, by analogy, those of living entities to more specific zoological issues.

Lamarck's chemistry, in particular, is seen as playing a predominant role in his conceptualization of both geology and meteorology. His conceptions of the nature of 'the crust of the earth' and of '(external) environment' (including what he formerly called 'climate') are looked upon as crucial to understanding his view on the possibility of delineating a field of living entities—*biologie*. There had been almost unanimous agreement in the scholarship on Lamarck's rejection of the 'new sciences' and in particular the concept of chemical analysis, an agreement shared by those who offered close readings, analyses, and interpretations of both his early and his later works, starting with Schiller and Hodge to the latest articles by Corsi. It is worth noting Schiller's emphasis on the emergence of 'a new science' in his 1978 book.

A different interpretation has been proffered by Angela Badinelli (2013), however. Badinelli offers an analysis of Lavoisier on the basis of which she reconstructs Lamarck's conceptualizations as moving gradually from his early position, combined with a conception of an essential molecule, toward his transformism, which also entailed his changing view of the relations between the inorganic and organic. Up to this point of Badinelli's argument concerning Lamarck's differing views on chemistry, it is very likely that other interpreters would largely concur. Yet, the crucial change in terms of Lamarck's attitude to the new chemistry occurred, according to Badinelli, mostly between 1802 and 1809 and was made evident in the *Philosophie zoologique* and further detailed in 1815, in the first volume of his *Histoire naturelle*. She argues that Lavoisier offered a novel perspective on living entities, a unifying and uniforming conception of both the organic and the inorganic, which led to a view of living entities as combustible bodies, and resulted in the discarding of the view of living systems as machines. Badinelli is clearly challenging Holmes' analysis on these issues (see esp. Holmes 1985 on Lavoisier and the Chemistry of Life). In his earlier phase, Lamarck differentiated between the principle of 'power of life' and 'matter.' In contrast, Barsanti (2000, 126–27), who argues for a holistic view of Lamarck's endeavor of *biologie*, already has a few statements indicating the direction of the importance of the 'new chemistry.' In Badinelli's interpretation, Lamarck's view of *biologie* evolved into a physical-chemical and even physiological science that was not dependent on mechanics and *accepted and even followed* Lavoisier's view that living entities (as well as non-living ones) were 'natural chemical systems' in a process of constant material transformation with their external environment. I believe that what is implied by this interpretation is that it was through the learning of Lavoisier's new viewpoint that Lamarck adopted the idea that compounding was capable of producing novelty.

The spectrum of positions on the meaning and significance of both '*milieux*' and '*milieux environnans*' is not very wide since there is general agreement on the crucial role of these 'environments' in impacting the course of action of the powers that are involved in shaping and molding living entities within a transformist framework. Also agreed upon is the notion of '*milieux*' as closely interwoven with that of 'change.' While elementary living entities are passively and directly exposed to the environment, the increasing complexity of living entities can be regarded as an index of the degree of their mediated and active interactions with their *milieux*, through 'needs' and 'habits.' Thus, *milieux*, together with the 'power of life,' are perceived as catering for, constructing, and nurturing variations. Most historians agree that the branching of the rising orders of living entities is related to the impact of these various interactions, and that their dynamics are underlain by the action of fluids inside and outside the living entities.

A note on 'organization' is called for here. There is general agreement on the centrality of the concept and its accompanying concept, 'function.' There is, however, disagreement on its deployment. The controversy has to do with respective positions on 'the physical' and 'the chemical' in Lamarck's oeuvre and their relationship with the notion of organization. The positions expressed cover a wide range. At one end, 'organization' and 'self-organization' are seen as rich notions that allowed a holistic mode of explanation that was alternative to the then prevalent mechanistic one, notions through which living entities were defined (on self-organization, Conry, 1981, 1994; Jordanova, 1989). At the other end, there is an array of interpretations that explicate 'organization' primarily as a central concept in Lamarck's theory of transformism, a concept that turned the naturalist's gaze from the external characteristics of living entities to their internal organization, as argued by Burkhardt (1995). Alternatively, it is seen as being used to draw hierarchical-organizational distinctions among living entities, while at the same time unifying all living entities within a single theoretical frame. 'The organic' is conceived as organizational, but its cause was physical, as exemplified by the motion of fluids, with the intricacies of the relations between organism and organization brought out by Cheung (2006) and Wolfe (2019).

All interpreters note and comment on the discrepancy between Lamarck's contemporaneous renown as a classifier and innovative taxonomist and what was regarded for a long time in historiography as his marginalization as an innovative naturalist. In the past three decades, however, a great deal of new work has appeared, which has re-instated Lamarck as a member belonging to multiple communities, inside as well as outside the more researched institutions such as the MuséumNHN and the Institut. The spreading of his ideas to many corners of Europe, their complex assimilation and deployment through the influence of the courses he taught at the Muséum has been reconstructed, and the various interlocutors of his '*dialogues manqués*' have been resurrected, especially by Corsi, but also some of the others.

Furthermore, recent historiographic work on Cuvier has complexified and brought into focus the controversies about the proper sites and methods for practicing natural history, considering not only the personal elements but also situating Lamarck within the political and cultural contexts of the Napoleonic and Restoration periods, briefly discussed above (Appel, 1987; Laurent, 1987; Outram, 1984; Rudwick, 2005, 2008; also Corsi, 2018).

The role of taxonomic work in Lamarck's *biologie* is discussed by most members of the group of contextualizing historians discussed above. They have pointed out that there are two ways to understand the term '*classification*,' one limited and the other more comprehensive and general. In its more limited version, it means sorting and dividing, i.e., classifying for practical purposes, in field botany; a classification that was looked upon as 'artificial,' rather than reflecting 'the natural order' within Lamarck's own discussions of method, e.g., in the discussions of Burlingame (1973) and Schiller (1969, 1971), both relating to Daudin ([1926] 1983); and Farber (1981) and Sloan (1995) on the division into 'natural' and 'artificial.' The second, broader way of understanding 'classification' is as a comprehensive *taxonomic enterprise*, which concentrated on how to move from the 'points of animaculation' that were constantly being produced through spontaneous generation, to living entities, first the simplest and lowest in the evolutionary sequence and then moving upward to those appearing later, which were more complex. In this version of taxonomic enterprise, Lamarck's *biologie* posited the contour of an 'ideal' as well as a delineation of a movement of complexification of living entities, a movement seen as 'reproductive-factual,' continuous, and, in principle, gradual. Thus, *biologie* proffered two taxonomic pictures, both linear and branching, and both were considered as rendering 'the order of nature,' with the issue of 'branching' dealt by Gould (1999a, b). Moreover, as Corsi has emphasized, this represented a mode of constituting diversity within the plurality of living entities.

The issues at the focus of present-day disagreements have to do with the question of whether one can actually view Lamarck's two activities of *classification* and *biologie* as separate or as intimately intertwined. If these two activities were indeed intertwined, then Lamarck's arguments on taxonomy were either posited as if they substantiated the theoretical claims at one stage of his work, or else these arguments exemplified such claims at a later stage of his work. Answering this question is complicated by the fact that Lamarck did not experiment, nor reconstruct experiments, nor dwelt at great length on accumulations of empirical-data comparisons.

I repeat the following comment: in one of the rare semi-autobiographical paragraphs in an opening course lecture, Lamarck stated (1817) that only gradually did he realize what a rare glimpse into the working of nature the investigation of invertebrates had offered him. He looked upon the distinction between invertebrates and vertebrates as one of his important original contributions, changing both the boundaries of, and the perspective on the realm of animals he was investigating, leading him to posit the invertebrates as 'model organisms.' As I argued earlier, in 1806 Lamarck introduced the evolutionary perspective as *foundational for classification* by defining living entities on the basis of the most general feature of nervous activity found *in the lowest forms of organization*. One could argue that Lamarck endeavored to turn *Histoire naturelle des animaux sans vertèbres* into an exposition of 'evolutionary systematics' (with a heavy emphasis on 'evolutionary') because he probably realized that the book would be of interest to a wide audience. The principal thrust of the general part was presented in the form of 'first principles' in which Lamarck tried to tie together method, model, and a rhetoric adequate to the task, the emphasis being on the fact that this framework worked adequately all along the evolutionary path (1815–1822, 1:23–4). It worked that way between organisms and their subsystems and faculties and their 'needs.' This had an underlining pointing at Cuvier's conception of 'nature,' intending to undermine Cuvier's definition for 'animal.' Lamarck's own exposition of the characteristics of a 'living body' harked back to the *Biologie* project. Yet, in the supplement to the *Histoire*, Lamarck admitted that the actual order, or 'the order of productions,' as he named it, was differently constructed than the 'ideal' one expounded in the book.

Lastly, most of the recent authors trace the appearance of Lamarck's project on *biologie* and date it to either 1800 or early 1802, as already suggested by Hodge (1971a, b), contrary to its much later dating by the first editor and publisher—Grassé (1944). Thus, there is general agreement on the *chronology* of the appearance of the term and the project related to it, i.e., *Biologie* (1800), '*biologie*' declared as 'terrestrial physics' in the final part of *Hydrogéologie* (1802a). There is also a consensus on Lamarck's later decision to discontinue the project, claiming reasons such as age, failing health, pre-occupation with other projects in *Recherches sur l'organisation des corps vivans* (1802b), and the repetition of that decision in the *Philosophie zoologique* (1809). Then, there was his very partial, somewhat declarative, return to the project, as indicated by his usage of *biologie* and its derivatives, such as '*biologiste*' in the first volume of *Histoire naturelle des animaux sans vertèbres* (1815). In further volumes of the oeuvre, he withdrew it, and there is only a limited number of explanations as to why this was so, these being either mostly internalist or in terms of his relations with the naturalist community. Corsi's contextualization of the *Biologie* project and in particular of Lamarck's rhetorical strategies and methodological choices throughout the period of 1800–1822 has added another dimension to our understanding. Corsi (2012) connects Lamarck's use of the term and the description of his project with the fast-changing political regimes and the fortunes of 'materialist' positions, of 'God,' and 'the Catholic Church' in the politics and culture of Napoleonic France and the early Restoration period. Corsi also points out that Lamarck used the same pattern when he explained (away) putting an end to his meteorology project in 1810.

1.5 Concluding Comment

As I have argued in this chapter, Lamarck lived in a period of transition, of shifting boundaries and fast-changing conceptual schemes. Two of the principal scientific institutions to which he belonged were initially conceived as sites for the interaction and mingling of scientific discourses and to some extent continued to be such sites while undergoing reorganization. To understand Lamarck, the picture needs to be refined by distinguishing developments in the first revolutionary decade and those in later periods, distinguishing between parallel developments, such as the mingling of discourses and the professionalization of the sciences, which took place side by side for a while, and by analyzing the political-economic-cultural context that made these possible. In some ways, Lamarck was a product of an earlier period, yet he offered an innovative conceptual scheme and deftly mixed different time-scales within that scheme. His ideas began to be more clearly formulated in his writings around the turn of the eighteenth century, although he used concepts, mechanisms, and methods that he laid down earlier, and that were derived from earlier modes of thinking. When he formulated his theory on a grand scale that encompassed all of living nature, the scientific field in particular and the cultural field at large were beginning to undergo a transformation that eventually hardened the disciplinary boundaries.

His audiences were to be found both in scientific and non-scientific communities, mostly outside the two institutions to which he belonged during that period—the Institut and the MuséumNHN. Combined with his idiosyncratic style of writing, this meant that there were many ways of reading and interpreting him during his lifetime. These were enabling conditions for the later dissemination of *severed* components of his enterprise and their incorporation into theories that were at times rather incompatible with his own, but it also ensured the many long-lasting transformations of some of his principal conceptions.

Lamarck's innovative endeavor of establishing transformism, a science of transforming living entities, received various explanations and interpretations that mostly centered around the natural-history and physical-science communities. I have argued that it is important to add the social discourses and practices of that time to this picture and tried to explicate some facets of his view on neuro-physiological psychology. Finally, I have offered a short periodization of the *longer-term historiography* and briefly indicated the principal foci of historical-conceptual agreements and controversies of *recent historiography* centered on various facets of Lamarck's endeavor and its possible sources, suggesting the possibility of viewing his taxonomic and his evolutionary enterprises as an integrated framework—his *biologie.*

Materials from the following articles have been used by permission of:

The MIT Press
2009 –" Interactions between social and biological thinking" – the case of Lamarck. Perspectives on Science 17/3: 237–306.

Springer Books: License Num. 5632340750689
2010 – Lamarck on Feelings: From Worms to Humans. In The Body as Object and Instrument of Knowledge, edited by C.T. Wolfe and O. Gal. pp. 211–39.

Routlegde /Taylor &Francis
2017 – 'Biologie' - Lamarck's endeavor of a science of living entities. In Philosophy of Biology before Biology, edited by Charles Wolfe and Cecilia Bognon-Küss. pp 170–187. London: Routledge.

References

Agulhon, M. (1979). *Marianne au combat: l'imagerie et la symbolique républicaine de 1789 à 1880*. Flammarion.

Appel, T. A. (1987). *The Cuvier-Geoffroy debate: French biology in the decades before Darwin*. Oxford University Press.

Baker, K. M. (1973). Politics and social science in eighteenth-century France: la Société de 1789. In F. Bosher (Ed.), *French government and society 1500–1850: Essays in memory of Alfred Cobban* (pp. 208–230). Athlone Press.

Baker, K. M. (1990). *Inventing the French revolution: Essays on French political culture in the eighteenth century*. Cambridge University Press.

Bandinelli, A. (2013). Le Origini Chimiche della Vita:legami tra la rivoluione de Lavoisier e la biologia di Lamarck. *Biblioteca di Nuncius, Studi e Testi, 71*, 133–200.

Barsanti, G. (1995). La naissance de la biologie: observations, theories, metaphysiques en France 1740-1810. In C. Blanckaert, J. L. Fischer, & R. Rey (Eds.), *Nature, histoire, Société: Essays en homage a` J. Roger* (pp. 197–227). Klinchsiech.

Barsanti, G. (1997). Lamarck et la naissance de la Biologie. In Goulven Laurant (Eds.), *Jean-Batiste Lamarck 1744–1829* (pp. 349–68). Editions du CTHS.

Barsanti, G. (2000). Lamarck: Taxonomy and theoretical biology. *Asclepio, 52*(2), 119–132.

Barthelmy-Madaule, M. (1982). *Lamarck, the mythical precursor: A study of the relations between science and ideology* (M. H. Shank, Trans.). MIT Press.

Ben David, J. ([1960] 1991). Roles and innovations in medicine. In Gad Freudenthal (Ed.), *Scientific growth by J. Ben David* (pp. 33–48). University of California Press.

Biener, Z., & Schliesser, E. (Eds.). (2014). *Newton and empiricism*. Oxford University Press.

Blanckaert, C. (2002). Le moment 'naturaliste' des sciences des homme. *Revue d'histoire des Sciences Humaines, 3*(2), 117–160.

Bowler, P. J. (1983). *The eclipse of Darwinism, anti-Darwinian evolution theories in the decades around 1900*. Johns Hopkins Press.

Bowler, P. J. (1988). *The non-Darwinian revolution: Reinterpreting a historical myth*. Johns Hopkins University Press.

Bret, P. (1991). L'emergence d'une histoire des sciences du moment revolutionnaire. In M. Vovelle (Ed.), *Recherches sur la Révolution* (pp. 169–176). La Decouverte.

Bret, P. (2002). *L'état, l'armée, la science: l'invention de la recherche publique en France (1760–1830)*. Presses Universitaires de Rennes.

Broman, T. (1998). The Habermasian public sphere and science in the enlightenment. *History of Science, 36*, 123–149.

Burkhardt, R. (1995). *The Spirit of system: Lamarck and evolutionary biology*. Harvard University Press.

Burlingame, L. J. (1973). Lamarck. In C. C. Gillispie (Ed.), *Dictionary of scientific biography* (Vol. 7, pp. 584–594). Scribner.

Butler, S. (1879). *Evolution, old and new; or, The theories of Buffon, Dr. Erasmus Darwin, and Lamarck, as compared with that of Charles Darwin*. Fifield.

Cabanis, P. J., & Georges. (1805). *Rapports du physique et du moral de l'homme* (2nd ed.). Crapart, Caille et Ravier.

Carrithers, D. (1995). The enlightenment science of society. In C. Fox, R. Porter, & R. Wokler (Eds.), *Inventing human science: Eighteenth century domains* (pp. 232–270). University of California Press.

Censer, J. R., & Hunt, L. A. (2001). *Liberty, equality, fraternity: Exploring the French revolution*. Pennsylvania State University.

Cheung, T. (2006). From the organism of a body to the body of an organism: Occurrence and meaning of the word 'organism' from the seventeenth to the nineteenth centuries. *The British Journal for the History of Science, 39*(3), 319–339.

Conry, Y. (1981). Une lecture Newtonienne de Lamarck, est elle possible? In J. Schiller (Ed.), *Lamarck et son temps, Lamarck et notre temps* (pp. 37–64). Vrin.
Conry, Y. (1994). Lamarck, penseur de frontiere. *Nuncius, 9*(2), 559–592.
Corsi, P. (1978). The importance of French Transformist ideas for the second volume of Lyell's principles of geology. *The British Journal for the History of Science, 11*(3), 221–244.
Corsi, P. (1982). Models and analogies for the reform of natural history. Features of the French debate 1790-1800. In W. Bernardi & A. La Vergata (Eds.), *Lazzaro Spallanzani e la biologia del Settecento: Teorie, esperimenti, istituzioni scientifiche: Atti del convegno, marzo 1981* (pp. 381–396). L.S. Olschki.
Corsi, P. (1988). *The age of Lamarck: Evolutionary theories in France 1790–1830* (J. Mandelbaum, Trans.). University of California Press.
Corsi, P. (1997a). Le Muséum et l'Europe. In C. Blanckaert, C. Cohen, P. Corsi, & J. L. Fischer (Eds.), *Le Muséum au premier sie'cle de son histoire* (pp. 635–647). Editions du Muséum d'histoire naturelle.
Corsi, P. (1997b). Les élèves de Lamarck: un projet de recherche. In G. Laurent (Ed.), *Jean-Baptiste Lamarck 1744–1829* (pp. 515–526). Editions de CTHS.
Corsi, P. (1998). A Devil's Chaplain's calling? *Journal of Victorian Culture, 3*(1), 129–137.
Corsi, P. (2005). Before Darwin: Transformist concepts in European natural history. *Journal of the History of Biology, 38*(1), 67–83.
Corsi, P. (2006). Biologie. In P. Corsi, J. Gayon, G. Gohau, & S. Tirard (Eds.), *Lamarck philosophe de la nature* (pp. 37–64). Presses Universitaires de France.
Corsi, P. (2012). Idola Tribus: Lamarck, politics and religion in the early nineteenth century. In S. Fasolo (Ed.), *The theory of evolution and its impact* (pp. 23–25). Springer.
Corsi, P. (2018). Systèmes de la nature and theories of life: Bridging the eighteenth and nineteenth centuries. *Republics of Letters, 6*, 1.
Cuvier, G. (1810). *Rapport historique sur les progrès des sciences naturelles depuis 1789, et sur leur état actuel....* l'Imprimerie impériale.
Cuvier, G. ([1815] 1989). *Rapport à l'Empereur sur le progrès des sciences, des lettres et des arts depuis 1789. 2. Chimie et sciences de la nature*. Edited by Yves Laissus. Belin.
Cuvier, G. (1860). *Éloges historiques: précédés de l'éloge de l'auteur par M. Flourens*. E. Ducrocq.
Damasio, A. R. (1999). *The feeling of what happens: Body and emotion in the making of consciousness*. Harcourt Brace.
Damasio, A., & Damasio, H. (2006). Minding the body. *Daedalus, 135*(3), 15–22.
Dantec, L., & Alexandre, F. (1903). *Les limites du connaissable: La vie et les phénomènes naturels*. Alcan.
Daston, L. (2000). The coming into being of scientific objects. In L. Daston (Ed.), *Biographies of scientific objects* (pp. 1–14). University of Chicago Press.
Daston, L. (2004). Attention and the values of nature in the enlightenment. In L. Daston & F. Vidal (Eds.), *The moral authority of nature* (pp. 100–126). Chicago University Press.
Daston, L. (2005). Fear and loathing of imagination in science. *Daedalus, 134*(4), 16–31.
Daston, L. (2007). Condorcet and the meaning of enlightenment". In *Proceedings of the British Academy, Volume 151, 2006 Lectures*.
Daston, L., & Sibum, O. (2003). Introduction: Scientific personae and their histories. *Science in Context, 16*(1–2), 1–8.
Daudin, H. (1926). *Cuvier et Lamarck. Les classes zoologiques et l'idée de série animale, De Linné à Lamarck. Méthodes de la classiacation et idée de série en botanique et en zoologie*. F. Alcan.
D'Hombres, N. (1989). *Les Savants en Révolution: 1789–1799*. Cité des sciences et de l'industrie.
Duchesneau, F. (1982). *La physiologie des lumie'res: Empirisme, mode'les et theories*. Martinus Nijhoff.
Edelman, G. M. (2006). The embodiment of mind. *Daedalus, 135*(3), 23–32.
Farber, P. L. (1981). Research traditions in 18th century natural history. In W. Bernardi & A. La Vergata (Eds.), *Lazzaro Spallanzani e la biologia del Settecento* (pp. 397–403). L.S. Olschki.

Furet, F., Ozouf, M., & with B. Baczko. (1988). *Dictionnaire critique de la révolution française*. Flammarion.
Giglioni, G. (2013). Jean-Baptiste Lamarck and the place of irritability in the history of life and death. In S. Normandin & C. T. Wolfe (Eds.), *Vitalism and the scientific image in post enlightenment life science, 1800–2010*. Springer.
Gillispie, C. C. (1956). The formation of Lamarck's evolutionary theory. *Archives internationals d'histoire des sciences, 9*, 323–338.
Gillespie, C. C. (1959). The Encyclopédie and the Jacobin Philosophy of Science: A Study in Ideas and Consequences. In M. Clagett (Ed.), *Critical problems in the history of science* (pp. 255–290). University of Wisconsin Press.
Gissis, S. B. (2009). Interactions between social and biological thinking: The case of Lamarck. *Perspectives on Science, 17*(39), 237–306.
Gissis, S. B. (2010). Lamarck on feelings: From Worms to humans. In C. T. Wolfe & O. Gal (Eds.), *The body as object and instrument of knowledge* (pp. 211–239). Springer.
Gissis, S. B. (2017). 'Biologie'- Lamarck's endeavor of a science of living entities. In C. Bognon-Küss & C. T. Wolfe (Eds.), *Philosophy of biology before biology* (pp. 170–187). Routledge.
Glass, B., Temkin, O., & Straus, W. L., Jr. (Eds.). (1959). *Forerunners of Darwin: 1745–1859*. Johns Hopkins Press.
Goetz, R. (1986). La 'Science de l'homme' des ideologues. In *Doctrines,sciences ou pratiques socials?* Colloque. Nancy: Presses universitaires de Nancy.
Goetz, R. (1993). *Destutt de Tracy*. Droz.
Gohau, G. (1997a). L'hydrogéologie et l'histoire de la géologie. In G. Laurent (Ed.), *Jean Baptiste Lamarck 1744–1829* (pp. 137–147). Editions de CTHS.
Gohau, G. (1997b). Le cadre mineral et l'evolution Lamarckiennne. In J. Schiller (Ed.), *Colloque international Lamarck* (pp. 31–46). Blanchard.
Gohau, G. (2006). Lamarck philosophe? In P. Corsi, J. Gayon, G. Gohau, & S. Tirard (Eds.), *Lamarck philosophe de la nature* (pp. 9–36). Presses Universitaires de France.
Goldstein, J. (2002). Mutations of the self in old regime and post-revolutionary France. In L. Daston (Ed.), *Biographies of scientific objects*. Chicago University Press.
Goldstein, J. (2005). *The post-revolutionary self: Politics and psyche in France, 1750–1850*. Harvard University Press.
Goodfield, J. (1960). *The growth of scientific physiology; physiological method and the mechanist-vitalist controversy*. Hutchinson.
Gould, S. J. (1999a). A division of worms: Jean-Baptiste Lamarck contribution to evolutionary theory, Part I. *Natural History* (February).
Gould, S. J. (1999b). Branching through a wormhole: Reclassifying the types of 'Worms, Part II. *Natural History* (March).
Grassé, P. P. (1944). La biologie: texte inedit de Lamarck. *Revue scientifique, 82*, 267–276.
Greene, J. C. (1959). *The death of Adam; Evolution and its impact on western thought*. Ames, Iowa State University Press.
Gross, M. (1979). The lessened locus of feelings: A transformation in French physiology in the early nineteenth century. *Journal of the History of Biology, 12*(2), 231–271.
Hamy, E. T. (1893). Les derniers jours du Jardin du Roi: et la fondation du Muséum d'Histoire naturelle. In E. T. Hamy (Ed.), *Centenaire de la fondation du Muséum d'Histoire naturelle* (pp. 1–162). Imprimerie Nationale.
Hamy, E. T. (1907). Les Débuts de Lamarck, matériaux inédits pour servir à sa biographie. *Paris: Bibiliothèque d'histoire scienti*ªque., *11*(1).
Heilbron, J. (1995). *The rise of social theory* (S. Gogol, Trans.). University of Minnesota Press.
Hesse, C. (1991). *Publishing and cultural politics in revolutionary Paris 1789–1810*. University of California Press.
Hodge, M. J. S. (1971a). Species in Lamarck. In J. Schiller (Ed.), *Colloque international Lamarck* (pp. 31–46). Blanchard.

Hodge, M. J. S. (1971b). Lamarck's science of living bodies. *The British Journal for the History of Science, 10*, 323–352.

Hodge, M. J. S. (1995). Lamarck: un grand changement de cadre conceptuel. In C. Blanckaert, J. L. Fischer, & R. Rey (Eds.), *Nature, Histoire, Société: essais en homage à J. Roger* (pp. 229–239). Klincksieck.

Holmes, F. L. (1985). *Lavoisier and the chemistry of life: An exploration of scientific creativity*. University of Wisconsin Press.

Hooykaas, R. (1959). *Natural law and divine miracle: A historical-critical study of the principle of uniformity in geology, biology, and theology*. Brill.

Hutton, F. W. (1899). *Darwinism and Lamarckism, Old and New; four lectures*. G. P. Putnam.

Jordanova, L. (1981). La psychologie naturaliste et le 'probleme des niveaux' la notion du sentiment intérieur chez Lamarck. In J. Schiller (Ed.), *Lamarck et son temps, Lamarck et notre temps* (pp. 69–80). Vrin.

Jordanova, L. (1989). Nature's powers: A reading of Lamarck's distinction between creation and production. In J. R. Moore (Ed.), *History, humanity and evolution* (pp. 71–98). Cambridge University Press.

Kaitaro, T. (2008). Can matter mark the hours? Eighteenth-century vitalist materialism and functional properties. *Science in Context, 21*(4), 581–592.

Lamarck, J.-B. (1794). *Rechèrches sur les causes des principaux faits physiques*. Chez Maradan.

Lamarck, J.-B. (1800). Biologie ou Considérations sur la nature, les facultés, les développemens et l'origine des corps vivans, edited by P. P. Grass. *Revue Scientifique, 82*, 267–276.

Lamarck, J.-B. (1801a). *Discours d'ouverture prononcé le 12 Floreal a l'an 8*. http://www.lamarck.cnrs.fr/index.php?lang=fr. 313.

Lamarck, J.-B. (1801b). *Système des animaux sans vertèbres; préceédée du discours d'ouverture du cours de zoologie, donn. dans le Muséum national d'histoire naturelle l'an 8 de la République*. Deterville.

Lamarck, J.-B. (1802–1806). *Mémoires sur les fossiles des environs de Paris*. Muséum national d'histoire naturelle.

Lamarck, J.-B. (1783–1808). *Encyclopédie méthodique botanique* (Preface, Introduction, Classes, Espèce, Caractère, Botanique). Panckoucke; Plomteux.

Lamarck, J.-B. (1809). *Philosophie Zoologique*. Dentu. http://www.lamarck.cnrs.fr/index.php?lang=fr.

Lamarck, J.-B. (1812). *Extrait du cours de zoologie sur les animaux sans vertebres*. D'Hautel & Gabon. http://www.lamarck.cnrs.fr/index.php?lang=fr.

Lamarck, J.-B. 1817. Articles from the various volumes of the *nouveau Dictionnaire d'histoire naturelle*: Faculté, Fonctions organiques, habitude, homme, Idée, Irritabilité, Inteligence, imagination, instinct, judgement, Météorologie, nature. Deterville.

Lamarck, J.-B. (1820). *Système analytique des connaissances positives de l'homme*. Chez l'auteur et Belin.

Lamarck, J.-B. (1815–1822). *Histoire Naturelle des animaux sans vertèbres*. Deterville. http://www.lamarck.cnrs.fr/index.php?lang=fr

Lamarck, J.-B. (1907). *Discours d'ouverture: an VIII, an X, an XI et 1806*. Bulletin scientifique de la France et de la Belgique.

Lamarck, J.-B. (1933). *The Lamarck manuscripts at Harvard* (William Morton Wheeler & Thomas Barbour, Ed.). Harvard University Press.

Lamarck, J.-B. 1972. *Inédits de Lamarck*; d'après les manuscripts conservés à la Bibliothèque centrale du Muséum national d'histoire naturelle de Paris (Max Vachon, Georges Rousseau, & Yves Laissus, Eds.). Masson.

Lamarck, J.-B. ([1802] 1986). *Recherches sur l'organisation des corps vivans: et particulièrement sur son origine, précédé du discours d'ouverture du cours de zoologie*. Fayard/Chez l'auteur, Maillard).

Lamarck, J.-B. (1990). *Les Introuvables* (D. Goujet, Ed.). Societe Francaise de Systematique.

Landrieu, M. ([1809] 1909). *Lamarck, le fondateur du transformisme; sa vie, son œuvre*, a l'occasion du centenaire de la 'Philosophie zoologique.' la Société zoologique de France.
Lang, A. (1889). Lamarck und Darwin: ein Beitrag über Geschichte der Entwicklungslehre, I-IV. *Kosmos, 132–142*, 243–250, 408–417, 510–533.
Laurent, G. (1987). *Paléontologie et évolution en Cuvier et Lamarck à Darwin*. Editions du Comité des travaux historiques et scientifiques.
Le Dantec, F. (1899). *Lamarckiens et Darwiniens, discussions de quelques theories sur la formation d'espe'ces*. F. Alcan.
Limoges, C. (1980). The development of the Muséum d'Histoire naturelle of Paris 1800–1914. In R. Fox & G. Weizs (Eds.), *The organization of science and technology in France 1808–1914* (pp. 211–240). Cambridge University Press.
Mayer, E. (1972). Lamarck revisited. *Journal of the History of Biology, 5*, 55–94.
Moravia, S. (1972). Philosophie et medicine en France a la fin du xviiie siècle. *Studies on Voltaire, 472*(89), 1089–1151.
Moravia, S. (1980). Enlightenment and the sciences of man. *History of Science, 18*, 247–268.
Omodeo, P. (1997). Le transformisme au xviiie siécle. In G. Laurent (Ed.), *Jean-Baptiste Lamarck 1744–182* (pp. 471–482). Editions de CTHS.
Osborne, M. (1992). Applied natural history and utilitarian ideals: Jacobin science at the Muséum d'histoire naturellle 1789–1870. In B. T. Ragan & E. A. Williams (Eds.), *Recreating authority in revolutionary France* (pp. 125–143). Rutgers University Press.
Outram, D. (1984). *Georges Cuvier: Vocation, science, and authority in post-revolutionary France*. Manchester University Press.
Packard, A. S. (1901). *Lamarck, the founder of evolution: His life and work*. Longmans, Green.
Perrier, E. (1884). *La philosophie zoologique avant Darwin*. Felix Alcan.
Perrier, E. (1888). *Le transformisme*. librarie J.B. Bailliére et fils.
Perrier, E. (1909). Jean de Lamarck. *Revue pedagogique, 55*(7).
Pickstone, J. V. (2007). Working knowledges before and After circa 1800: Practices and disciplines in the history of science, technology, and medicine. *Isis, 98*(3), 489–516.
Reddy, W. M. (2001). *The navigation of feeling: A framework for the history of emotions*. Cambridge University Press.
Reil, H.-P. (2005). *Vitalizing nature in the enlightenment*. University of California Press.
Rey, R. (1994). Naissance de la biologie et redistribution des savoirs. *Revue de Synthese, 1*(2), 167–197.
Riskin, J. (2002). *Science in the age of sensibility: The sentimental empiricists of the French enlightenment*. University of Chicago Press.
Roger, J, Ed. (1979a). Les Néo-Lamarckiens Français. *Revue de Synthese, 95/96*, 279–469. (J. Roger, J.C. Sillard, A. Lagarde, Cl. Blanckaert, M. Viré, G. Gohau, A. Diara, K. Wellman, J.L. Fischer).
Roger, J. (1979b). Presentation. In *Les néo-lamarckiens français* (pp. 95–96), edited by Jacques Roger. Special issue of *Revue de synthèse* 3ie'me série.
Royer, C. (1870). Remarques sur le transformisme. *Bulletins de la Société d'anthropologie de Paris*. II Série. *5*, 265–317.
Rudwick, M. J. S. (2005). *Bursting the limits of time: The reconstruction of geohistory in the age of revolution*. University of Chicago Press.
Rudwick, M. J. S. (2008). *Worlds before Adam: The reconstruction of geohistory in the age of reform*. University of Chicago Press.
Rudwick, M. J. S. (2009). Biblical flood and geological deluge: The amicable dissociation of geology and genesis. *Geological Society Special Publication, 310*(1), 103–110.
Schiller, J. (1969). Physiologie et classification dans l'oeuvre de Lamarck. *Histoire et biologie, 2*, 35–57.
Schiller, J. (1971). L'echelle des êtres et la serie chez Lamarck. In J. Schiller (Ed.), *Colloque international Lamarck* (pp. 87–103). Blanchard.

Sewell, W. H. (1987). Le citoyen/la citoyenne activity passivity and the revolutionary concept of citizenship. In C. Lucas (Ed.), *The French revolution and the creation of modern culture* (Vol. 2, pp. 105–124). Pergamon Press.

Sloan, P. (1995). The gaze of natural history. In C. Fox, R. Porter, & R. Wokler (Eds.), *Inventing human sciences: Eighteenth century domains* (pp. 112–151). University of California Press.

Staum, M. (1980). *Cabanis: Enlightenment and medical philosophy in the French revolution*. Princeton University Press.

Staum, M. (1996). *Minerva's message: Stabilizing the French revolution*. McGill-Queen's University Press..

Stedman Jones, G. (2002). The new social history. In C. Jones & D. Wahrman (Eds.), *The age of cultural revolutions: Britain and France 1750–1820* (pp. 94–136). University of California Press.

Terrall, M. (2007). Speculation and experiment in enlightenment life sciences. In S. Mueller-Wille & H.-J. Rheinberger (Eds.), *Heredity produced: At the crossroads of biology, politics and culture 1500–1870* (pp. 253–276). MIT Press.

Tirard, S. (2006). Générations spontanées. In P. Corsi, J. Gayon, G. Gohau, & S. Tirard (Eds.), *Lamarck philosophe de la nature*. Presses Universitaires de France.

Vidal, F. (2006). *Les sciences de l'âme* (pp. xvi–xviii). Honoré-Champion.

Vovelle, M. (1985). *La mentalité révolutionnaire: société et mentalités sous la Révolution française*. Editions sociales.

Wagner, A. (1908). *Geschichte des Lamarckismus: als Einführung in die psycho-biologische Bewegung der Gegenwart*. Franckh.

Wahrman, D. (2004). *The making of the modern self: Identity and culture in 18th century England*. Yale University Press.

Williams, E. A. (2002). *The physical and the moral: Anthropology, physiology and philosophical medicine in France 1750–1850*. Cambridge University Press.

Wokler, R. (1995). The enlightenment science of politics. In C. Fox, R. Portert, & R. Wokler (Eds.), *Inventing human sciences: Eighteenth century domains* (pp. 323–346). University of California Press.

Wolfe, C. T. (2014). On the role of Newtonian analogies in eighteenth century life science: Vitalism and provisionally inexplicable explicative devices. In Z. Biener & E. Schliesser (Eds.), *Newton and empiricism* (pp. 223–261). Oxford University Press.

Wolfe, C. T. (2019). *La philosophie de la biologie avant la biologie: Une histoire du vitalisme*. Classiques Garnier.

Young, R. M. (1970). *Mind, brain and adaptation in the nineteenth century: Cerebral localization and its biological context from gall to Ferrier*. Oxford University Press.

Young, R. M. (1973). The role of psychology in the nineteenth-century evolutionary debate. In M. Henle, J. Jaynes, & J. J. Sullivan (Eds.), *Historical conceptions of psychology* (pp. 180–204). Springer.

Chapter 2
Herbert Spencer: The Tripartite Model

2.1 Introduction

In this chapter, I propose that Herbert Spencer's transfer and appropriation from Lamarckism enabled him to embark on the endeavor to constitute novel fields dealing with aspects of the social and the psychological as scientific ones. I will discuss his work insofar as it was relevant to the founding and elaboration of the conceptual and methodological frameworks of these fields, which had a deep impact on their formation all over Europe, the USA, and much beyond. His evolutionary Lamarckism emerged in several ways: determinism and plasticity elaborated through organism–environment, biological heredity/cultural inheritance relationships, hierarchical order, the challenge of parallelism, and individuals and collectivities.

In recent decades, much work has been done on Spencer and his impact both within British and European societies, as well as globally.[1] Thus, my aim is circumscribed. I will deal with the central themes of his work, which had the greatest impact, and given these recent studies, my own contextualizing will be minimal (Fig. 2.1).

Spencer's work spanned many fields: philosophy, psychology, biology, sociology, political theory, education, and ethics, and from the early 1850s until his death in 1903, he was struggling to refashion them as components of a system of human

[1] I have used three detailed groups of interpretation since the 1970s: *one* that focused on Spencer's concepts of 'evolution' and 'progress' and their relation to those of his contemporaries, including Darwin and somewhat earlier writers; a *second* that grappled with Spencer's multi-disciplinary network of concepts, the centrality of 'evolution' and 'complexity'; and a *third* that situated Spencer within historical, philosophical, scientific, and biographical contexts and contemporaneous controversies, and offered insights on the primacy of the ethical, the social, or the scientific in his work.

S. B. Gissis, *Lamarckism and the Emergence of 'Scientific' Social Sciences in Nineteenth-Century Britain and France*, History, Philosophy and Theory of the Life Sciences 36, https://doi.org/10.1007/978-3-031-52756-2_2

Fig. 2.1 Young Spencer, date unknown, portrait (1820–1903). Engraving Geo E. Perine, New York. Source: Wikipedia, public domain

sciences. He did so by placing those components onto an evolutionary continuum that situated human behavior within living nature. Since Spencer dedicated most of his work to the elaboration of such a system, the toll of repetition-cum-variation would be evident to any reader. This evolutionism was not merely a one-way transfer of concepts and models from the life sciences into the social sciences. Spencer's thoughts on how social institutions formed and re-formed gave him new ways to talk about nature. The application of the term 'naturalism' to the positions expressed in his work has been a topic of ongoing controversy, but I use it as a (self) identity actor's category (see Dawson & Lightman, 2014).

By his own testimony, Spencer learned his Lamarckism as a young man by reading Charles Lyell's presentation and refutation of it in the *Principles of Geology* (1830–1833, vol. 2), though one can assume that he had some familiarity with evolutionary ideas before that, given his father's interests and intellectual activities in Derby, the diffusion of Robert Chambers' *Vestiges of Creation* (1844), and his own testimony of reading Rymer Jones, William Carpenter, Henri Milne-Edwards, and others. Since the mid-1840s, Spencer lived in London and was close to some of the radical figures of the 1830s (Desmond, 1989). One can surmise that he was familiar with extant versions of Lamarckism, catastrophism, and other in-between contemporary narratives of evolution, as well as with the natural theology arguments against them. Upon reading his essays prior to the first edition of *Principles of Psychology* (*Princ.Psycho.*i, 1855) and immediately following it, but before 1859, it seems that he was committed to a Lamarckian version of evolution (Bowler, 2015b *v*. Francis, 2007). In his own evolutionizing of the main mechanism for solving problems—methodological, epistemological, and structural—Spencer made use of certain presuppositions of Lamarckism.

Spencer's work was carried out while participating in contemporaneous controversies, using the conceptual tools available then and forging new ones that were widely used and criticized within the next half-century. A number of questions kept recurring: what were the structures of dynamic interactions between organisms, between organisms and their environments, between various parts and functions of organisms? They pertained not only to the relationship between an individual as an individual or as a member of a group, but also to how such escalating interactions could result in evolutionary developmental systems that became more complex. A review on a late edition of his collected essays stated the following: "In 1855 – four years before Mr. Darwin published his *Origin of Species* – Mr. Spencer set forth that doctrine of evolution to the systematic application of which to the various phenomena of nature, and especially of organic life, his larger treatises and the greater part of his subsequent part life have been devoted" (*The Saturday Review*, April 11, 1891). The gradual constitution of Spencer's foundational framework and its initial application are found in his principal works written before and between the publications of the *Princ.Psycho.*i and of the first edition of the *Principles of Biology* (*Princ.Bio.*i, 1864–1867; in particular, Spencer, 1851, 1852a, b, 1853, 1857a, b, 1859). A preliminary basic frame of analysis began to emerge during the 1850s relating to biological, social, and psychological development. It came to the fore again in two essays in 1860a: "Bain on the Emotions" and "The Social Organism" ([1863] 1891, 1:241–264, 265–307). In *First Principles* (*First.Princ.*, 1862), there were further attempts at extending it universally. In the *Princ.Bio.*i, this universal extension was found to be counterproductive, and Spencer put constraints that gave the model a specific plasticity, relevant to the particular entities it was applied to, while a sociological point of view was already beginning to emerge. By the time Spencer was writing the many volumes of his *Principles of Sociology* (*Princ.Soc.*) in 1874–1896, with volume I revised and enlarged in 1876 and 1885, the struggles to establish 'social science,' also called 'sociology,' as an autonomous discipline were widespread in Western Europe and in the USA.

Spencer's work exemplifies a particular mode of transfer. While Lamarck (1809) and Charles Darwin ([1859] 1964) channeled their approaches from the non-human to the human (even though it has been argued that in Darwin 'the question of man' was present since the beginning), Spencer initially addressed questions regarding human individuals and human sociality. Applying human models to the non-human gave him valuable insights into the interdependence of parts and wholes and the potentialities of complex relations of independence, partial or full interdependence, among them, implicitly assuming the distinction between systems whose parts were defined independently and parts whose place in the system defined them. His work served in his time as a paradigm for a biological-social approach toward constituting social science, irrespective of whether Spencer started with ethical-social motivations or the other way around.

2.2 Spencer's Conceptual Framework

> But I had undertaken to set forth a general theory of Evolution as exhibited throughout all orders of existence ... On the whole the result seems to have shown that the attempt was not unwarranted.
>
> In one respect, indeed, I had, as an outsider, studying the phenomena of organic life as phenomena of Evolution at large, a certain kind of advantage over specialists, dealing after the ordinary manner with their respective separate subjects—plant-life and animal-life. The man of science who limits himself to a department, is apt to overlook, or else not sufficiently to appreciate, those most general truths which the phenomena he studies display in common with other groups of phenomena ... thus conducting to a more philosophical conception of biological facts. (Herbert Spencer, 1904, 2:103–4)

Spencer was groping toward an evolutionary framework that would provide a rich, detailed account of the complexity of the living world and its biological, psychological (mental), and social diversity and give meaning to structure, functionality, adaptability, intentionality, and other aspects of life as these evolved. All these were characterized by interactions, openness, and directionality and were time-and-history bound. Concurrently, he wanted this evolutionary framework to encompass the non-organic, the subject matter of the physical sciences, and yet to make this exceedingly complicated (and unrealistic) endeavor 'scientific,' i.e., in his eyes—'Newtonian.'Please add here (Gissis, 2017).

In the mid-1850s, Spencer crystallized his view that Evolution (with a capital 'e') was the all-encompassing framework of living nature, a general process and principle whose main feature on all levels was a growing complexity of relationships between environments and organisms and their evolving effects. Spencer enunciated this radically innovative stance when introducing evolution to the analysis of the mental apparatus in the *Princ.Psycho*.i. There, he articulated what I consider to be his basic epistemological and methodological frame of analysis of living nature, namely, *the changing interrelations of the changing interactions between organisms and the environment, and the changing effects they induce.*

Two resources contributing to his theoretical framework and its unresolved inner tensions are discernible. For one, the Newtonian tradition in the physical sciences served at the time both as a basic conceptual scheme and as a paradigmatic model of how to do science, though Baconian science with its emphasis on empiricism had its adherents. Two features were relevant: the search was for universal immutable laws of nature, and the nature of the basic entities—the particles—which had no history and were simple, discrete, and unchanging. Another significant source was organic chemistry, for whose practitioners, atoms were not conceived as the ultimate ontological units, but rather as components of their models. There was a search for descriptions and theoretical explanations in terms of forms or *structures* and *functions*, which paralleled the biological approach. I believe that one reason for the unresolved inner tension stemmed from the conflicts between these contemporaneous conceptual schemes in the physical sciences and Spencer's evolutionary theoretical framework.

2.2.1 Spencer's 'Fundamental Principle'

Given his basic frame of analysis, the notion of complexity played a central role in Spencer's system-building, as stated by John Tyndall in his BAAS Address in Belfast: "There are two obvious factors to be … taken into account – the creature and the medium in which it lives, or as it is often expressed, the organism and its environment. Mr. Spencer's fundamental principle is, that between these two factors there is incessant interaction … Life he defines to be 'a continuous adjustment of internal relations to external relations'" (Tyndall, 1892). I call this the Spencerian *tripartite relationship: organism, environment, interaction*, i.e., organism, environment, and their mutual interactions and the ensuing effects of these interactions.

This *tripartite relationship* (rather than the more prevalent binary perception of his thinking) became the central element of all of Spencer's thought, applied to the mental, social, and anthropological realms. Furthermore, the role played by 'organism' in this tripartite could be fulfilled by individual organisms and by collectivities. The interactions changed, resulting in greater complexity over time, and complexification was also the measure of progress. Spencer smuggled in another assumption which he claimed was based in physics: the instability of the homogeneous. It meant wherever there was an action, there were at least two effects brought about by that single action. In turn, those effects sparked still more actions. Because of this, later states were always more complicated than earlier ones. Thus, evolution led to complexity, as the simple (homogeneous) led to a differentiated and specialized (heterogeneous) result. He called it '*the movement from homogeneity to heterogeneity*.' It came to stand for 'complexity' and vice versa, arguing that states of heterogeneity produced webs of inter-dependencies, rather than relations marked by dominance and subordination.

Spencer deployed Karl Ernst von Baer's concept of directional development of the embryo from the homogeneous to the heterogeneous, since embryogenesis offered insights into the general processes of development, including openness, constraints, and limits of development (Churchill, 1991). It also indicated an irreversibility of the process: organisms could not revert to an original state from a more evolved, heterogeneous one. The embryonic model also made it possible to talk about the micro-level heterogeneity (see Gould, 1977; Adam, 1988; Hopwood, 2009). Spencer consistently refrained from discussing an originating unit of life either in *Princ.Bio*.i or in *Princ.Soc*. Frederic Harrison (1905, 102), a contemporaneous sociologist, referred in his criticism to the significance of embryology to Spencer's developmental explanations, claiming that Spencer dealt with beginnings—primitive man, embryos, invertebrates—and with the mature forms such as their contemporary society, but did not provide for the actual process of development. The relations among the tripartite elements of the model manifested growing heterogeneity, greater complexity from earlier to later states. The connections between 'the internal and the external' and the relations between physiological and mental states in *Princ.Psycho*.i were constituted according to the same model of growing complexity.

2.2.2 Environment

In both *Princ.Psycho.*i and *Princ.Bio.*i, Spencer often called these integral features of the model 'correspondence' (on the inaptness of the term, see Spencer, 1855, 373), which disappeared once the frame was fully worked out. The 'physiological division of labor' (Milne-Edwards, 1837, 1851, translated 1851, 1856, respectively), i.e., the combination of continued distributed actions among localized parts (later on, cells) to maintain an organism and their morphological differentiation and thus specialization, was seen as explicating the processes whereby complexity increased. 'Use and disuse,' the behavioral aspect of organisms adapting to their environments, led to 'inheritance of acquired modifications,' providing for both change and trans-generational continuity. These features of the mechanisms exemplified the tripartite relationship.

Given the then-existing level of knowledge and the available technical and conceptual tools, Spencer's model could not be mathematized, nor quantified. At the time, both Spencer and Thomas H. Huxley, when discussing the methodological aspects of physical laws versus biological laws, were critical of mathematics as a tool for describing change in living nature (Desmond, 1997, 2:377). Nevertheless, Spencer (e.g., 1855, 368, 381, 382) believed that by rejecting the metaphysics of isolatable, unchanging, objects and/or entities, the model incorporated a fundamental intuition expressed by some contemporaneous naturalists and philosophers regarding living nature and social nature, and that it emphasized the evolvement of an ever-changing, increasingly complex order within these realms.

In his earlier writings and still in *Princ.Psycho.*i, Spencer, in a Lamarckian fashion, attributed to organisms an internal drive—though, according to Corsi (1988), Lamarck did not believe in an internal drive as an innate teleological force, but as the result of physical-chemical action in an endless variety of changing circumstances. Such drive would account for the presumed responsiveness of organisms to challenges from their environment and thus for the necessarily growing complexification. In *Princ.Bio.*i (but most explicitly in 1855, 1:405–8), Spencer tried to rid his theoretical framework of any allusion to such an internal drive, for reasons described by his biographers as social, scientific, personal-psychological, and/or political (Turner, 1985; Peel, [1971] 1992; Francis, 2007).

There were thus a number of extant modes of viewing the relations between living entities and the conditions, circumstances, surroundings, and milieu, in which they were located. For one, 'environment' was depicted as passive since any mechanisms of change as well as those of stable continuity were contained within the organism and worked from the inside out (e.g., John Knox as discussed in Desmond, 1989, especially 73–4; Richards, 1989). Second, 'organism' was posited as passive, subjected to acting external forces, mostly physical but also social. Third, an attempt to produce a combination of the internal and the external, whereby the organism (often, considered as an individual, but not necessarily so) could be interpreted as responding differentially to 'the external' (Barsanti, 2000).

In his erudite article, Pearce (2010) argues that unifying the extant modes into one term, namely 'environment,' and positing organism and environment as dichotomous constituted Spencer's critical innovation in the British context. I disagree both on the issue of historical beginnings and on the attribution of significance to the passage from 'circumstances' to 'environment.' Our disagreements boil down to a *dichotomous view* of 'organism–environment' versus a *tripartite model*, which I put forward here.[2]

For Spencer, organisms could be considered only within some environment. In order for 'environment' to have any reference, there had to be an 'organism/environment/evolving effects' frame of analysis. Environment was considered as the sum of all—physical, chemical, biological, social—external conditions and circumstances, in which organisms found themselves, including the existence of other organisms. Did Spencer consider 'environment' in a global manner, as existing 'out there' irrespective of organismal perspectives on it—in a manner analogous to the way one would assume the independent existence of the external world—or as a specific environment for specific organisms? I believe the latter was the case. Spencer's treatment of nutrition as a process of environment–organism relations is an illuminating example. In *Princ.Bio.*i, he considered nourishment an environmental action that sparked internal processes of physiological differentiation, promoting complexity of the organism (1864–1867, 2:378–83).

Spencer commented in various places that the effects of the environment on the lowest (in terms of evolution) and on lower organisms were limited and would become more diversified and complex as the organisms themselves became more complex. Conversely, organisms impacted differentially on their environment (e.g., 1864–1867, 1:426): the lowest organisms did so to a minimal degree and the most complex organisms maximally, in accordance with the hierarchies of individuation. One can think of the effect of an amoeba on its environment, compared to that of a beaver colony, compared to a human twentieth-century town. Yet, the boundary between the environment and organisms did not signify zones of internal cooperation versus zones of external competition. The environment was seen as an environment for an organism, and thus local and not rigidly circumscribed, and there was no way to conceive of organisms independently of their environment, nor of environment-less organisms. Therefore, some of an organism's changes, conceived as changes of characteristics rather than as accumulated variations, could be looked upon as being passed on trans-generationally, or as 'embedded' in their environment and thus passed on to their organism's environment trans-generationally by virtue of the changed organisms and Spencer's generalized mechanisms of heredity. None of the elements of the tripartite relationship was viewed as totally passive, with the frame of analysis both reciprocal and looping.

[2] I am grateful to Trevor Pearce for letting me read two chapters from the MS of his book 2020 on pragmatism.

Spencer posited that the changing states of organisms over time, by virtue of their interactions (which were also changing), reached differing states of balanced adaptation with their respective environments. He thought these reflected active processes taking place both inside organisms and within the tripartite relationship, named '*equilibrations*' or '*moving equilibrium*.'[3] Homeostasis, as conceived by Claude Bernard, was the opposite. Spencer thought equilibrium would be reached, whereas Bernard insisted on the maintenance of such states. Nonetheless, one can see Spencer moving haltingly toward such a Bernardian conception. These equilibrations had possible alternative trajectories in the future. Since the reciprocal effects were not fully predetermined, there was latitude; the framework contained a measure of openness toward the future. This allowed for 'emergence'—of new structures and new means of interaction, conspicuous in Spencer's sociology.

Many of Spencer's contemporaries believed in a general tendency to decomposition (Thomson, 1862; Smith & Wise, 1989). Peel ([1971] 1992) and Bowler (1989, 2015a) have argued that British Victorian thinkers exhibited a tension stemming from the adoption of the 'progress' discourse and the growing significance attributed to the laws of thermodynamics. William Thomson's formulation of the second law of thermodynamics concerning the heat death of the universe was soon interpreted as also applying to organisms (Tattersal, 2014), while Tyndall's emphasis was on the first law. By contrast, in *First Princ.*, Spencer strove to provide universal growing complexity *and* decomposition (or *dissolution*) in a uniform manner in all realms.

His commitment to a tripartite frame and to complexification implied a hierarchical view of individuation and of living entities at large. It was *hierarchical* in that the progressive development Spencer was committed to required stable component entities (on a specifiable timescale) that were also endowed with a measure of *plasticity* that would allow for some change in structure, properties, or functions. Examples would be 'evolution of the embryo' at the component level of organisms, polymorphism at the level of individual organisms, and varieties at the collectivity level. This prescription served as a common ground that, according to Spencer, would allow for continuity between the physiological, psychological, and social forms of living and within the field of psychology between the simplest activities and those of reasoning and inference.

Spencer's theory of complexity did not match his practice. He declared that methodologically one should start from complex systems to investigate hypotheses regarding simple ones, and similarly, in the social realm, one should proceed from the present to the past (1855, 711, 1873, 109, 320–22). Yet, in both *Princ.Bio.*i and *Princ.Soc.*, Spencer's path of analysis (in fact, a reductive one) started from the simple, considered the most elementary, and followed its compounding, re-compounding, and re-re-compounding, to produce multiplicity, variety, and complexity.

[3] Discussed, for example, in 1862, 440–48; 1864–1867, 1:432–42, also 443–63 (on natural selection as an 'indirect equilibration'); see also 1864–1867, 2:221–390.

2.2.3 Short Comment on Darwin's Natural Selection

Spencer had an ambivalent attitude to Darwin's 'natural selection' in the *Origin*, which he both accepted and partially rejected. The significant distinction relevant here is related to their differing views of biological individuals/organisms. Darwin's Natural Selection as well as Sexual Selection applied to *varied* individuals. Though members of a species, individuals/organisms were a population, an aggregate of variations in the statistical sense. Ancestry and genealogy, as well as diversity, played a more significant role in Darwin's view, and the historicity of living nature was more marked. Spencer looked upon individuals, or a subgroup of individuals of a species, especially individuals that made up a lower-level species in a particular environment, as if they were *identical*, in a Lamarckian manner, so that any particular changes and their consequences would also be identical. He held that only for some descriptive and analytic limited purposes 'populations' were made up of slightly differing individuals. Under changing external circumstances, some such deviating group might survive better than the average, and their descendants would form a larger sector of the next generation (*Princ.Bio*.i is where the expression 'survival of the fittest' first appeared, 1864–1867, 1:444).

Spencer thought, however, that none of the evolutionary dynamics, neither his nor Darwin's, sufficed to singly explain living nature. For Spencer, natural selection did its evolutionary work *on top of* the dynamics of the tripartite model, solely as a secondary factor that produced fitness through changes that were either accidental or a result of the direct adaptation affected by the mechanisms of the tripartite relationship. Yet, natural selection could speed up the transfer of such changes, which worked equally on all individuals under the same circumstances (though in some contexts individuals were diverse enough not to be similarly affected). Spencer was reluctant to relate human progress to this mechanism. He explicitly argued that structural changes could be satisfactorily explained solely through inherited functional modification brought about through the mechanism of use/disuse.

Consequently, natural selection was relegated to being a negative factor, eliminating unfit individuals, and operating as a constraint on the explanatory system since for Spencer it explained survival but not the appearance of novelty. Furthermore, it could perhaps be instrumental, Spencer argued similarly to Alfred R. Wallace, in explaining change in lower forms, but certainly not in humans. Spencer held this position more or less consistently throughout his life. In "The factors of organic evolution" (1886), he denied any role to natural selection in the development of humans, i.e., emphasizing instead civilization—growth of intelligence and its usage of resources and human emotive capacities. This position was explicit in the exchange of essays with August Weismann, and in the chapters added to the 1898 edition of *Principles of Biology*. Furthermore, in *Princ.Bio*.i, Spencer refused to accept Darwin's notion of spontaneous variation because of its contemporaneous association with randomness, seen as undermining causal explanation. In my view, these vacillations had to do with the double-sided anchoring of the human within living nature.

2.3 Lamarckism Transferred

2.3.1 Individuals and Collectivity: The Centrality of Lamarckian Evolutionism

By the mid-nineteenth century, the variety of living entities was considered under the term 'organisms.' Controversies focused mostly on the lowest boundary of the living, the continuity/discontinuity of the living and non-living, and the origin of life.

2.3.1.1 General Framework of Individuation in Principle of Biology i

*Princ.Bio.*i opens with chapters deploying contemporaneous chemo-physical findings for descriptions and analyses of the boundaries between what Spencer called 'matter,' 'organic matter,' and 'organisms.' It is worthwhile to read them as a background to his deliberations on how to explain functionality and adaptability as relational, while keeping a semblance of the procedures of the reductive physical sciences. His choice of the tripartite model indicated that he chose to problematize that reductive program.

Spencer published the volumes of *Princ.Bio.*i between 1864 and 1867. I succinctly analyze the Spencerian entwinement of evolution—with *Lamarckian features*—with notions of individuality and collectivity in biology, and the centrality of hierarchical thinking there. Though Spencer read at least the 30 odd pages on Lamarck in Lyell's *Principles of Geology*, and though one can find in *Princ.Bio.*i a direct critique of Lamarck, it is important to recall that disconnected components of a grand system have an impact also through their appearance within other systems, in other fields of knowledge, or in socio-political contexts. As Corsi (2005) warns us, in the first half of the nineteenth century, 'Lamarckian' indicated a wide variety of theoretical or philosophical options often loosely, or not at all, related to Lamarck. For Lamarck and for some of his physiologist contemporaries, functional/behavioral changes in the organism were measured by the level of organization of the nervous system.

The evolvement of individuation was presented in *Princ.Bio.*i as a broad-ranging panorama. Each facet was delineated through the widest range of living entities, from single and multi-cellular organisms to plants to vertebrates. In the theoretically inclined chapters, Spencer argued that whatever held for biological evolution seamlessly suited the cosmic-universalized evolution presented in *First Princ.* Yet, the use of non-biological *social analogies* appeared only in crucial parts of these chapters, as a means to elucidate a chain of reasoning without which one could not fully understand the mechanisms of biological individuation. As their historical source, all social analogies used contemporary perceptions of the processes of British industrialization, urbanization, with Manchester and Birmingham serving as model cities (see, e.g., de Tocqueville, [1833]1988), the growth of transportation and communication nets, and the attendant change in class division. Recall that young Spencer intermittently worked in the 1840s as an engineer/surveyor in the train system. Thus, Spencer (1864–1867, 1:192–99) discussed a hypothetical example of the iron-ship industry.

While the mere concept of the individual was a given for Spencer (1852–1870 works), his concern focused on how to explain the evolutionary relationships among the diversity of individuals. The tripartite model was used to analyze how new organic functions and forms evolved, were maintained, and reproduced, and thus how there was diversity. The process of adaptation was first exemplified in evolutionarily very simple living entities, at the boundaries of being organisms. These boundaries signified minimal conditions for individuality and were the site of interaction and thus adaptation amid environmental changes. These processes of evolving, entailing plasticity, distinguished organisms from non-living entities. "Each organism exhibits within a short space of time, a series of changes which, when supposed to occupy a period indefinitely great, and to go on in various ways, instead of one way, gives us a tolerably clear conception of organic evolution in general. In an individual development we have compressed into a comparatively infinitesimal space, a series of metamorphoses equally vast with those which the hypothesis of evolution assumes to have taken place during those immeasurable epochs that the earth's crust tells us of … the evolution of the most complex organism out of the simplest" (1864–1867, 1:349–50).

Functions as tools of adaptation characterized living entities, even though there seemed to be an emphasis on structure and form in *Princ.Bio*.i. Functional specializations were expressed in the tripartite model at the behavioral—use and disuse—level, and there were several time-scales involved (e.g., of the individual and of the components of the environment). Spencer intended the process to be reciprocal, looping. Yet, he did not posit an active environment-creating organism, e.g., such as suggested by Lewontin's *The triple helix* (2000).

2.3.1.2 Individuals and Hierarchies

For Spencer, *an individual was a consequence of evolutionary processes*, not a foundational unit in a description of living nature. Stability did not inhere in particular components of biological individuals, but in the sameness of mechanisms and patterns and the continuity of their workings. Viewed from Spencer's evolutionary perspective, the entities of his wide-ranging investigations manifested—to put it anachronistically—extensive phenotypic plasticity, tangled hierarchies, complicated classifications, and controversial developmental histories. Yet, despite the overwhelming diversity, he attempted to arrive at a notion of *biological* individuality, to offer a systematic review of examples across the gamut of living nature, so as to anchor his theorizing in an interpretation of well-accepted empirical findings. In *Princ.Bio*.i, Spencer presented the evolvement of individuation as a broad-ranging time-dependent panorama, implying that biological individuality could only be understood by observing how it evolved. The style of the descriptions was as if they were field reports or at times concrete experimental reports. One could view the book as a collection of evolutionary-developmental narrations by virtue of which some were also stories of individuation. Although the developmental narrative related to groups of living entities (rather than to species), it treated the members of the group as homogeneous as far as the mechanisms of individuation were concerned. Change, which was seen as activating individuation, was explained through the tripartite model. An amalgamation of material components, understood as a process of becoming part of something more complex, was placed at one end of the range; at the other end, intelligence, reasoning, information exchange, and storage as the integration of change were referred to (e.g., 1864–1867, 1:62–3). There was a repertoire of patterns that was constitutive of the process of individuation of organisms (e.g., rhizopods, elaborated upon by Huxley, [1852] 1892; Spencer, 1904, chaps 35, 37).

In *Princ.Bio.*i, Spencer started with a *liminal individuality* (in fact, the single-cell organism) in which differentiation from the environment operated at the lowest degree of bounding, together with minimum coherence of its matter and bare definiteness of form. These two features—material bounding and shape/form—were the most basic conditions for being a (liminal) living individual (e.g., 1864–1867, 2:78). In addition, they possessed material living tissue that exhibited continuity with that of lower forms.[4] Further development of individuality followed a path, generally charted by the tripartite model (Fig. 2.2).

Spencer comments, p. 80* that in spite of the engraver's emphasis of the surrounding curves instead of vague linear representation, the homogenous sarcode is primary. "Its inaccuracy curiously illustrates the vitiation of evidence"

Homogeneous states, assumed as inherently unstable, developed into more stable heterogeneous ones, by virtue of temporary balanced states inside the organism, and temporary equilibria between the organism and its specific environment. Spencer characterized this process of temporal evolution as a 'moving equilibrium.'[5] In order to discuss living systems within an evolutionary framework, Spencer had to account for continued change over time and concurrently for stability and for emergence, i.e., to construct mechanisms that would allow for change and stability as compatible notions within the description of the system. The nebular hypothesis (1858) was an example wherein a mechanism—gravitational attraction—was adduced to give rise evolutionarily to the observed stability and equilibrium of the solar system. A principal mechanism Spencer posited for organisms was the emergence of 'physiological division of labor' inside the organism, or among externally, spatially, or otherwise related organisms. Thus, individuality developed under particular sets of environmental pressures. "There is, indeed, as already implied, no definition of individuality that is unobjectionable. All we can do is to make the best

[4] In a footnote related to an illustration of a particular protozoa from William Carpenter's Researches on the foraminifera (1861), Spencer commented on the craftwork that sharpened and made definite what should have been blurred! (1864–1867, 2:80n*).

[5] E.g., 1864–1867, 1:184–200, 'adaptation' in the discussion of a social analogy, 193–99, and 432–62.

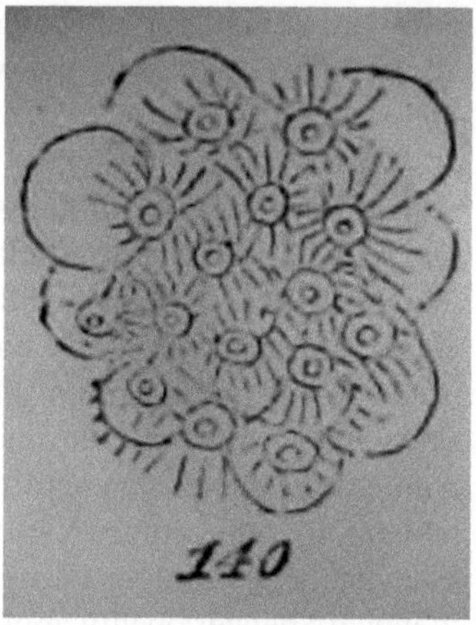

Fig. 2.2 Spencer, an illustration of *Sphaerozoum punctatum* (after William Carpenter). *Principles of Biology*, vol. II, p. 80. Source: Wellcome collection, London

practicable compromise … A biological individual is any concrete whole having a structure which enables it, when placed in appropriate conditions, to continuously adjust its internal relations to external relations, so as to maintain the equilibrium of its functions" (1864–1867, 1:206–7). Also, "the *ensemble* of vital phenomena presented by the organic world as a whole, cannot be properly dealt with apart from the *ensemble* of vital phenomena presented by each organism, in the course of its growth, development, and decay. The interpretation of either implies the interpretation of the other; since the two are in reality parts of one process" (1864–1867, 1:475). In other words, the interactions between an organism and its facets of its environment matter. Different kinds of organisms would react differently. They would be concurrently 'in the same environment' but in fact in a differing one, according to their own action and reaction capabilities (e.g., a sea environment with both cephalopods and crustaceans).

Spencer's 'individuation narrative' depended on four assumptions applied to organisms regarded singly, though they were members of groups (ordinarily species groups, e.g., the annelids): the tripartite model; change as development in organisms; development as emergence of functions; and inheritance of the effects of 'modifications of functions.'

Spencer's contemporaries found it difficult to offer an agreed-upon definition of biological individuality once they departed from higher vertebrate organisms, easily identifiable as distinct and separate individuals. The difficulty centered on numerous unicellular organisms, many multi-cellular organisms, particularly colonies of marine organisms, most of the invertebrates, organisms that could seem to be a loose combination/compound of repetitively identical, functionally independent, units, a plurality of organisms whose partial integration only extended to some of the functions deemed necessary for existence, and symbionts, etc. For example, Spencer compared a polyp and a polyzoon and their reaction to touch from the outside (1864–1867, 2:367–68). In the polyp there was hardly any functional differentiation and the connections among semi-independent body parts were very slight; thus its reaction was very localized. In the polyzoon there was functional differentiation of parts for outside pressure and a higher degree of coordination that allowed for reactions of the whole organism to outside touch on any of its remote body parts.

*Princ.Bio.*i was presented as an investigation of living nature at large. It contained four pregnant human-social explications of physiological mechanisms and one on the divergence and classification of living entities, discussed in analogy to human languages. The penultimate chapter of *Princ.Bio.* applied to humans the explications, tensions, and antagonisms within organisms (the whole range, from processes of individuation discussed above to reproduction), as falling under the same general mechanisms and repertoire of patterns of change as other organisms. Numerous examples were primarily drawn from those human groups deemed 'problematic' by mid-century Victorians—e.g., the Irish, Boor peasants on the Cape, and Fuegians. The relations between reproduction and individuation were the one topic to which Spencer openly applied contemporaneous views on utility, profit, and loss, thereby offering a 'calculus' on the biological economy of individual organisms. The last chapter of the book also unfolded a biologically driven utopian vision for individual humans. In all cases, Spencer attempted to elucidate mechanisms of development whose effects were studied through changing relationships among parts of the organism over time, explicating the stability of change of the overall organismal framework: the looping/feedback effect of the division of labor, differing tempo, pace of change, and specialization among organs, the notion of accelerated processes of change as a speeded-up recapitulation process whose early stages were obliterated. In a somewhat parallel manner, just as Darwin neutralized the need for a designer, so did Spencer. He described the dynamics and its mechanisms as providing for both social and biological emergence of self-organization and regulation. In his view, this made the need for centralized control and regulation superfluous. I join Elwick's view (2003) on Spencer's politics on that aspect. Yet, I argue about the level of his evolutionary dynamics and mechanisms and suggest regarding the descriptions of biological individuals as consequent upon them. Looking at the features characterizing the variety of individuals in *Princ.Psycho.*i and *Princ.Bio.*i, one can sort them into various categories of criteria of individuality. These categories exposed contemporaneous difficulties in providing an encompassing definition of individuality: minimal material conditions; further constraints—formal and spatial; and conditions for compounding and complexifying parts into wholes.

Within Spencer's worldview, the accumulation of features in a particular individual indicated that individual's position in the *hierarchy of complexity*. There was a difference between the use of 'heterogeneity' at the macro-evolutionary level and at a micro-evolutionary one: parts of *Princ.Psycho.*i and parts of the first and most of the second volume of *Princ.Bio.*i were dedicated to detailed descriptions of these degrees of individuality across the range of living nature. Another use of heterogeneity was among individuals in a given population—which was Darwin's use. Spencer was equivocal.

Spencer produced a view of organic nature as made up of nested hierarchies that reflected levels of complexity rather than a simple unilinear hierarchy. The components of a higher-level system could themselves be elements found at lower levels.

An example would be organisms with forms of reproduction that encompassed multiple life cycles and rhythms or single ones with both differences and similarities when looked at and compared inter-generationally (e.g., Nyhart & Lidgard, 2011). It is noteworthy that for Spencer these hierarchies were feature-dependent, therefore considered pragmatic and not 'part of nature.' *Nature as nested* was emphasized also by assuming that wider frameworks served as environments for more limited ones, as new, differing, types of individual entities gradually arose. Yet, these new types of individuals were not seen as marking evolutionary transitions; therefore, they did not require new evolutionary-developmental mechanisms. The variety of individual types was viewed within a historical continuum and methodological continuity (in a couple of places, Spencer tentatively offered to look upon some species as such a new type of individual, e.g., 1864–1867, 1:458 ff, vol. 2). Spencerian biological individuality was graded (and not relativized), since there was a continuum—from its lowest, i.e., liminal individuality to its uppermost abstract boundary. The Spencer–Bain principle, which argued that sensations, feelings, and reason were on a continuum, was also perceived as a continuum among levels of organization. It seems that contrary to Darwin (and Wallace), considerations of speciation were secondary, and phyletic evolution would be the result of individual changes.

Spencer struggled to delimit the range of the concept of biological individuality. The phenomena he encountered in his attempt to both describe and analyze living entities, societies, and 'the mental/mind' did not seem to him to fit any available definition of 'individual,' however narrowly analyzed. He noted that many of these exhibited modes of collective existences. *Perceiving any living entity as an individual for some purposes and as a collectivity for others would have greatly simplified his work.* The categories of the criteria of individuality point out why this was so: the minimal-conditions list fitted, more or less, the conventional constraints on what is a common-sense concrete individual, and the further-constraints list was its elaboration. Both the whole–part list and the graded-relations list included modes of accommodating a plurality of units and modes of collective existence, all within a fully or partially hypothesized framework.

I suggest that Spencer was acutely aware of the tensions produced by the constraints of his framework of individuality. Therefore, I look upon both the categories of organizational grades and of levels of individuals as constituting his conceptual gesture toward a hybrid notion of a *collective individuality*, for example, when he offered a short thought-experiment whereby instead of generations of individuals one would discuss a single continuous individual (1864–1867, 1:435–36). That conceptual framework contained multiple modes of relating and hierarchies of organizational complexity of relations. It encompassed a collectivity of diverse kinds of individuals, differentially related to each other and to the collectivities whose members they were. I find implicit support for this suggestion in the role of collectivities in the construction of a repertoire of mental activity patterns in *Princ.Psycho.*i and in the construction of the history of 'evolutionized individuals' as 'wholes' in the second volume of *Princ.Bio.*i. Somewhat similarly to his argument in *Princ.Psycho.*i and in its second, revised edition (*Princ.Psycho.*ii—1870/1872) concerning the role of collectivity, Spencer argued in various places in *Princ.Bio.*i that ancestral paths of heredity, particularly in higher (complex) individuals, affected the way they were modified by contemporaneous action of evolutionary mechanisms. A similar impact of collective past generations was employed in *Principles of Sociology* (*Princ. Soc.*—1874–1896), whose basic evolutionary unit was a collectivity, presented as a special kind of complex multi-level individual, the 'super-organism,' while in "The Social Organism" (1860b), societies and organisms similarly grew by an increase of structural complexity and by an increase of the mutual interdependence of their parts.

In his controversy with Weismann in the 1890s, Spencer defended a cluster of ideas—not just heritable adaptations—namely, the intertwinement of development and evolution, reflected also in entities below and above the level of 'ordinary' organisms, arguing for a different mechanistic concept of evolution that would account for features of living nature important for many nineteenth-century biologists: interactions and interrelations of organisms and environments, directionality, complexification, and emergence.

Thus, in major fields of his theorizing, biology, psychology, and sociology, Spencer held a hybrid notion of individuality that could be viewed either as 'collective individuality' or as 'collectivity of individuals.' He made explicit his underlying assumption, crucial for understanding his totalizing endeavor, that it was conceptually impossible to separate and isolate individuality from collectivity. This allowed him to use either individuals or hybrid individuals or collectivities in the role of the 'organism' within his basic frame of analysis, and thus stretch the extension both of the tripartite model and of possible interactions. It fitted well with his agenda in the *Principles of Ethics* to explicate the deep roots of human morality through evolution, offering a reconstruction of the history of the evolvement of living entities as an evolutionary development of moral actions: starting with lower life forms and ending with humans.

2.3.2 Biological Heredity and Cultural Inheritance in the Two Editions of the Principles of Psychology 1855, 1870/72

> If the doctrine of evolution is true, the inevitable implication is that mind can be understood only by observing how mind is evolved. (Spencer, 1872, 1:291)

Throughout, I ask: what was the 'work' of evolutionary Lamarckism in *Princ. Psycho.*i and *Princ.Psycho.*ii, and what was its 'cost'? Given that Spencer's essays just before and after the first edition of *Principles of Psychology* dealt with aspects of progress and evolution, the question arises as to why he decided on psychology as his project, for which there is no answer in his own testimony. I view it as an attempt to lay down the foundational assumptions that would enable him to pursue his enterprise of refashioning/evolutionizing the contemporary human sciences.[6] Furthermore, some of the mechanisms posited in both editions, such as those cater-

[6] I disagree with Renwick (2015a, b) who regards *First Princ.* as the first and foremost thrust.

ing to the relations between biological heredity and social-psychological inheritance, were widely deployed in the late nineteenth-century formation of the social sciences.

*Princ.Psycho.*i is foundational in the following sense: it dealt with human nature, human understanding, and the requirements for humans 'to associate together'—the ever-recurring themes of eighteenth and early nineteenth-century philosophers. Yet, parts dealing with philosophy were based on physiology, and parts dealing with psychology were based on an evolutionized physiology. The basic analytical units, the mechanisms, the generalizations, and the laws, all would have to apply to living organisms in general in order to apply to humans, i.e., to human perception and consciousness, to purposeful, planned as well as to unconscious and spontaneous human activities; to thought, feelings, desires, and to both individual and social behavior. Thus, Spencer 'naturalized' human behavior by putting it on a par with that of any living organism—to constitute a new psychology meant to lay the foundations for sociology, anthropology, politics, and ethics. In his *Autobiography*, Spencer noted that "Psychology underlies sociology; and there had to be specified a number of those more special truths of psychology which have to be handed in to sociology as part of its data" (1904, 2:240). Similarly, in a letter from 1855: "After dealing with general psychology, it became requisite to enter upon the special psychology of man in preparation for sociology. Certain traits of human nature are presupposed by the ability to live in *the associated state*" (1908, 2:3—emphasis added). This is at a time of deep controversies on the questions of 'mind' and 'life.'

The very term 'psychology' became an accepted designation for the field in Britain only in the 1840s. J.S. Mill discussed it in his *Logic*, and *The Journal of Psychological Medicine and Mental Pathology* was founded in the late 1840s. This does not preclude books with that term in their titles from being published there much earlier. Any position taken was suffused with implicit ideological assumptions and directed not just toward a limited circle but to the public sphere at large. Somewhat later, the writing of men-of-science was turning more internally-professionally, also affecting Spencer's shrinking readership in the late 1880s and 1890s (Csiszar, 2018)

I put forward that in both editions of *Princ.Psycho.*, Spencer's innovative stance stemmed from a particular Lamarckian evolutionizing of an amalgam of tenets from the then current mental and physiological theories of psychology. It enabled him to intentionally blur the demarcation between biological heredity and social—cultural—psychological inheritance, in order to posit hybrid categories – dealing neither solely with the physiological nor solely with the mental. Furthermore, this edifice could be coherently sustained if the underpinning of the blurring-hybridizing became collectivist, while the rhetoric—political, ideological, and scientific—remained individualistic, thus providing legitimation, effecting acceptance for a new science.

The period between the two editions was eventful on all fronts in Britain, with significant political repercussions, impacting laws, the economy, and the electorate. Also, a number of books that brought a new perspective on social institutions and on diverging human capabilities, such as in British anthropology, were published, whose resonance could be found in Darwin's *Descent*, just as in Spencer's works of that period.[7] Spencer rarely referred to his sources, but sifting through his socio-anthropological writings might indicate indirect participation in the then-ongoing controversies. Note also that Lubbock was, like Spencer himself, a member of the X Club. A large number of new periodicals appeared, attuned to diverse audiences, sounding the major controversies, when the written article gradually became as significant as the oral lecture (and its publication). Numerous journals either published articles or reviews on Spencer's work or dealt with issues similar to those discussed in his books. An important factor was the canceling of the tax on paper in 1860. Publications could be sorted into two kinds: those intended to open new vistas of knowledge to larger audiences and work of amateurs; and those publications whose effort was to establish professional distinction (Dawson et al., 2015). Therefore, *Princ.Psycho.*i and *Princ.Psycho.*ii were not singular events but part of an ongoing activity in which the issues discussed in Spencer's books formed one thread in a fabric in which writing on things scientific, psychological, social, and biological were intimately interwoven, as manifested in the sequence of major writings by Bain, Darwin, Lyell, Spencer, and Wallace, as well as Lewes, between the 1850s and the mid-1870s.

Spencer published *Princ.Psycho.*i in 1855 (670 pp.); *Princ.Psycho.*ii (1800 pp.) was first issued as sections to subscribers and then as a book in 1870/72.[8] To my mind, *Princ.Psycho.* exemplified his deployment of a Lamarckian evolutionary position. He had first written the biological-evolutionary part of *Princ.Psycho.*i, yet chose to open the book with a general epistemological-philosophical one. The first volume of *Princ.Psycho.*ii was devoted to the neural-physiological-psychological aspects, given the contemporaneous debate on the dependability of perception and the constraints on scientific knowledge (as in his "The Universal Postulate," 1853). This may indicate not only the readership's evolving familiarity with notions of evolution but also Spencer's thrust to legitimate the new field as a science. Retrospectively, some forty years later, Spencer said about *Princ.Psycho.*i: "At the time when it was published, extremely few were prepared even to entertain its fundamental conception, much less to agree with it" (1904, 1:468); and about *Princ. Psycho.*ii: "At that date … an evolutionary view of the Mind was foreign to the ideas of the time, and voted absurd … Naturally therefore after the publication of the *Origin of Species* had caused the current of public opinion to sway the other way, a more sympathetic reception was to be counted upon" (1904, 2:220).[9]

[7] Between the appearance of *Princ.Psycho.*i and *Princ.Psycho.*ii: Maine, *Ancient Law* (1861); Lubbock, *Prehistoric Times* (1865); McLennan, *Primitive Marriage* (1865); Tylor, *Researches into the Early History of Mankind* (1865); Lubbock (ed.), *The Origin of Civilization and the Primitive Condition of Man* (1870); Maine, *Village Communities in the East and West* (1871); Tylor, *Primitive Culture* (1871).

[8] The third edition 1880 was identical to the second, except for a 50-page penultimate chapter of "congruities," separately paginated.

[9] Also, an 1867 memorandum to the same effect (1908, 1:184).

Psychology was constituted by Spencer as a 'relational science' that dealt with the *relationship* of relations among external phenomena *with* those among internal phenomena. This position was somewhat differently articulated in the more 'biological-evolutionary' parts of the 1855 edition. In both editions, *each category* used by Spencer in the description, analysis, and explanation of psychology became constituted as a hybrid one—neither solely mental in the traditional sense nor solely physiological.

Using physiology as the 'foundational science' in his psychology, Spencer allotted an important role to the investigation of sensory–motor connections, otherwise formulated as muscular–neural, discussed at length in the biological sections of *Princ.Psycho*.i. *Princ.Psycho*.ii opened with a discussion of the nature of motion. Spencer accepted there the claim that self-mobility could serve as a criterion for an organism's position on the evolutionary scale and be used as a paradigmatic case in the search for the basic units of analysis. Adopting an evolutionary framework enabled a novel repertoire of models and metaphors that emphasized interconnections and relatedness. More importantly, it enabled Spencer to evade a parallelist position (such as Bain's) through *bridge-constituting hybrids* made up of physiological states *and* mental ones *and* their effects, i.e., interactions/relations between relations, which were second-level relations within a functional-dynamical frame. These hybrids then served as basic units and as the conceptual foundation of the new psychology. There seemed to be a difference between foundational positions and rhetorical caution, given the religious and institutional difficulties of being considered 'a materialist.' 'Surface-parallelists' could actually be monists, dualists, and even epiphenomenalists.

In *Princ.Psycho*.i, Spencer resorted to the empiricist method implied in the associationist assumption, to which he subscribed (Young, 1968), of choosing basic (traditionally discrete) units, while all other components of the system were regarded as combinations and compounding of them. In contradistinction to his predecessors, the evolutionary mechanisms he adopted operated on these basic units and their compounding, resulting in processes of complexification. He adopted a methodology of continuities (sections on "Philosophical analysis" and on "Biological synthesis" of the 1855 edition, similarly in the 1870/72 edition), for example, between man and nature; between states with and without consciousness; between automated activity and 'intelligence'; mode of thought employed by a scientist and the elementary forms of irritability in organic nature. In *Princ.Psycho*.ii, written after *Princ. Bio*.i, these continuities were already summed up under 'From homogeneity to heterogeneity.' The most conspicuous example, relevant to contemporaneous discussions in anthropology was the positing of a continuity between 'primitive thought' and abstract thought, such as in science and in religion. Similarly, both perceptions and conceptions of children and 'undeveloped races' were posited as illustrations in his depiction of progress in discussing the concepts of 'self' and 'personal identity.' His metaphors and analogies were quite similar in both editions and had two principal sources: the railway networks in Britain and the anthropological narratives on non-European 'races.' The first was more pronounced in 1855, the second in 1870, perhaps as an offshoot of one of Spencer's little-known enterprises, *Descriptive*

Sociology (1873–81). Both sources of repertoires were looked upon as systems in processes of expansion, growth, or complexification.

In both editions, Spencer used a specific rhetoric in talking about psychology: the rhetoric of constituting a new field of knowledge and diverse tactics of legitimation. There was, though, a marked difference. In *Princ.Psycho.*i, he presented the evolutionary adaptational mechanisms as providing explanations and frameworks for the whole gamut of 'mental phenomena,' including the production of culture and science (e.g., 1855, "General Synthesis," chaps 10ff). Psychology gained its legitimation as a science through physiology (by analogy, as Comte had attempted for sociology in relation to biology), where in the simplest/lowest forms of organisms, aspects that were later treated in psychology and physiology as separate, were practically united. In *Princ.Psycho.*ii (1870/1872, 1:97–128), Spencer discussed the empirically problematic nature of the subject matter and method of psychology and criticized its means of investigation—'self-observation of feelings'— especially where consciousness was concerned (the much despised 'introspection' of old). Yet, he argued that psychology was constituted in relation to physiology because its actual subject matter was neither solely phenomena of consciousness ('mind') nor solely neural phenomena ('brain'). In *Princ.Psycho.*ii, the scientific status of psychology was discussed in two distinct ways: as a science of 'superstructure' (meta-science) dealing with relational (hybrid) units and as a concrete science dealing with the subjective. This reflected the need to accommodate the hybrid units of psychology within the prevalent classifications of substrates into which they did not fit. Concurrently, direct subjectivity of the phenomena had to be accounted for: "the correlated phenomena of the environment are as essential to every psychological idea as are the correlated phenomena of the organism … the propositions of biology when they imply the environment at all imply almost exclusively its few general and constant phenomena (which may be ignored) … whereas the propositions of psychology refer to its multitudinous, special and ever varying phenomena … [that] cannot be left out of consideration" (1870/1872, 1:134–35).

The main feature of the evolutionary mechanism in *Princ.Psycho.*i was perceived as *adaptation* between organisms and their environment (including 'social environment'), while throughout 'development' and 'evolution' were used interchangeably, 'development' much more often, and 'evolution' did not appear in adjectival form. The term 'development,' first used in 1852 essays, implicitly criticizing Chambers (*Vestiges of Creation*, 1844), disappeared in *Princ.Psycho.*ii. In *Princ.Psycho.*i, 'adaptation' was only applied to complex organisms, and inheritance of acquired patterns of action was its main mechanism with two predominant features: a gradually increasing complexity and its ordering in an evolutionary system. This system had to keep 'a state of equilibrium' at every evolving stage of complexity (named 'consensus' when transposed to psychology). Schabas (1997) finds it a common feature of economic and psychological theorizing during this period, but I think that in Spencer's case it is a somewhat superficial similarity, as the 'balance' was an evolutionary time-directed growing complexity rather than a mechanistic one. Yet, there was also physics there, as later on Spencer insisted that his process was also linked to physical (mechanics) formulae. In *Princ.Psycho.*ii, there was an increased emphasis on the neural aspect—'internal states' and 'external states' were often discussed as 'neural and muscular states' but also as 'neural states' and as 'feelings.' Evolutionary processes at large, and in particular those with which psychology dealt, were characterized as a passage or movement from homogeneity to heterogeneity, with 'feeling' considered the basic, lowest, unit of consciousness.

In *Princ.Psycho.*i, Spencer radically argued for a developmental view of various procedures of cognition such as classifying, reasoning (e.g., induction and deduction), and language itself, deploying 'use – repetition – habit' and even 'habituation' as an important explanatory tool.[10] These allowed the reasoner, the speaker, to perform significant shortcuts because stages in the process 'merged' and 'sank to the unconscious,' and the 'shorter version' became the ordinarily functioning one. One of his more detailed examples was an analysis of how a child learned to classify and to read (1855, 205–7). "When learning to read the child has to class each individual letter by a distinct mental act … By continued practice these processes become more and more abbreviated and unconscious. Presently the power is reached of classing by one act a whole group of such symbols – a word. And eventually an entire cluster of such words is taken in at a glance … their having become automatic or organic" (206–7).

Spencer argued that this 'merging and sinking,' whereby the activity of those parts of reasoning, classifying, estimating, etc., were converted into 'automated' ones, started at early infancy and thus became 'organic.' This, in a nutshell, was Spencer's epistemological/developmental answer to both the Lockean and the Kantian theories of understanding. He extended the same mechanism to the perception of space and time, arguing that these were actually classifying modes of experiencing relatedness of objects via motion, which by virtue of endless compared repetitions produced the abstracted 'relativity of position,' i.e., perceptions of space and of time. His use of the sense of touch signaled that he too, like other associationists then, was still under Berkeley's influence on this matter. Thus, conceptions of space and time were not the condition to experiencing but rather its developmental product. Consciousness itself was defined as an emergent consequence of experiencing changes, rather than its precondition. Consequently, an infant's perception would differ from that of a child, and a child's from that of an adult: "being inherited by the infant in a proximate form, is progressively modified by the daily activities that accompany development, until it reaches complete form." There were, Spencer argued, concurrently modifications that had to do with the overall adjustment of the organism on the basis of the 'ancestral one' (1870/1872, 2/1:193, fn.). The sense of necessity of seeing the world one way rather than another was consequent upon the 'shortcut mechanism' of 'merging and sinking.' The inability to conceive of an alternative denoted the necessity that arose through the evolutionary developmental process, and in that sense was a universal postulate.

---

[10] When discussing 'repetition' in Spencer's mental mechanism, Francis (2007) relates it to Spencer's empiricism, while I would argue that it is part and parcel of his being a Lamarckist, rather than a proponent of neo-Lamarckism.

Spencer further criticized the phrenological faculties-divisions of the brain/mind and their hierarchical implications and suggested the de-composition of each and every 'faculty.' In the first, 'neural' volume of *Princ.Psycho.*ii, Spencer explained in great detail how experiencing produced neural changes that had a hereditary-evolutionary collectivist aspect to them. This was beautifully expressed in a critical comment on *Princ.Psycho.*ii by Douglas Spalding in *Nature*, February 20, 1873, "the doctrine that the brain and nervous system is an organized register of the experiences of past generations, that consequently the intelligence and character of individuals and of races depend much more on this, on the experiences of their ancestors, than on their individual experiences" (299). Or, as Spencer put it in *Princ.Psycho.*i: "Every one of the countless connections among the fibres of the cerebral masses, answers to some permanent connection of phenomena in the experience of the race" (1855, 581), and then went on to say "These pre-established internal relations [i.e., the nervous molds], though independent of the experiences of the individual, are not independent of experiences in general; but that they have been established by the accumulated experiences of preceding organisms. The corollary from the general argument that has been elaborated, is, that the brain represents an infinitude of experiences received during the evolution of life in general: the most uniform and frequent of which, have been successively bequeathed, *principal and interest*" (583—emphasis added; see also 539–53). Note the economic metaphor 'principal and interest' in order to emphasize successive additions. In Spencer's view (and of some contemporaries), these molds fulfilled the function of the Kantian transcendental apparatus and yet were themselves a product of experiencing.

Instead of talking about the development of humans, Spencer suggested to consider a graded continuity from the lowest forms of irritability and contractibility in organic nature to the complex processes of an 'advanced man of science,' all of them under the same mechanisms. The 'vital' and the 'mental' were thus undivided: "no line of demarcation can be drawn between them otherwise than arbitrarily" (1855, 349)—being unified in the lowest forms of organic life. Deploying the basic units is meant to reconstruct the developmental path as an evolutionary one.

The developmental mechanism discussed above gained its full import as the mechanism that explained reflex action. The latter was defined as the basic/primary hybrid unit, as a simple one, the compounding of which would be called 'instinct.' In the contemporaneous context, 'reflex action' and 'instinct,' unless discussed from a purely neurological perspective, were difficult notions to explain both for those committed to innate ideas and for those who were not, particularly so if one did not assume some kind of divine power. Spencer claimed that if one regarded his mechanism of accumulative-repetitive experience as developmental and evolutionary, then it could explicate both. As such the results of its action were to be passed on in *hereditary transmission,* i.e., *a biological heredity of psychological patterns of experiencing acquired through the life of an individual.* Spencer took over some of Laycock's reflex features, e.g., it being a mediator between organism and environment, its automatic character, and its pervasiveness at all levels of the neural system (following Marshal Hall). In Laycock, this appeared within the discussion of addiction (alcoholism), as throughout the nineteenth-century, discussions of addiction were a locus of a 'mixed discourse' on heredity, biological-psychological.[11]

Spencer was aware that this explanation would not suffice, given that it had to provide for the universally necessary and permanent character of both instinct and reflex action. Moreover, there were individual idiosyncrasies, diversity of environments, and thus of experiences and other contingencies to account for. Spencer solved the problem by resorting to collectivist assumptions, which became a principal underpinning of the whole system, holding its multilayered strands together. Experience and behavior became manifestations of collective hereditary transmission, working equally on all possible generations of individuals of the relevant group. What was transmitted was not solely a psychological-cultural pattern. Rather, in line with the notion of hybrid units, a changed biological pattern, neural in this case, was hereditarily transmitted. This biological pattern, called 'tendency' (or 'latency'), jointly provided the members of the collectivity with modified competencies through which their experiences would be organized. This 'tendency' molded the physical and psychological behavior of the individual (Spencer, 1855, 526–38). In the cases of instinct and reflex action, the molding that originally might have been more plastic became inflexible through biological transmission of gradually hardened patterns. Presumably, this inflexibility had its adaptational uses as a response to specific needs that produced the pattern in the first place.

The causal model used was the tripartite model. Spencer did not conceive of causes as being solely 'one-way directed,' or as formulated by Mill in his *Logic* as a complete disconnection between cause and its effect, entailing the impossibility of reversing the direction of the causal action. Rather, Spencer argued for a model more like a feedback or an interaction mechanism, but with a difference. Whatever was affected by causal action would then by virtue of it undergoing changes causally affect change in 'the cause.' The order of cause and effect was turned into an order of effects that are also causes vis-à-vis their 'primary' causes. For example, the perception and cognition of an individual would be affected by the collective patterns inherited; his/her experiencing mediated through these patterns would cumulatively with those of other individuals and then causally affect change in the collective patterns themselves, which would then cause changes in the experiencing patterns of the following generations of individuals. Spencer posited here a powerful tool for the investigation of the social-psychological along a timeline, a tool that took seriously the ordinary human experience of communication and interaction. A significant feature of this 'spiral of causes,' 'which were effects which were causes,' was an alternation between the biological and the cultural, and between individuals and collectivity along a transgenerational line.

Spencer went to great lengths in explaining the necessary level of development of the neural system in order for such a pattern to be formed and to be transmitted. He also drew a fine distinction between automated cognition-behavior that

[11] See also the two appendices at the end of vol. I of *Princ.Psycho.*ii "On the actions of anaesthetics and narcotics," "Consciousness under chloroform" (Spencer 1870/1872, 631–40).

developed over the lifespan of individuals and automated cognition-behavior that was so 'readily-organized' at birth. Once this form of hereditary transmission was established as an explanatory tool, it could be used to explain more flexible patterns of behavior too. Conversely, (hereditary) instincts could, according to Spencer, become a component of some more complex activity, in which case the isolated repetitive character would become distorted, experience and action less automated, until it lost both its character and its adaptational role. The latter process seems to be an extrapolation of the concept of reflex as a simple automated unconscious event, stretching it to cover conscious and complex ones. Purposive behavior could thus be completely physiologized. The thrust of Spencer's argument concerning instinct and reflex action meant that every mental procedure had to be in some sense a learned one, thus further substantiating his principal presupposition on the graded continuity of 'processes of the mind' and the uniform mechanisms operating on it. Human adaptation was distinguished by the ability to learn and pass on gathered and organized information to other members of the collectivity. This process was accompanied by a parallel gradual extension of the relevant environment (Spencer, 1855, 394–412, see also 1870/1872, 1:350–69).

A further consequence was that 'feelings,' considered 'inborn or innate sentiments' in the contemporaneous philosophical discourse, had to be explained by the same mechanism of experience–repetition–habit–etc., with transgenerational biological-hereditary transmission, moving from the level of individual heredity to the level of collective heredity. Central among these were 'will' and 'sympathy.' In *Princ.Psycho.*ii, Spencer devoted a whole section to such sentiments and a special chapter to "Sociality and Sympathy."

The decision to treat 'will' as a natural continuation of that argument was in many ways a rather courageous act. One can hardly think of more significant concepts than 'will' and 'reason' in Victorian ethos and culture (for both believers and non-believers). The individual, a centerpiece of the modern empirically based social and biological sciences, was regarded in Victorian Britain as a foundational epistemological and methodological category, perceived on the one hand as the ultimate unit to be explained and on the other as that entity in terms of which all there is was to be described, analyzed, and explained. In their diverse interpretations, 'will' and 'reason' encompassed the moral code and the 'self-forming' ideals of the 'individuals' comprising the ascending British middle class. In this case, Spencer's analysis undermined the separateness and distinctness of the particular chain called the 'ego'—but at the same time performed an act of salvation by anchoring it, once again, in collective experience and biological heredity. Instead of the traditional 'will' and 'freedom of will,' Spencer suggested to his audiences the laws of growing adaptation.

Experience and learning processes, he argued, very gradually formed and fashioned biological-psychological competences—the mechanism that operated via a collective biological heredity. In their turn, these competencies were the enabling preconditions for further individual accumulation of experience and thus for learning and communication. Note the role of the collective in the underpinning of a science, psychology, which purported to deal with the processes of individuals.

Thus, the de-composition of sensationalist notions, of the utilitarian worldview, of Kantian transcendental epistemology, of certain facets of associationism, and of faculty psychology, all depended on a specific mode of biologizing/Lamarckian evolutionizing.

*Princ.Psycho.*ii similarly deployed these arguments but also added the role of the evolutionary model of natural selection and significantly elaborated on the socio-anthropological implications of evolutionizing psychology. Concurrently with his work on *Princ.Psycho.*ii, Spencer developed the schematic tables of *Descriptive Sociology*, and immediately after finishing *Princ.Psycho.*ii embarked on writing a short version of his views on sociology—*The Study of Sociology* (*SoSoc.*, 1872–1873). 'Progress' was conceived as either another facet of biological evolution or consequent upon its applicability to human biological evolution. By evolutionizing psychology, Spencer gained hitherto untapped resources for psychological analysis, which provided numerous examples and illustrations. These were the materials authored by social and anthropological thinkers (believers in 'progress' in either its plastic or more rigid version). Just as a Lamarckian evolutionist might use examples for the process of complexification from 'organic nature,' Spencer made use of examples from 'social nature' as described by contemporaneous social thinkers and particularly by anthropologists. This was probably the context of his decision to devote the whole last section in *Princ.Psycho.*ii to the societal aspects of his biologizing/evolutionizing psychology.

I think that the most illuminating discussion to be found in the biological part of *Princ.Psycho.*ii, volume I, was the one on "Pleasures and Pains" that concluded the second section, "The inductions of psychology." A highly critical comment in Lewes' review of *Princ.Psycho.*i posited pleasure and pain as the missing assumptions from which one should derive moral sentiments (Lewes, 1855, Life and Mind review article, last sentence). Even in *Princ.Psycho.*ii, Spencer avoided doing just that.

The sub-section, "Pleasures and pains," had Lamarckian aspects combining both evolutionary models as an alternative to utilitarian models, especially J.S. Mill's. Spencer began, claiming "the phenomena of Pleasure and Pain are perhaps the most obscure and involved which psychology includes" (1870/1872, 1:273). Both feelings were explained as components of the adaptational strategy of the organism. Throughout the evolutionary process, Spencer argued, pain indicated a perturbation to the adaptational equilibrium between the organism and its surroundings, while pleasure indicated a growing adjustment. Spencer did not explain the development of such gauges on the basis of hereditarily transmitted biological transformation, but as a result of natural selection: the benefit which accrued thereupon would be for the collectivity ('race') due to its long-term survival. Spencer tried, rhetorically at least, to escape from the collectivist underpinning of that mechanism, formulating it in a way more congenial to Darwinians. Thus, he mentioned 'ancestral races of creatures' rather than specifically human evolution. However, more interestingly, he talked about ancestral individuals (e.g., 194), or 'a long list of individuals,' when discussing von Baer's recapitulation model and its relevance (e.g., 220–21, fn.).

The possibility of regaining individual 'equilibrium,' i.e., adaptation to one's surroundings in such contexts, was hampered, in Spencer's view, because of the effect of what he called 'direct equilibration,' since the Lamarckian mechanism was (too) slow in complex organisms experiencing multifactorial disturbances. This was due to the vicious circle created by the reciprocity between modification of the social environment and the resulting individual modifications, especially in cultures that either advocated interfering with natural selection, or lauded pain and suffering. Did Spencer thereby reject utilitarian ideals? No, if we attend to his rhetoric. Yes, if we draw the implications of this position that rejected the possibility of aggregate social equilibrium. And yet, no, if we note that by the apparent use of natural selection Spencer hoped to stay on the individual-adaptation level without resorting to a hidden collectivist-laden mechanism. A more general estimate can be reached when the interplay between the two evolutionary mechanisms is elucidated. The Corollaries section illustrated Spencer's view of the applicability of his general psychological assumptions to human sciences. By juxtaposing his 'psychological analysis' with the one found in these chapters, one gains a sharper understanding of those features that made Lamarckism attractive to people in the human sciences in the latter part of the nineteenth century and of the ways in which appropriating Spencerian Lamarckism could have been an important vehicle for them. In the very short introduction to the Corollaries section, Spencer made it clear that the question: "What makes people stick together in society?" was a basic one and could be partially answered by evolutionizing psychology. In his discussion there, just as in other sections of *Princ.Psycho.*ii and to a lesser extent in *Princ.Psycho.*i, evolution could be exemplified by looking at human development, at anthropology and history. "We will thus measure the leading traits of intellectual development as it affects and is affected by civilisation" (1870/1872, 2/2:520); "let us compare the mental activities of the child, the savage and the civilised man in his various grades of culture" (2/2:518); "when from the comparatively judicial intellect of the civilized, we pass to the intellect of the uncivilised man … Between the thought of men and women" (1:581); "the minds of inferior human races cannot respond" (1855, 464); "the argument from the history of civilisation, or from the comparison of different existing human races" (573).

The Corollaries section was the most political one in both editions since Spencer's positions on policies and economies of war were bluntly stated there. All in all, in "Corollaries" and "Pleasures and Pains," he offered a grim view of modern 'militant colonial' society and gave a first approximation to what, 20 years later, would turn into two versions of binary sociological typology: that of Durkheim and that of Tönnies.

Thereafter, Spencer set the narrative framework of the following chapters by weaving together social and economic wellbeing as the enabling conditions for the intellectual development of the collectivity and of the individual. He added a significant layer to that narrative by correlating social evolution with the ability to express emotions.

In both editions, in so far as Spencer employed Lamarckian assumptions, there was an intentional effort to blur the distinction between biological heredity and cultural inheritance. The hereditary mechanism resorted to delineating an alternating-rhythm pattern between the two. An important feature was the choice of 'basic units,' which were neither solely mental nor solely physiological, and whose process of recombining and amalgamating was a hybrid one too. Human competencies or 'tendencies' or 'latencies' were neither solely biological nor solely mental-cultural. They were pre-organized molds, inherited as neural modifications that functioned as neural-psychological patterns for organizing experience. At the same time, they were open to further functional–adaptational modifications by that experiencing, which in its turn would be biologically inherited; heredity was conceptualized on the borderline between the individual and the collectivity (the 'species,' the 'race'). I would further add that in the period between the late 1840s and early 1870s, or until Galton's *Hereditary Genius* (1869) sank in, evolutionizing psychology was intended to discuss physiology and heredity in conjunction. In the more general context of 'progress,' it meant that racialization was applied to imperial policies and to social-class determination as well as to human developmental potentialities from 'a psychological point of view.' In this interim period, British middle-class individuals could gain a measure of immortality as their experiencing became that of the collectivity *jointly* and the collective patterns became part of each individual *singly*. Moreover, behavioral patterns—habits—of the individuals were conceived as shapers of the cultural history of the collectivity. But, it is doubtful whether this could have been a way to overcome the self-interested discreteness attributed to individuals in most of the political and economic theories of the time.

The late 1870s could be viewed as a fairly clear dividing line between those who would investigate the brain only and those who would discuss categories of thought culturally and philosophically only. The mixing of languages, of metaphors and analogies—philosophical, medical, biological, general-scientific—that Spencer used would become more difficult later on. Thus, in more than one sense, a comparison of Spencer's two editions affords today's reader an insider's gaze into the passage from one phase to the other and not just in terms of evolutionary models used, or conceptions of heredity.

Early twentieth-century historians of psychology did not find much to say in favor of *Princ.Psycho.*, starting with the early-1920s renderings by E.G. Boring, G.S. Brett, and H.C. Warren, as neo-Lamarckism was already rather controversial. Historians' main effort was directed toward establishing a genealogy of precursors, founders, and followers that would prove psychology evolving to become a fully-fledged experimental science, clearly demarcated from other psychologically introspective enterprises. The change in appreciating *Princ.Psycho.* had to do with the renewed interest in the history of evolutionary thought, and particularly various readings of Darwinism, but also with the process turning natural history into a subfield of evolutionized biology (R.M. Young, Roger Smith, Peel). The next wave started in the early 1980s, with L.S. Jacyna, C.U. Smith, with R. Richards somewhat later. Since then, one can find in some of the textbook histories of psychology a chapter or half a chapter devoted to Spencer and evolution within the nineteenth-century gallery, usually tucked between Gall and Combe and Bain (with Danziger as an exception).

2.3.3 Biological Heredity and Cultural Inheritance in Principles of Sociology

I propose to look at the kind of work the entanglement of evolutionary concepts did for Spencer in the construction of his sociology, a sociology in which both 'individual' and 'collectivity,' i.e., 'the social,' were the foci. Though Spencer's evolutionism impacted human sciences already with the publication of *Princ.Psycho.*i and "The Social Organism" essay, the actual publication of *SoSoc.* and the earlier volumes of *Principles of Sociology* (*Princ.Soc.*) put forth a wide-range synthesis. The diverse reactions to this enterprise rippled through social thinking of the era and contributed to the ongoing process of crystallization of sociologies all over Western Europe and the USA. This was true whether following Spencer's path with local variations, or opposing him while struggling with the impact of his work, or else steering a strongly non-biological course. Recently, some works tracing his impact have been published (e.g., Renwick, 2015b; Pearce, 2020). I have chosen a certain corpus as the critical one for the modes in which Spencer conceptualized the entanglement of individuality and collectivity while evolutionizing it in a Lamarckian mode: "The Social Organism" (1860b), *SoSoc.* (1873), *Princ.Soc.* (1874–1896), and some shorter 1890s texts.

Although Spencer posited himself as an individualist in matters political, his basic *evolutionary* unit in Princ.Soc. is *the collectivity, presented as a special kind of individual.* The methodological tension in the background can be seen in the exemplifying material between 'science-like facts' and historical and anthropological material. The biological analogy of society to organism was helpful to Spencer in preventing outright teleological explanations. Instead, he deployed concepts such as 'function,' 'harmony,' and 'equilibrations,' where the latter implied an in-built feedback or looping mechanism for system maintenance. In this manner, whatever appeared to the members of the collectivity as goals and ends could be looked upon as a consequence of the way the feedback mechanisms were arranged. Thus D'Arcy Thomson said (pro-oemion of his Spencer lecture in 1913): "his [Spencer's] *mechanical outlook* on phenomena" (4—emphasis added); "he draws a broad picture, confessedly of a mechanical kind" (7); "came to the study of biology in the spirit and with the equipment of an engineer" (11), while Josiah Royce wrote in his book on Spencer: "Mill said that Bentham's disciples learned from the master the art of breaking up a subject into its constituent parts. Spencer's assistants learned from him the more difficult art of building up the disintegrated elements into an *organic whole*" (Royce, 1904, 195–96—emphasis added).

Spencer first considered the discipline of sociology and its constituting elements in his 1860 essay "The Social Organism." In an act of naturalizing, he asserted that "society is a growth and not a manufacture" (1891, 1:269, also 266). Societies and organisms were considered to be distinct from all other entities through two fundamental features: the continual increase in complexity of their structure and the functional interdependence of their component parts. Though individuals and societies were mutually formed and fashioned, society was likened to an individual organism; more precisely, hierarchies of differing societies were likened to hierarchies of differing species of organisms. Societies and organisms were both perceived to grow by an increase of structural complexity and of the mutual interdependence of parts, resulting in the 'whole' being more stable and longer lasting than its component parts. Spencer posited human societies in contradistinction to individual organisms, which had to have some bounding material delimiting them from their environment, as well as a definite shape, and possessed continuity in material living tissue.

Human individuals of societies were seen as discrete materially, though not necessarily functionally, as mobile, endowed with sensation, feeling, and consciousness, though differentially so. Does it follow that society was 'an organism plus' or that a biological organism was 'a society *manqué*'? This is debated in *Princ.Soc.*

The guiding thread in "The Social Organism," an analogy between individual biological organism and human society, continued to inform Spencer's work on social phenomena until the very end. Analogies are based on assumed similarities, which are deemed relevant to the argument they implicitly make. They are but one mode of 'transfer.' In the latter part of the nineteenth century, evolutionized biology played a vital role for a number of sociologists in Europe and the USA, molding their programs, methods, practices, and even their self-image as scientists. This analogical and metaphorical transfer/appropriation/engagement established similarity and suitability, but also a difference between the biological and the 'non-biological.' It seems that the character of this transfer/appropriation changed between the 1870s and the 1890s so that sociologists would present their emerging field as fundamentally similar and yet as uniquely distinct.

Spencer mapped or projected a generalized enhanced notion of 'biological organisms' on human society. It was done to enhance certain projected features that would enable him to claim that *both* societies and biological organisms were different from anything else. Thus, he could argue that at his chosen level of abstraction *society was an organism* (later, a 'super-organism'). These projected features were hierarchically graded structural complexification, growing functional division of labor, gradual growth of internal structures of coordination, and evolvement of internal structures for the regulation and control of component units. Concurrently, Spencer invested a marked effort in minimizing other features in the analogy in a manner that allowed him to argue that society was in fact distinct from an organism. It enabled him to call for a new mode of description, explanation, and argumentation when dealing with societies. This new mode would constitute a new discipline, sociology, that would deal with sociocultural evolution. One of the most striking differences between his biology and sociology was in the modes of classifying societies as compared to classifying organisms. Spencer classified societies solely by their degree of organization and by the predominant features of their overall institutional systems. The significant feature was the impact of the institutional systems on the modes of life of the component units in their interactions with these institutions. Thus, he posited 'industrial' versus 'military' societies. A conspicuous distinguishing feature of 'the human social' was a degree of *plasticity* (e.g., 1874–1896, 1:549, 571), defying the modes of classification explicated in *Princ.Bio.* Classifying by complexity of structure, implied in the case of social organisms, classifying by (an implicit ladder of) culture.

Throughout the first volume of *Princ.Soc.*, Spencer invested much work to prove that the analogy between society and organism was an almost perfect fit, and exceptions could either be bent or held up to emphasize the ruling analogy even further. Noted at the time was Albion Small (1894) defending this analogy against contemporaneous critique. Most of the biological examples were taken from unicellular and multicellular invertebrates.

Yet, at the end of his "Qualifications and Summary" in *Princ.Soc.*, Spencer argued that although some principles of organization might be common to both orders, the social and the biological, the social organism could not be compared to any single biological organism, positing the analogy as a scaffolding (1874–1896, 1:592–93), even claimed as an ad-hoc argument (592, fn.). Thus: "Here let it once more be asserted that there exist no analogies … the social organism, discrete instead of concrete, asymmetrical instead of symmetrical, sensitive in all its units instead of having a single sensitive centre, is not comparable to any particular type of individual organism, animal or vegetal." I believe that Spencer deployed the analogy in the way metaphors are often used, namely, to project an affinity in some chosen features while at the same time creating a distance. That constituted distance opened up space for an analysis of societies as *sui generis* and thus for an analysis of the relations between individuals and collectivities within a non-biological conceptual framework.

SoSoc. published as a series of articles in 1872 in *Contemporary Review*, a widely read journal, and in 1873 as a book by Henry S. King, was Spencer's first explicit statement of the need for a separate discipline. It was a juggling act in which Spencer tried to maintain the centrality of society, of collectivity, and concurrently preserve a wide space for the actions expressive of moral autonomy for its components—human individuals. For a few years from 1876 onward, the book was adopted by Sumner as a textbook in his sociology course at Yale (Hofstadter, 1955; Francis, 2007, 189, 381n2) and by Dewey in his courses during the late 1880s–early 1890s (Pearce, 2020, 252, 292). Furthermore, Pearce (2020, 335) argues that around the turn of the twentieth century Spencer's works, especially *Princ.Soc.* and *Principles of Ethics* were required reading, object of philosophy courses at Harvard, Michigan, Yale, Western Reserve, Minnesota, and the University of the South.

SoSoc. presented the parallelist analogy combined with the tripartite model and its consequences. Spencer discussed the nature of social phenomena from a methodological reflexive angle. He believed they were inaccessible either through perception or through introspection. He adumbrated the difficulties presented by the in-built involvement of researchers in their subject matter, with the researchers' biases to choose 'the exceptional and extraordinary' as their topic rather than the quotidian, with the difficulties of how to deal with the so-called 'self-evident,' and more generally, the opacity of social phenomena to perception and the difficulties in understanding alien social systems. Thus, the book elaborated on the researcher's need 'to unlearn.' As yet another facet of this methodological reflexivity, Spencer claimed that the natural sciences named abstract-concrete, physic and chemistry, were even less adequate as models for investigating the social, i.e., for dealing with "the entangled plexus of co-operating causes" (1873, 320–21).

In *Princ.Soc*, unlike *SoSoc.*, Spencer dealt with the actual workings of the social, and sociology was presented as reformist. The generalized introductory discussions of each topic were based on biological concepts previously expounded in *Princ.Bio.* and transposed unto the social realm, thereby producing concepts whose unique features were constituted as mere local exceptions. At the very beginning of *Princ. Soc.*, Spencer claimed that there were degrees of sociality (1874–1896, 1:5–6). From this perspective, sociality seemed to be dependent on the characteristics of the individual elements of the aggregate (9). Yet, the list of fundamental features of sociality found in the animal world at large characterized the collectivity rather than its individual units. Examples of that could be the internal coordination in the collectivity, generalized control coupled with the existence of sanctions and their extent, their collective recognition, and in higher mammals the evolution of '*social sentiments*.' For Spencer, one of the principal factors in enhancing 'social sentiments' was social life itself, primarily its '*density*.' The relation between the size of the collectivity and its 'density' was supposed to be a specific feature unique to human collectivities. 'Size' was related to divisions of labor that transformed a multiplicity of units, here the human aggregate, into a whole (1874–1896, 1:453). Consequently, 'social density' was presumably expressed through the intensification of human relations, specified cooperation through communicative means—linguistic, emotional, and intellectual (460). Cooperation was seen as an important factor in enhancing social plasticity. 'Density' as a feature of human societies referred not just to the number of people per space-unit but also to the number and quality of interrelations and interactions among them. The primary use of 'density' in *Princ.Bio.* was in analogous sites, e.g., molecular movement and density (1864–1867, 1:14, 2:521), population density, as well as in *First Princ.* and in essays dealing with the physical aspect of evolution.[12]

The 'continuity' of the social organism was also differently construed, through including all living components of the environment and all *material products* of the activities of the collectivity, as well as its '*mental products*,' defined as the cultural inheritance from past generations, e.g., gods (1874–1896, 1/1:305–6) as society's components. These constructed together 'social nature,' a mode of material contiguity, claiming to cancel the traditional separation of natural and artificial. Yet, Spencer did not provide any detailed analysis of material culture or of technology (e.g., in contradistinction to Marx, who also included the products within the analysis of society), though railways loomed large in his examples and analogies.

In *Princ.Soc.*, social institutions—domestic, ceremonial, political, ecclesiastical, professional, industrial—produced environmental frameworks for other institutions and for the human 'individual units.' Perhaps one could argue that for Spencer the constitution of sociology as 'scientific' entailed making its field of phenomena part

[12] Altruistic sentiments are beyond the scope of the discussion here, having to do with the construction of human reproduction as an exceptional case.

of a generalized nature, i.e., of living nature, and of 'social nature' in order to stretch the bounds of what was to be included in 'society.' This 'social nature' was presumed to obey the same laws of evolution. It allowed him to use 'survivals' and 'fossils' (in their contemporaneous biological connotation) to reconstruct social evolutionary continuity rather than historical chronology and continuity between knowledge of nature and that of social nature/societies.[13]

Similarly to *Princ.Psycho.* and *Princ.Bio.*, the first volume of *Princ.Soc.* brought forth 'the data of sociology' after it had defined its topic: the study of the most advanced form of an evolutionized organism, i.e., the 'super-organism.' These data turned out to be descriptions, analyses, and explanations of 'primitive man' regarded as 'an elementary form' evolutionarily. Francis (2007) rightly emphasizes the difference between the view of the savages/primitives in *Princ.Psycho.* and in *Princ. Soc.* In the latter, 'primitive man' was seen as pre-social but not devoid of emotions and virtues, and rather the other way around in the former. Thus, to my mind, one may consider these chapters as Spencer's methodological parallel of the chapters in *Princ.Bio.* devoted to evolutionarily lower forms of living entities. It resonated with the parallel drawn and pointed to in *Princ.Psycho.* between the child and the 'primitive,' frequently found in the nineteenth-century discourse of 'Progress cum Empire,' even though both Spencer and Wallace criticized British imperialism, albeit for differing reasons. Another facet of the same was the role Spencer allotted to habit, custom, and tradition in such societies. He contrasted a high degree of plasticity and late individual and cultural maturation attributed to complex societies, with the rigidity and early maturation of individuals attributed to societies considered simple, low in the hierarchy. This dichotomous contrast resonated with analogous dichotomies in *Princ.Bio.* Custom and habit in simple societies, tradition in modern societies, conveyed the power of the collectivity to mold and fashion its 'individual units.' According to Spencer, a capacity for absorbing new impressions existed only in the evolutionarily highest organisms, and only in them did individual experiences play a significant role in shaping behavior.

The enterprise of *Descriptive Sociology of Groups of Sociological Facts* is a case in point. Between 1867 and 1881, Spencer invested much effort and money to bring out a series of volumes, eight of which were actually published starting in 1873, whose topics were diverse 'races'/ancient nations. Under Spencer's guidelines (who considered them rich troves for sociologists), historical and anthropological materials were organized to proffer a comparative universal study, covering 72 societies. Some contemporaneous reviewers applauded the gigantic endeavor, while others criticized it severely, e.g., the tables provided some schematic inter-cultural comparisons, but not any historical sequence (*Saturday Review*, August 8, 1874).

Spencer's position on 'race' may briefly be summarized as follows. On the one hand, he severely criticized British imperialism because maintaining it necessitated a military society regime. At times, he argued for the moral equality and perhaps even superiority of 'the savages' over the civilized, defending the rights of 'the

[13] There Albion Small (1905, chap. 7) questioned Spencer's theorizing of social change.

primitive peoples' under British imperial rule. Spencer was a believer in eventual social (utopic) progress, also related to the potentially remedying power and impact of the environment. On the other hand, he took part in the Victorian theorizing on hierarchies of races, not necessarily linear ones, in their relation to cognitive, emotional, and behavioral capacities and potentialities. He assumed Anglo-Saxon superiority, yet analogized 'the poor' to these 'low races.' He profusely used examples based on such hierarchies and their implications in his evolutionary psychology and sociology, e.g., the sociocultural dependence of growth of brain size in 'the savage' and in 'the civilized.' His lecture essay "Comparative Psychology of Man" ([1876] 1891) explicated his and contemporaries' positions through the use of 'complexity' (mental), 'plasticity' (biological, experiential, and cognitive), the role of 'social environment' in inculcating 'habit.' Finally, he indicated the implied reversion/retrogression or degeneration of those inside such race hierarchies, using transgenerational inheritance mechanism. Much of his work could be read as manifestations of *racial environmental determinism* since the fate of 'lower races' was inevitable. Was there a contradiction? The evolution of societies and tribes was described in analogy to biology, to the development of organisms. Though there was the abstract possibility of looking upon non-western societies as stages in such evolution, in most of his writings the evolution—biological, social, cognitive—of such societies was an in-principle arrested one. Thus, Spencer criticized British imperialism very sharply (as a military-society regime), defended the rights of non-western peoples under its rule, and held on to the Victorian views on these peoples and their arrested evolutionary potentialities.

Spencer's views on growing complexity stood in stark contrast to the views derived from the second law of thermodynamics and in contrast to the tendency of many contemporaries to uphold notions of decomposition as a necessary consequence of that. His notion of 'dissolution' in *First.Princ.* was supposed to provide for both development and constraint, but it also expressed a certain ambivalence toward (inevitable) progress, marked later more explicitly, in the appendix to the sixth edition of that book. There, progress was slow and given to regressions (particularly under the impact of the environment). In *Princ.Soc.*, the notion of not interfering with the course of evolution might in fact have served not to signify that humans have no control over the process, but rather that their control was expressed through intentional non-interference. Right from his mid-early work, he adopted the notion of equilibrium in order to bound evolution.[14] Discussing relative equilibrium allowed him to turn 'dissolution' too into a relative/temporary one, rather than the doomed apocalyptic vision of some of his contemporaries.

Spencer's basic model in discussing 'social nature' remained the tripartite model. Environments were defined very broadly, and societal interactions were given a special and predominant role. His analysis of the tripartite relationship for the social had two facets to it: one focused on the 'super-organism,' i.e., on human society, on collectivity, and its environment, including the social; the other on human

[14] In "Filiation of Ideas (1899), Spencer said that the idea came after a talk with Tyndall in 1858.

individuals, the individual component-units of the social aggregate, and their environments. The latter included organic, physical, climatic, geographic, and social components. Society was graded according to 'complexity of structure.' Human individuals were graded according to a scale based on the evolution of their emotions and on the evolutionary status of the society and civilization they were part of. The correlations between the two foci of Spencer's sociology were such that components that made up elementary social groupings embodied liminal forms of human individuals. Through one perspective one could look upon the human individuals that made up a society only because they were its components whose existence and perseverance were its telos (as a goal-guided system). Through the other perspective, one could look upon collectivities, societies, as the foundational units of the super-organic evolution. But note that Spencer did not look for an 'absolutely originary' unit in *Princ.Soc.*, nor did he do that in *Princ.Bio.* The dynamics of social evolution he was trying to construct could not be captured unless both perspectives were addressed.

Most of Spencer's sociological work, and there are five volumes of *Princ.Soc.* alone, was a detailed elaboration based on historical and anthropological investigations carried out and written up mostly by earlier nineteenth-century authors or by contemporaneous ones. It was organized so as to delineate, illustrate, and explicate institutional patterns of cultural evolutionary dynamics, and its overall emphasis was on the societal perspective. Spencer constantly stressed the special importance of the directionality of time which imbued his evolutionary narrative with a distinctive style. Yet, it was surprisingly devoid of historical sense. He proffered a rather sophisticated notion of culture (relative to his contemporaries), e.g., "the consensus of functions becomes closer as evolution advances … in developed aggregates of both kinds, that combination of actions which constitutes the life of the whole, makes possible the component actions which constitute the lives of the parts" (1874–1896, 1/2:487). The social evolutionary narrative was rich in exemplifying case studies and contained some striking and apt insights on the working of contemporaneous Western imperial-colonial societies. Paradoxically, one could characterize *Princ.S*oc. as consisting of both 'thick' and 'thin' (highly abstract) descriptions.

Lamarckian evolutionized biology was perceived as relying on a partially observable past, which could be reconstructed by supplementary hypotheses. The inheritance of acquired characteristics was such a hypothesis, which presumably would supply an exemplification of a mechanism of the present as well as of the dynamics from the past to the future. The actual passage from the organic to the super-organic was depicted as emerging, "*yet it cannot be held inevitable in each particular society, or even probable*" (1874–1896, 1/1:97—emphasis added).

The impact of collective past generations, which was forcefully present in *Princ. Psycho.* in the notion of collectivities inhering within individuals through the back-and-forth mechanism between biological heredity and cultural/psychological inheritance of acquired characteristics, was similarly employed in *Princ.Soc.* Spencer conceived of 'development' biologically, as broader than heredity since it included both the transmission and the characteristics that had been realized through transmission. There was an element of contingency in Spencer's evolutionary view since evolution depended on the impact of the environment. He emphasized this facet so as to try to remove the attribution of an inner drive from his notion of progress. 'Social nature' was constituted as a 'dense' framework of interactions and interrelations between past adapted and transformed sociocultural products and present experiences. Like many nineteenth-century social thinkers (e.g., Marx, Durkheim), Spencer distinguished between surface phenomena and deep structures, where the latter were supposed to indicate the real state of things. On the surface, the 'units' of human individuals seemed to be the bearers of evolution, who transformed the collective evolutionary framework, but the deep-structure analysis would reveal them as predominantly transformed trans-generationally by that very framework. In some contexts, Spencer suggested that the aggregate nature of society ontologically pointed to human individuals as the foundational entities, which in turn would mean that the individual perspective was the one to adopt epistemologically. Yet, in the major part of his discussions, the collectivity, whether very elementary or quite advanced, was at least as significant. Moreover, the human individual could not be perceived nor explicated unless in conjunction with, and as a component unit of, a collectivity. This is the utopian dimension of Spencer's sociological work—the changed ethical quality of life. The social enabling conditions for it were very high degrees of mutual interactions and cooperation, which served as indices of social evolution. Although in Spencer's narrative, the changes would seemingly happen to individuals and concerned their happiness, these changes could only become socially meaningful when the collectivity at large was affected through a biological–cultural back-and-forth civilizing process. What made possible the alternation between the cultural and the biological in Spencer's sociology and made room for psychology as a unique science of the individual was the role assumed by the collectivity.

2.4 Concluding Comment

In this chapter, I offer an interpretation of Spencer's works based on the argument that his basic epistemological and methodological frame of analysis of living nature was the tripartite model of changing interrelations of the changing interactions between organisms and the environment, and the changing effects they induce. Spencer invested a great deal in evolutionizing the mind–body–world relations, offering evolutionary dynamics interlaced with notions of plasticity and hierarchy. Thus, he constructed a path of back-and-forth movement between the biological and the 'mental/behavioral/social,' providing a useful model for ways to extend plasticity to the psychological and the social realms. He put forward a hybrid notion of individuality, which could be viewed either as 'collective individuality' or as 'collectivity of individuals.' For him, a collective individuality contained multiple modes of relating and *hierarchies* of organizational complexity within its framework. Alternatively, the collectivity of individuals contained diverse kinds of individuals differentially related to each other and to the collectivities whose members they were. The collectivity could be looked upon as a special kind of individual. The first was found to be more useful in biology, while the second turned out to be so in sociology. Thus, there was a clear conceptual impossibility of isolating/disengaging individuality from collectivity in any of these fields.

Another set of problems that stands out is Spencer's attempt to account for the immeasurable diversity that resulted, in his view, from rather minimal beginnings – both in the field of biological entities and that of social individuals. To help understand the overwhelming *diversity*, he took note of the different *time scales* operating.

Even though Spencer held a fairly deterministic view of evolutionary laws, laws in which the environment always played an all-important role, his mechanisms for change included an in-principle contingency. The plasticity of organisms, the existing diversity, and its multiple hierarchies and contingency of change became key elements when trying to account for the enormous diversity in the biological world. The same was true for societies as super-organisms: their plasticity, diversity, and their ever-growing complexity—the latter an inherent feature of all living beings—together with that element of contingency were key factors in accounting for the growth and stability of societies.

Given the above, I would argue that Spencer was deploying Lamarckian mechanisms in an attempt to convey within nineteenth-century conceptual schemes certain insights on diversity, plasticity, hierarchy, and complexity, and on individuals and collectivities. However, these conceptual schemes were rather inhospitable to such views. His endeavors to use these relations and these features for the construction of a synthetic all-encompassing evolutionary worldview are the reason for the limited renewed interest in his work.

Materials from the following articles have been used by permission of:

University of Chicago Press:

2017 – "Spencer's Evolutionary Entanglement: From Liminal Individuals to Implicit Collectivities" in Biological Individuality. Edited by Lynn Nyhart and Scott Lidgrad, 158–183. Chicago: University of Chicago Press.

References

Adam, A. E. (1988). *Spontaneous generation in the 1870S: Victorian scientific naturalism and its relationship to medicine* (PhD). Sheffield City Polytechnic http://shura.shu.ac.uk/2695/.

Barsanti, G. (2000). Lamarck: Taxonomy and theoretical biology. *Asclepio, 52*(2), 119–132.

Bowler, P. J. (1989). *The invention of progress: The Victorians and the past*. Johns Hopkins University Press.

Bowler, P. J. (2015a). From agnosticism to rationalism: Evolutionary biologists, the rationalist press association, and early twentieth-century scientific naturalism. In B. Lightman & G. Dawson (Eds.), *Victorian scientific naturalism: Community, identity, continuity* (pp. 309–326). Chicago University Press.

Bowler, P. J. (2015b). Herbert Spencer and Lamarckism. In M. Francis & M. W. Taylor (Eds.), *Herbert Spencer legacies* (pp. 203–241). Routledge.

Carpenter, W. B. (1861). *Researches on the foraminifera*. Taylor and Francis.
Churchill, F. B. (1991). The rise of classical descriptive embryology. In S. Gilbert (Ed.), *Conceptual history of modern embryology* (pp. 1–29). Plenum.
Corsi, P. (1988). *The age of Lamarck: Evolutionary theories in France 1790–1830* (J. Mandelbaum, Trans.). University of California Press.
Corsi, P. (2005). Before Darwin: Transformist concepts in European natural history. *Journal of the History of Biology, 38*(1), 67–83.
Csiszar, A. (2018). *The scientific authorship and politics of knowledge in the nineteenth century*. Chicago University Press.
Darwin, C. ([1859] 1964). *On the origin of species*. A facsimile of the First Edition. Harvard University Press.
Dawson, G., & Lightman, B. (Eds.). (2014). *Victorian scientific naturalism: Community, identity, continuity*. Chicago University Press.
Dawson, G., Lintott, C., & Shuttleworth, S. (2015). Constructing scientific communities: Citizen science in the nineteenth and twenty first centuries. *Journal of Victorian Culture, 20*(2), 246–254.
de Tocqueville, A. (1988 [1833]). *Journeys to England and Ireland* (J. P. Mayer, K. P. Mayer, G. Lawrence, Eds.). Taylor & Francis.
Desmond, A. J. (1989). *The politics of evolution: Morphology, medicine and reform in radical London*. The University of Chicago Press.
Desmond, A. J. ([1994] 1997]). *Huxley: From Devil's disciple to evolution's high priest*. Addison-Wesley.
Elwick, J. (2003). Herbert Spencer and the disunity of the social organism. *History of Science, 41*, 35–72.
Francis, M. (2007). *Herbert Spencer and the invention of modern life*. Acumen.
Gissis, S. B. (2017). Spencer's evolutionary entanglement: From liminal individuals to implicit collectivities. In L. Nyhart & S. Lidgrad (Eds.), *Biological individuality* (pp. 158–183). University of Chicago Press.
Gould, S. J. (1977). *Ontogeny and phylogeny*. Belknap Press of Harvard University Press.
Harrison, F. (1905). *The Herbert Spencer lecture*. Clarendon Press.
Hofstadter, R. (1955). *Social Darwinism in America*. Beacon Press.
Hopwood, N. (2009). Embryology. In P. J. Bowler & J. V. Pickstone (Eds.), *The Cambridge history of science* (Vol. 6, pp. 285–315). Cambridge University Press.
Huxley, T. H. (1892). On animal individuality [1852]. In M. Foster & E. Ray Lankester (Eds.), *The scientific memoirs of Thomas Henri Huxley* (Vol. 1, pp. 146–151). Macmillan.
Lamarck, J.-B. (1809). *Philosophie Zoologique*. Dentu. http://www.lamarck.cnrs.fr/index.php?lang=fr.
Lewes, G. H. (1855, November 3). Life and mind, 3rd review essay on Spencer's Principles of Psychology. *The Leader*.
Lewontin, R. (2000). *The triple helix: Gene, organism, environment*. Harvard University Press.
Milne-Edwards, H. (1837). *Elémens de zoologie: Ou Leçons sur l'anatomie, la physiologie, la classification et les moeurs des animaux*. H. Dumont.
Milne-Edwards, H. (1851). *Introduction á la zoologie générale, ou considérations sur les tendences de la nature dans la constitution du règne animal, premiere partie*. Masson.
Nyhart, L., & Lidgard, S. (2011). Individuals at the center of biology: Rudolf Leuckart's Polymorphismus Der Individuen and the ongoing narrative of parts and wholes. *Journal of the History of Biology, 44*(3), 373–443.
Pearce, T. (2010). From 'circumstances' to 'environment': Herbert Spencer and the origins of the idea of organism-environment interaction. *Studies in History and Philosophy of the Biological and the Biomedical Sciences, 41*, 241–252.
Pearce, T. (2020). *Pragmatism's evolution: Organism and environment in American philosophy*. University of Chicago Press.
Peel, J. D. Y. ([1971] 1992). *Herbert Spencer: The evolution of a sociologist*. Gregg Revivals.

Renwick, C. (2015a). Biology, social science and history: interdisciplinarity in three directions. *Palgrave Communications, 1*, 1–5.
Renwick, C. (2015b). Herbert Spencer, biology, and the social sciences in Britain. In M. Francis & M. W. Taylor (Eds.), *Herbert Spencer: Legacies*. Routledge.
Richards, E. (1989). The 'moral anatomy' of Robert Knox: The interplay between biological and social thought in Victorian scientific naturalism. *Journal of the History of Biology, 22*(3), 373–436.
Royce, J. (1904). *Herbert Spencer: An estimate and review*. Fox, Duffiled & Co.
Schabas, M. (1997). Victorian economists and the science of mind. In B. Lightman (Ed.), *Victorian science in context* (pp. 72–93). Chicago University Press.
Small, A. (1894). Organic concept of society. *The Annals of the American Academy of Political and Social Science, 5*, 740–746.
Small, A. (1905). *General sociology: An exposition of the main development of sociological theory from Spencer to Ratzenhofer*. Chicago University Press.
Smith, C., & Wise, N. (1989). *Energy and empire: A biographical study of Lord Kelvin*. Cambridge University Press.
Spencer, H. (1851). *Social statics: Or, the conditions essential to human happiness specified and the first of them developed*. J. Chapman.
Spencer, H. (1852a). A theory of population, deduced from the general law of animal fertility. *The Westminster Review, 57*(April), 468–501.
Spencer, H. (1852b). The development hypothesis. *The Leader, 3*(March 20), 280–281. (Reprinted in *Essays Scientific, Political and Speculative, I*, 1–7).
Spencer, H. (1853). The universal postulate. *The Westminster Review, 60*(October 1), 513–550.
Spencer, H. (1855). *The principles of psychology*. Longman, Brown, Green and Longman.
Spencer, H. (1857a). Progress, its law and cause. *The Westminster Review, 67*, 445–85 (Reprinted in *Essays Scientific, Political and Speculative, 1*, 8–62).
Spencer, H. (1857b). The ultimate laws of physiology/transcendental physiology. *National Review, 5*(October), 332–55 (Reprinted in *Essays Scientific, Political and Speculative, 1*, 63–107).
Spencer, H. (1858). Recent astronomy and the nebular hypothesis/the nebular hypothesis. *The Westminster Review, 70*(July 27), 185–225 (Reprinted in *Essays Scientific, Political and Speculative, 1*, 108–91.
Spencer, H. (1859). The Laws of organic form. *British and Foreign Medico-Chirurgical Review, 45*(January), 189–202.
Spencer, H. (1860a). Bain on emotions and the will. *The British and Foreign Medico-Chirurgical Review, 49*(January), 42–52. (Reprinted in *Essays Scientific, Political and Speculative, 1*, 241–64).
Spencer, H. (1860b). The social organism. *The Westminster Review, 17*, 90–121 (Reprinted in *Essays Scientific, Political and Speculative, 1*, 265–307).
Spencer, H. (1862). *First principles*. Williams and Norgate.
Spencer, H. ([1864] 1867). *The principles of biology*. Williams and Norgate.
Spencer, H. ([1870/1872]. *The principles of psychology*, 2nd edn. Longman, Brown, Green and Longman.
Spencer, H. (1874 [1873]). *The study of sociology*. Appleton.
Spencer, H. (1886). The factors of organic evolution. *Nineteenth Century, 19*(April–May), 570–89, 749–70 (Reprinted in *Essays Scientific, Political and Speculative, 1*, 389–466).
Spencer, H. (1891). *Essays scientific, political and speculative* (Vol. I). Williams & Norgate.
Spencer, H. (1891 [1876]). The comparative psychology of man. *Essays Scientific, Political and Speculative*, 351–370. Williams and Norgate.
Spencer, H. (1892–93). *The principles of ethics* (2 vols.). D. Appelton & Co.
Spencer, H. (1893). The inadequacy of 'natural selection,' I and II. *Contemporary Review, 63*, 153–66, 439–456.
Spencer, H. (1894). Weismannism once more. *Contemporary Review, 63*, 592–608.
Spencer, H. ([1874] 1896). *Principles of sociology*. Williams and Norgate.

Spencer, H. (1899). The filiation of ideas. In David Duncan (Ed.), *The life and letters of Herbert Spencer*, Appendix B (pp. 533–76).

Spencer, H. (1904). *An autobiography* (Vol. 2). D. Appleton & Co.

Spencer, H. (1908). *Life and letters of Herbert Spencer* (David Duncan, Ed.). Methuen.

Tattersall, M. (2014). Thermal degeneration: Thermodynamics and the heat-death of the universe in Victorian science, philosophy, and culture. In M. Härmänmaa & C. Nissen (Eds.), *Decadence, degeneration and the end: Studies in the European fin de Siècle* (pp. 17–34). Palgrave Macmillan.

Thomson, W. (1862). On the Age of the Sun's Heat. *Macmillan's Magazine, 5*, 288–293.

Turner, J. H. (1985). *Herbert Spencer: A renewed appreciation*. Sage.

Tyndall, J. (1892). "The Belfast Address" [1874]. *Fragments of science: A series of detached essays, addresses and reviews*. *2*, 135–201. Appleton.

Young, R. M. (1968). Association of Ideas. In P. P. Wiener (Ed.), *Dictionary of the history of ideas*. Charles Scribner's Sons.

Chapter 3
Interlude: The Cluster of Plasticity and the Impact of Its Transfer

> Plasticity, then, in the wide sense of the word, means the possession of a structure weak enough to yield to an influence, but strong enough not to yield all at once … Organic matter, especially nervous tissue, seems endowed with a very extraordinary degree of plasticity of this sort … the phenomena of habit in living beings are due to the plasticity of the organic materials of which their bodies are composed. (William James, *Principles of Psychology*, 1890, 105)

There is no single agreed-upon answer to the question of when the term 'plasticity' came into use in biology—perhaps sometime in the last third of the nineteenth century. In spite of that quote,·a review of Pigliucci's *Phenotypic Plasticity* by Fuller (2003) claims the Swedish geneticist Herman Nilsson-Ehle was the first in 1914 to describe the effect of the environment on an organism's phenotype. Yet, the phenomena it was referring to, the cluster of notions related to them in the biological, neurological, social, and mental fields, had already been the focus of study much earlier, in the natural realm with the human sphere as an indivisible part thereof, and in the human realm as separate and particular. The materiality of organs such as the brain was posited through diverse metaphors as plastic (e.g., Diderot on the brain as 'living wax,' Wolfe, 2016a, b). In this short interlude, I offer a succinct comment on the relations among the components of the Cluster of Plasticity (CP) within diverse SL/NL frames in the nineteenth century. I propose that it played a significant *foreground* role in the work of Lamarck and of Spencer and remained an important *background* factor in the work of the later founders I discuss. I succinctly discuss the concepts involved, their historical use within the Cluster of Plasticity (CP), and the historiographical discussions about them.[1]

[1] Besides the primary materials of the various chapters I have directly used the following secondary sources on *the historical angle*: Allen, 1985; Amundson, 1996; Berlucchi and Buchtel, 2009; Bowler, 1983, 1988; Churchill, 2007; Geroulanos, 2014; Gould, 1977, 1992; Griffiths, 2003; Laubichler, 2007, 2009; Müller-Wille and Rheinberger, 2012; Nicoglou, 2018; Osborne, 1994, 2019; Papineau, 2007, Pearce, 2020; Simpson, 1953; Wolfe, 2016a, b; Young, 1970/1990.

S. B. Gissis, *Lamarckism and the Emergence of 'Scientific' Social Sciences in Nineteenth-Century Britain and France*, History, Philosophy and Theory of the Life Sciences 36, https://doi.org/10.1007/978-3-031-52756-2_3

I suggest that throughout the period discussed in this book, CP served for contemporaries as a '*conceptual package deal*' so that deploying any of its component concepts necessarily implied assuming the others. CP with plasticity at its core included: tropes directly related to 'environment/milieu,' e.g., environmental change, effect of/interaction with/adaptation to environment; use and disuse; learning; and habit. Throughout the long nineteenth century, CP had to do with having the ability to change and/or to cause change under specific environmental circumstances and within specific relations to the environment, i.e., to be malleable and/or to 'sculpt'/form/construct, whether at a particular stage of the organ's or organism's—the living system's—developmental process or throughout its life span (in anachronistic terms, both developmental and phenotypic plasticity). The CP was elaborated upon particularly in conjunction with facets of behavior, but also of 'inheritance (of modifications)'—thus both with questions of development and of heredity, modes of continuous similarities, trans-generational transmission. CP has played a renewed, albeit ambivalent, role in recent discussions in developmental and evolutionary biology, in genetics and epigenetics, in the neurosciences, as well as in philosophy of biology in general.

In my view, CP was central as a conceptual enabling condition in the transfer from and engagement with the SL/NL evolutionary biological in the period and the places discussed in the various chapters. Therefore, I delineate the ways components of the CP were entangled with the emergence of 'scientific' social-science disciplines that had recourse to evolutionary biological repertoires, especially SL/NL ones.

Questions and distinctions implicitly or explicitly posed within late-eighteenth-century discussions on the malleability of living matter, such as colonial and acclimatization projects (Osborne, 1994, 2019), were further elaborated in the discourses of the second half of the century, particularly on organisms. The concepts of CP were supposed to name/describe/analyze relations in nature—i.e., between parts of nature, specifically relations, including feedback loops, between parts of organisms, parts and wholes among organisms, among these and parts of nature defined as 'environment' within those relations, but also, as regards humans, parts defined as 'social environment' (e.g., Spencer).

A variety of components, often deemed as structural, limiting, non-malleable (but also stable, unaffected, and non-perturbed by changes in the environment) at the cellular, physiological, and behavioral levels as perceived then, were regarded as necessary for balancing plasticity with 'resilience'—understood primarily as the ability to keep continued stability/stable-identity/stable-functioning in the face of and amidst changes. These constraints were particularly conspicuous in discussions of plasticity of the nervous system (as innateness was often attributed to such structural constraints), and especially the brain (before the work of Ramon Y. Cajal 1913–1914) and its changes in relation to experiencing and to 'external (environmental) elements.' This was also related to the trend away from the significant role attributed to conscious introspection in the late eighteenth century. It was also relevant to the later ambivalent attitude toward introspection, as in Spencer, to its complex roles in Jackson and in Ribot, and its paradoxical inversion in Wundt. Vidal and

Ortega (2017) offer a scathing critique of the contemporary non-critical use of 'neuroplasticity.' In contrast to this dichotomous tradition, Bateson and Gluckman (2011) suggest modifying the meaning of plasticity and present-day robustness and argue further that plasticity and robustness are complementary.

On the face of it, being active or being passive could be attributed to any 'player' or 'party' within such a frame of relations. However, it is my view that within the SL/NL frames the relations depicted were mostly between *active parties*, whether these were posited as individuals or as collectivities. That activity was not attributed to the 'ontic,' i.e., metaphysical characters of 'living matter' in itself, but rather to the parties of the *interactive* relationships within a transformist/evolutionary conceptual scheme. Organisms were perceived to be self-organizing, though differentially, relative to their degree of organizational complexity, i.e., taking part in their own organizational development, maintaining and regulating their activities, including their response to changes in their environment.

CP was understood by its nineteenth-century users to express a contemporaneous anti-reductionist stance (thus as actors' category) and was deployed to explain evolutionary novelties, rising complexity, expression of organismal characteristics, and their transmission. The variety of modes of deploying CP within SL/NL frames resulted at times in a blurring of distinctions between internal and external causation.

I suggest that Spencer's significant use of Lamarckian tools and their implied assumptions of plasticity-cum-agency were widely influential, especially his deployment of plasticity through the tripartite model of the changing interrelations of the changing interactions between organism and environment, the changing effects they induce, and the transgenerational inheritance of these acquired modifications. Spencer assumed that the consequences of such organismal-biological plasticity could also be inherited as changes in behavioral patterns and vice versa. Thereby, he constructed a path of back-and-forth movement between the biological and the 'mental/behavioral/social,' providing a useful model for ways to extend CP to the psychological and the social realms.

'Learning' as well as 'habit' in both biological and social connotations played a significant role within the late-nineteenth century CP. These facets were crucial in the transfer from the biological to the social, in particular as CP was perceived—when positing it anachronistically, in present-day idiom—as such that through the permeability of organisms–environment a route to evolutionary change was marked by phenotypic adjustment/accommodation to *stressful* environmental conditions, and these effects were then transmitted to subsequent generations. Consequently, what affected the present could then affect change in the future (e.g., serve as cues from the transmitted past). I have tried in the various chapters to point to the potentially 'double-edged sword' of this use of CP—opening up paths of plastic reform of both individuals and collectivities and vice versa—of transgenerational environmental determinism.

A generalized notion of recapitulation, i.e., the conceptualizing of the entangled relations between ontogeny and phylogeny, was often deployed in plasticity discussions. Spencer followed von Baer, while the French neo-Lamarckists mostly followed Haeckel's different models and assumptions. Gould (1977, 1992) argues that not only NL but also American orthogeneticists and some Neo-Darwinists found Haeckel's biogenetic law, which implied a directed evolution, a useful generalization on the process of phylogeny construction. The analysis of such 'descent relationships' turned on the sequence of developmental stages of the individual ontogeny—with the embryo analogy—as reflecting the earlier history in producing phylogeny. Thus, it required a 'universal acceleration of development' (Gould, 1992), while it manifested continuity of similar traits trans-generationally. Bowler (1996) suggests relating the proliferation of this notion during the relevant period to conceptions of genealogical trees of life, as well as the rise of natural history museums.

Within the protracted and multilayered historical evolvement of the entangled modes of perceiving evolution, development, and heredity, the SL/NL frames helped for a while to postpone and even temporarily reject the Galtonian and then Weismannian theories of heredity. Although historically more complicated, it seemed to many contemporary actors *at the time* that whoever argued for the separation of nature and nurture could be perceived as implying a separation between evolution and heredity on the one hand and development and plasticity on the other. Though very different from each other, these theories shared the assumption that adaptive changes acquired during any stage in the lifetime of an organism were not heritable. The implication of this assumption for CP views and positions was an undermining of the models and mechanisms that served as enabling conceptual conditions for its particular transfer between the biological and the social-psychological. The ascent of this assumption was considered by some contemporaries as a crucial factor, though not the only one, in the conceptual separation between 'nature' and 'nurture,' between organisms and their environment, especially in as far as these interactive relations were perceived to be relevant to the production of both novelty and continuity in evolution. These separations had also the effect of cancelling out the 'progress' aspect of evolutionary change, significant especially for Spencer and Ribot, in particular in their socio-cultural transfer, including the change of norms, of social institutions.

In light of J.M. Baldwin's relevancy to understanding the role of the CP in the latter part of the nineteenth century, let me briefly comment on the present-day return of interest. None of the main figures I deal with in the book related to Baldwin, though Binet in his later period did. Such contemporaneous discussions of the role of behavior and learning had somewhat similar end-results, yet explanatory mechanisms differed significantly and so did their later implications. I shall not discuss interpretive controversies that accompany the renewed interest, e.g., in contradistinction to mechanisms discussed by Waddington, in relation to present notions as niche construction, epigenetics, and even the role of interaction with environment on behavioral-plasticity of whole organisms, etc.[2] Baldwin is perceived

[2] In particular the interpretation offered by Godfrey-Smith in Weber and Depew, 2003, chap. 3. See also Ceccarelli, 2019; Depew, 2003; Green, 2014; Griffiths, 2003, 2007; Loison, 2018, 2020; Nicoglou, 2018; Pearce, 2020; Scheiner, 2014; Sommer, 2016.

as conceptualizing a non-Lamarckian yet 'heritable' plasticity—individually acquired adaptive response/traits to be accommodated within a Darwinian frame, with components of learning by imitation and of social heredity. Thus, learning became congenital, emphasizing psychological behavioral plasticity, not through NL but rather by natural selection. Perhaps Baldwin's efforts could be understood within the framework of the endeavor to find a way to embed plasticity, and the inheritance of its effect, within the frame of 'individual–society–environment' in the late nineteenth century. This endeavor had diverse formulations within the Darwinian non-NL, American NL, Pragmatism of that period, the embryological and zoological traditions, etc. Of significance was the multi-faceted endeavor, *weltanschauung*-motivated, to defend, explain, and shape the mental dimension and its evolutionary change within a frame rivaling that of Weismann so that such a frame would incorporate social learning and its inheritance.

In the last decades, studies of Baldwin's work on these issues have thrown light on those entangled and contradictory efforts and brought the novelty of present-day concepts of plasticity into sharp focus. An important facet of this mini-renaissance has been the attention paid to the behavioral aspect of plasticity, i.e., learning, and its loss, especially to two concepts of Baldwin—'organic selection' and 'social selection/social heredity.' This comes out especially in the detailed interpretation offered by Griffiths (2003), e.g.: "At the population level Baldwin Effect would give the impression that the Lamarckian inheritance of acquired characters was taking place, but this impression would be an illusion – no individual organism would actually inherit an acquired character from a parent. Baldwin's account of social heredity and biology–culture co-evolution in important respects resembles the modern ideas of epigenetic inheritance and niche construction" (193–4). Griffiths' thrust in that article is that the Baldwin Effect was residual to Baldwin's principal interest in the nature of the interaction between social and biological heredity, and Griffiths' interest in Baldwin was probably related to the present-day plasticity *problématique*. This was aptly summarized in Buklijas (2018, 180): "larger transformation in biological science towards characterizing the organism as interconnected, plastic, permeable, and responsive to changes in its surrounding."

Historical studies of the late nineteenth century suggest differing contextualization of that period. Green (2014) stresses the social context of immigration to the USA from Europe and of the migrating blacks inside the USA from the South to the North in the wake of the Civil War and its aftermath. Others point to the rising tensions between NL, neoDarwinism, and the late-nineteenth-century perspective on the re-use of the old labels, especially of epigenesis versus pre-formation as labels for new positions on these controversial issues. Yet, others draw sophisticated distinctions; e.g., Loison (2018) draws in detail the important distinction between Baldwin's position and Waddington's radicality.

Let me end by quoting the last paragraph of "A Program for Biology," composed in 2006 by Levins and Lewontin, which implies the Cluster of Plasticity, that is, implying organism–environment, history, and the significance of biological–social entanglement: "We ask three questions: why are things the way they are instead of a little bit different, (the question of homeostasis, self-regulation, and stability); why are things the way they are instead of very different (the question of evolution, history, and development); and what is the relevance of the rest of the world?" (Levins & Lewontin, 2006, 335).

References

Allen, G. E. (1985). Heredity under an embryological paradigm: The case of genetics and embryology. *Biological Bulletin, 168*, 107–121.

Amundson, R. (1996). Historical development of the concept of adaptation. In M. R. Rose & G. V. Lauder (Eds.), *Adaptation* (pp. 11–53). Academic Press.

Bateson, P., & Gluckman, P. (2011). *Plasticity, robustness, development and evolution*. Cambridge University Press.

Berlucchi, G., & Buchtel, H. A. (2009). Neuronal plasticity: Historical roots and evolution of meaning. *Experimental Brain Research, 192*, 307–319.

Bowler, P. J. (1983). *The eclipse of Darwinism, anti-Darwinian evolution theories in the decades around 1900*. Johns Hopkins Press.

Bowler, P. J. (1988). *The non-Darwinian revolution: Reinterpreting a historical myth*. Johns Hopkins University Press.

Bowler, P. J. (1996). *Life's splendid Drama: Evolutionary biology and the reconstruction of life's ancestry, 1860–1940*. University of Chicago Press.

Buklijas, T. (2018). Histories and meanings of epigenetics. In J. Maurizio Meloni, D. F. Cromby, et al. (Eds.), *The Palgrave handbook of biology and society* (pp. 167–188). Palgrave Macmillan.

Ceccarelli, D. (2019). Between social and biological heredity: Cope and Baldwin on evolution, inheritance, and mind. *Journal of the History of Biology*, 161–194.

Churchill, F. B. (2007). Living with the biogenic law: A reappraisal. In M. D. Laubichler & J. Maeinschein (Eds.), *From embryology to Evo Devo* (pp. 37–82). MIT Press.

Depew, D. J. (2003). Baldwin and his many effects. In B. H. Weber & D. J. Depew (Eds.), *Evolution and learning: The Baldwin effect reconsidered* (pp. 3–32). The MIT Press.

Fuller, T. (2003). The integrative biology of phenotypic plasticity. *Biology and Philosophy, 18*, 381–389.

Geroulanos, S. (2014). The plastic self and the prescription of psychology; ethno-psychology, crowd psychology and Psychotechnics 1890-1920. *Republics of Letters, 3*, 2. http://arcade.stanford.edu/rofl/plastic-self-and-prescription-psychology.

Gould, S. J. (1977). *Ontogeny and phylogeny*. Belknap Press of Harvard University Press.

Gould, S. J. (1992). Ontogeny and phylogeny revisited and reunited. *BioEssays, 14*(4), 275–279.

Green, C. D. (2014). James Mark Baldwin, the Baldwin effect, organic selection, and the American 'immigrant crisis' at the turn of the twentieth century. In G. Barker, E. Desjardins, & T. Pearce (Eds.), *Entangled life: Organism and environment in the biological and social sciences*. Springer.

Griffiths, P. (2003). Beyond the Baldwin effect: James Mark Baldwin's 'social heredity,' epigenetic inheritance, and niche construction. In B. H. Weber & D. J. Depew (Eds.), *Evolution and learning: The Baldwin effect reconsidered* (pp. 193–218). The MIT Press.

Griffiths, P. (2007). The Baldwin effect and genetic assimilation. In P. Carruthers, S. Laurence, & S. Stich (Eds.), *The innate mind: Culture and cognition* (Vol. 2). Oxford University Press.

James, W. (1890). *Principles of psychology*. Henry Holt and Co.

Laubichler, M. D. (2007). Does history recapitulate itself? Epistemological reflections on the origins of evolutionary developmental biology. In M. D. Laubichler & J. Maeinschein (Eds.), *From Embryoplogy to Evo Devo* (pp. 13–34). MIT Press.

Laubichler, M. D. (2009). Form and function in Evo-Devo: Historical and conceptual reflections. In M. D. Laubichler & J. Maienschein (Eds.), *Form and function in developmental evolution*. Cambridge University Press.

Levins, R., & Lewontin, R. (2006). A program for biology. *Biological Theory, 1*(4), 333–335.

Loison, L. (2018). Lamarckism and epigenetic inheritance: A clarification. *Biology and Philosophy, 33*, 29.

Loison, L. (2020). Disentangling genetic assimilation from the Baldwin effect. A philosophical perspective. *Paradigmi, 38*(3), 441–462.

Müller-Wille, S., & Rheinberger, H. J. (2012). *A cultural history of heredity*. Chicago University Press.

Nicoglou, A. (2018). The concept of plasticity in the history of the nature–nurture debate in the early twentieth century. In J. Maurizio Meloni, D. F. Cromby, et al. (Eds.), *The Palgrave handbook of biology and society*. Palgrave Macmillan.

Osborne, M. O. (1994). *Nature, the exotic, and the science of French colonialism*. Indiana University Press.

Osborne, M. A. (2019). *The emergence of tropical medicine in France*. University of Chicago Press.

Papineau, D. (2007). The Baldwin effect and genetic assimilation reply to Griffiths. In Peter Carruthers, Stephen Laurence, and Stephen Stich (Eds.), *The innate mind*, vol. 2: *Culture and cognition*. Oxford University Press.

Pearce, T. (2020). *Pragmatism's evolution: Organism and environment in American philosophy*. University of Chicago Press.

Scheiner, S. M. (2014). The Baldwin effect: Neglected and misunderstood. *American Naturalist, 184*, ii–iii.

Simpson, G. G. (1953). The Baldwin effect. *Evolution, 7*(2), 110–117.

Sommer, M. (2016). *History within: The science, culture, and politics of bones, organisms and molecules*. University of Chicago Press.

Vidal, F., & Ortega, F. (2017). *Being brains: Making the cerebral subject*. Fordham University Press.

Wolfe, C. T. (2016a). *Materialism: A historico-philosophical introduction*. Springer.

Wolfe, C. T. (2016b). Materialism and the 'soft substance of the brain': Diderot and plasticity. *British Journal for the History of Philosophy, 24*(5), 963–982.

Young, R. M. (1970). *Mind, brain and adaptation in the nineteenth century: Cerebral localization and its biological context from gall to Ferrier*. Oxford University Press.

Chapter 4
John Hughlings Jackson: A Clinical Scientist

4.1 Introduction

I have chosen to proceed from the discussion of Spencer's effort to evolutionize the human sciences and thus also to establish 'evolutionary psychology,' to a discussion of the work of John Hughlings Jackson, since certain facets of Jackson's conceptualizing are highly relevant to the discussion of transfer/appropriation of varieties of Lamarckism in fields outside biology. In Jackson's work, both the theorizing and the classification of neurological pathologies manifested a specific evolutionary view of the nervous system, its structure, and its functioning in health and disease. His work thus formed a significant link in the transfer-network of Spencerian-Lamarckian/neo-Lamarckian (SL/NL) models and mechanisms that proved to be so generative for the emerging social sciences of the late nineteenth century, for psychology in particular. In *A History of Experimental Psychology*, Edwin G. Boring adds in his footnotes to British Psychology: "John Hughlings-Jackson affected psychology by way of his *doctrine of the evolutionary strata* of the brain *and the mind*" (Boring 1929, 502—emphasis added); a rather extraordinary credit by a historian of psychology, who was educated by Titchener at Cornell and had little appreciation for the 'non-experimental' as defined by Titchener and himself (Fig. 4.1).

Though developing a clinical methodology to deal with individuals, Jackson's overall purpose was to formulate a universalized concept of diseases of the nervous system and their sophisticated taxonomies as processes, especially of epilepsies and aphasias. This would give the field not only diagnostic tools and technologies but also a scientific disciplinary status, and thus construct a 'universalized individual' and his/her 'universalized brain,' for which other researchers then proceeded to produce *localizing* structural-functional maps and atlases. Yet, 'the universalized epileptic' could not have epistemological robustness for Jackson without assuming the 'collectivity of epileptic patients' beyond the specificity of patients, the difference, locality, and contingency of individual case studies. Moreover, within his

S. B. Gissis, *Lamarckism and the Emergence of 'Scientific' Social Sciences in Nineteenth-Century Britain and France*, History, Philosophy and Theory of the Life Sciences 36, https://doi.org/10.1007/978-3-031-52756-2_4

Fig. 4.1 John Hughlings Jackson portrait (1835–1911). Frontispiece from *Selected writings of John Hughlings Jackson,* Hodder & Stouhgton, 1931. Source: Wellcome collection

reconstruction of the evolutionary-developmental path of the nervous system, 'collectivities' played a role somewhat similar to the one conceived by Spencer. Likewise, Jackson held that certain 'physiological constants' of the human nervous system: these (Kantian) neural molds of perception, such as perception of time, space, and mass, evolved from adaptive physiological activities and were inherited trans-generationally.[1] His work had directly influenced discussions on automatism in France with which he was familiar. In recent years, a body of interpretive studies on Jackson has been accumulating.[2] I will not deal with the role of reflex theory in Jackson's work, in which he continued on Laycock's path, nor with issues related to the two hemispheres theory. I assume that he held on to both of them (as argued in Harrington, 1987, chaps 7–8).

[1] Already in the 1850s, the 'transcendental faculties' were also biological phenomena for Lewes, Martineau, Spencer, and the other members of that circle of intellectuals.

[2] The principal ones: Chirimuuta, 2019; Critchley & Critchley, 1998; Dewhurst, 1982; Greenblatt, 1970, 1999; Harrington, 1987, 1991; Jacyna, 1981, 1982, 2000, 2008, 2011; Lekka, 2015; López Piñero, 2010; Riese, 1950, 1954, 1956, 1965; Smith, 1982a, b, 1993, 2012; Smith, 1992, 2016, 2020; York & Steinberg, 1993, 1994, 2002, 2006, 2013; Young, 1970; certain facets discussed in Casper, 2014, 2016; Finn, 2012; Gere, 2011, 2013; Greenblatt, 1970, 1999; Head, 1926; Lorch Marjorie Perelman, 2004, 2008; Oppenheim, 1991; Shorvon, 2014; Star, 1989, 1992; Sterling, 2010; Swash, 1989; Temkin, 1994; Wallesch, 1989. Philosophical-ideological context: Daston, 1978; Danziger, 1982, 1990; Radick, 2007 (esp. chaps 1–2); a section on Hughlings Jackson in *Brain* 134, no. 10 (2011), Occasional Papers, 3106–26.

My focus here is on Jackson's evolutionizing of the nervous system and on the impact of SL assumptions on his theorizing and clinical practice, as relevant to my general argument. My argument here is two-pronged:

(a) I look at Jackson as *a clinical scientist* who created and formed this scientific-persona role (e.g., Algazi, 2016; Daston & Sibum, 2003), when medical specialization in Britain was just beginning. Thus, he fashioned his clinical activity in constant interaction with his scientific-neurological endeavor, and vice versa. His search for an epistemological foundation for neurology and the philosophical implications thereof stemmed from the need for such core interaction within the contemporaneous context. It resulted in opening up a new investigative space, a new scientific-medical field—neurology—discarding both anatomy and soul psychology as explanatory tools.
(b) Regarding Jackson as a clinical scientist reveals that for this endeavor he constructed meticulous modes of following his patients: data, record keeping, autopsy, and laboratory. Such effort could only take place within a framework that was fundamentally collective, even though structured as if each physician was on his own. His epistemological search, enhanced by his evolving observational regime, led him to draw on Spencer's notions of evolutionary hierarchies and apply them to the nervous system. Yet, he rejected Spencer's particular version of it. *Jackson uncoupled the Spencerian evolutionary direct relation/alignment between complexity and organization.* His view on the potential and actual possibility of 'emergence,' his complicated position on mind–brain–body and his 'methodological parallelism' were shaped by and related to that uncoupling.

Neurology and scientific psychology have ever since followed Jackson's version rather than Spencer's in their clinical and theoretical evolutionary explications, though Jackson's work was rarely explicitly quoted by his direct contemporaries outside Britain, except for Ribot, and somewhat later Freud. That changed from the end of the first decade of the twentieth century onward. For example, in a central German textbook on neurology from the early 1930s, Jackson was quoted almost as many times as Jean-Martin Charcot (Wallesch, 1989/2013). The decade when Jackson began his medical activity, the 1860s, and the two following decades were crucial in the investigation of the brain/mind and their relationship. The cluster of disciplines regarded as having these as their scientific object were forming their distinct and separate paths—psychiatry, psychology, neurology, physiology (and philosophy). This process brought about significant changes in the character of contemporary institutions related to these fields and their practices, as well as the establishment of new ones. Thus, it is highly relevant to understand the medical institutions and practices that were informing, shaping, and constituting the context of his work and being shaped by it.

The first section discusses Jackson's endeavor to constitute *clinical work as scientific practice* and the role of *medical cases* in that. The second section briefly introduces medical-contextual layers deemed relevant to it in the second half of nineteenth-century Britain: the emergence of specialization in the British medical field, institutions for 'nervous diseases,' and evolving procedures of data recording 'nervous-disease' patients. The third section discusses the interlacing of Spencerian-Lamarckian evolutionary models and mechanisms in Jackson's work as they were applied to the nervous system. A succinct discussion of localization, with regard to epilepsies and aphasias, follows. Thereafter I deal with a cluster of philosophical issues of concern to such a clinical scientist: the relations of brain, mind, consciousness, the unconscious, and 'self/person'—as explicated by Jackson from his particular evolutionary perspective.

4.2 Jackson's Clinical Work as Scientific Practice

Jackson was well acquainted with work done outside Britain. In his medical case studies and other articles, he extensively quoted from the work and experimental findings of both French- and German-speaking neurologists and corresponded with some of the French ones. I look at his clinical work within the context of British medicine as a mode and form of scientific practice.

Other scholars have already explicated Jackson's gradual development of methodical modes of investigating symptoms in order to constitute a disease, epilepsy in this case, and its classifications (in particular, Lekka, 2015; Shorvon, 2014; Star, 1989; Sterling, 2010). The significant feature is that Jackson's work was based on a close follow-up on the path of the patient's nervous-system-state as revealed through his/her body, relying on increasingly detailed and standardized forms of inquiry at Queen Square Hospital, carried out by physicians, general and specialized, and support staff. Jackson also relied on information on the diverse phenomena directly related to the disease gleaned from friends and family as well as the patient him/herself, and recorded in these forms. Contrary to asylum reports, which detailed patients' behavior and their attitudes toward the staff etc., these physical examinations came to include, besides the patients' history and some family medical history, most body parts, in particular eyes, face muscles, the four limbs, heart, lungs, liver, senses, reflexes, weight, blood, urine, fever, cognitive-mental state, as well as very detailed daily tables on acute states of disease and their stages and progress.

Research has been undertaken comparing the various institutional frameworks for the treatment of epilepsy and their differing features, the changing divisions of problem-fields throughout the last third of the nineteenth century between neurology, physiology, psychology, and psychiatry, and their differing paths in Germany, France, and Britain (Danziger, 1982; Daston, 1978; Rylance, 2000, chap. 3; Star, 1989, 1992; Wassmann, 2005).

Jackson started his work at several institutions: a general hospital (London Hospital); a charity asylum, West Riding Pauper Lunatic Asylum, at that time run by James Crichton-Browne; and a specialized hospital, in Queen Square. The large majority of his charity-hospital patients were poor, low-class people. For a while, the Asylum (West Riding) afforded him, and others, a unique opportunity for research, including autopsies, and publication in its *Reports*. In comparison to other dominant figures in Queen Square Hospital, such as William Gowers, Jackson stood out in his consistent avoidance of moral judgment or social assumptions regarding the behaviors of the epileptic patients before, during, or after an acute state. In an article published in the *West Riding Journal* in 1875, in an issue devoted to epilepsy, Jackson described temporary mental disturbances in the wake of epileptic paroxysm, emphasizing the automatic character of actions carried out during such states. He strongly conveyed that the patient could not/should not be considered responsible, neither morally nor legally, for such actions or their results. Furthermore, he argued that in many cases the patient would seemingly continue an action initiated before the oncoming epileptic episode, but it would be carried out in a wildly absurd manner, totally unconscious. Apparently, Jackson was primarily interested in studying and assessing particular cases, in order to reconstruct a range of organic causes, to exemplify medical-clinical standards, to estimate the pathological effect, so as to help his patients to 'return.' Such 'return' meant to achieve a meaningful reduction of epileptic seizures, to improve the patients' general physical state, and to allow those who had been working before to return to work.

Jackson's long-term professional aim seems to have been a combined neurologization and localization of the disease, which—with further internal classification—would result in the establishment of a standard mode of representing it. The process of constituting epilepsy (and to some extent aphasias) as a disease of the nervous-system, a robust scientific object, supported by presumably cross findings in ('experimental') physiology, pathology (brain autopsies), neurological surgery, and neurology itself. Concurrently, the person suffering from what became defined and classified as a particular form of epilepsy was constituted as a nervous-disease-epileptic-patient. This, in comparison with and in contrast to any 'disease of the mind,' was highly instrumental in constituting neurology as both a clinical specialization and a disciplinary field. Contrary to accepted convention at the time, Jackson did not conceptualize epilepsy as a 'social disease.' His descriptions and analyses did not include elements of the common myth of epileptic violence, sexual aggression, etc. Furthermore, he seemingly listened to at least some of his patients' descriptions of their experiencing of particular states and actually used their terminology. On the other hand, contrary to asylum physicians, Jackson avoided including a plethora of biographical details, while insisting on the inclusion of the empirical findings of the close inspection and follow-up. These were used in order to establish structural-functional-material etiology.

These practices stood out in comparison to the late Victorian discourse on the connections between epilepsy and hereditary diseases, degeneration, human regression, criminality, moral insanity, etc. Patients said to be suffering from epilepsy were often included collectively in the 'residuum'/degeneration discourse of the time (Harris, 1992)—following medical-psychiatrist/alienist understanding of heredity (Prosper Lucas, Benedict A. Morel, Valentin Magnan, then Henry Maudsley), of mental disturbances/pathologies, and of degeneration as a transgenerational incrementally accumulated organic transfer of loss of level of organization and atavistic jumps even without any precise formulation of hereditary laws. These discourses have been looked upon as expressing the dynamically changing contemporaneous bourgeois/middle-class anxieties concerning social class/social order, the representation of social deviance (e.g., in biological criminology), its treatment by the state institutions, e.g., courts, doctors, and schools (Chamberlain & Gilman, 1985; Nye, 1984; Pick, 1989).

The evolutionary mechanisms and models Jackson adopted organized the clinical and theoretical field in a particular way. Yet, this occurred within a medical-institutional context of gradual change, alongside the rise of alternative ways of conceiving notions of human evolutionary development and of heredity. There was also the formation of webs of alliances among neuro-physiologists, evolutionized-biology scientists, clinicians, pathologists, and surgeons, all trying to converge in producing a unified, universalized brain model. Such webs were manifested during the anti-vivisection trial of David Ferrier. Moreover, one should add a more general medicalization of the public discourse. Within this context Jackson did not consider 'dissolution' or epilepsy as a 'moral regression' nor as a 'regression to less-than-human' on the evolutionary continuum between humans and animals—that is, not in the way insanity was presented by Darwin, namely as loss of emotional restraint parallel to that of the great apes, as manifested in the text and choice of photographs and drawings in Darwin's *The Expression of the Emotions in Man and Animals* (some of which were supplied by Crichton-Browne from his asylum photograph-collection), nor as regression to an earlier, less human hereditary mold as found in Maudsley's and George H. Savage's work. Nor did Jackson draw any analogy between the epileptic individual and 'degenerate civilizations,' or construct the image of the 'epileptic character,' as Gowers (1880) did. Finally, contrary to many of his contemporaries, Jackson did not see epilepsy as a hereditary and irreversible 'degenerative process' that was worsening over the course of time and spiraling to all branches of the family-genealogy. Thus, within the context of Victorian notions of heredity, Jackson's evolutionary view and work did not belong in important ways, clinically and theoretically, to either the 'degeneration' trend as Maudsley's or the rising eugenical concerns.

Having drawn this distinction, it is important to note that other parallel dichotomies hovered in the background at the time, those between health and illness, normal and abnormal, since detailed observation and measurement implied a norm, even if not necessarily normativity, within a context where 'progress' and 'evolution' were directional and often merged together in contemporaneous use. As Jackson wrote in 1879: "There are two kinds of classification of disease – one scientific, generally called theoretical, for the advancement of knowledge; one empirical or clinical, for practice. In the scientific or theoretical, so far as is possible, we classify diseases as they are departures from healthy states – as they show abnormalities of structure, function and nutrition. Empirically or practically, we classify them as *they approach certain types*" (1958, 1:276). This perhaps allows for a partial understanding of why and how an empirical bodily perspective on epilepsy enabled Jackson to arrive at new insights and organic perspectives on the physiology of the brain and the nervous system, also leading to an inversion of Spencer's relation between complexity and organization. This raises the question of whether Jackson also conceived of a distinction between normal and abnormal brains, healthy and diseased ones, rather than various states of the brain. I would argue that the two were not merged in Jackson's discussion, rather it would not be far-fetched to see Jackson's effort to turn clusters of symptoms into 'epilepsy,' a well-defined scientific-medical object, as analogous to Ludwik Fleck's analysis and description ([1935] 1979) of how 'a disease' (syphilis) is established.

Though Jackson rarely discussed etiology beyond the physical boundaries of the individual, the indirect ways, in which 'the social' at large played a role in his conceptualization of the etiology of disease and its therapeutics proved Jackson as a non-believer in 'the creed of degeneration.' The 'social' came up in the constitutive effects on individuals' lifespan, development, and participation in the social institutions they inhabited. For example, in 1883, Jackson conducted a survey on nervous diseases among children, where causality was attributed to school pressures and the sanitary state of schools in poor areas. Until the end of the nineteenth century, epileptic children were not included in either the 1893 general Education Act for ages 5–13, or in the later addition for children with special needs (Oppenheim, 1991, 244).

4.2.1 *The Uses of Medical Cases*

Positing Jackson's clinical work as 'scientific practice,' I propose to look into his use of 'medical cases' in a circumscribed attempt to discern their role in constructing a scientific field, neurology. For that purpose, medical cases are defined as those published within a medical publication framework, rather than patients' files or physicians' case-notes. The historical-sociological literature on present-day medical cases and/or medical records is huge, a sizable part of which has been written under the impact of Foucault's work, as well as by historians and philosophers of science. They discuss a very wide variety of aspects and layers of the medical case/record/patient history, especially the contemporary one.

What sort of labor did medical cases do for Jackson? Were they instances of a conceptual 'class' that they exemplified? Or more of an agent, with at times complex properties? How could they enable generalization, even universality? For the discussion, I adopt the characterization of *the medical case as an epistemic genre*, especially as elaborated by historian Gianna Pomata (2010, 2013, 2014) in her reconstruction of the constitution of its modern version in the sixteenth century and its stabilization by the late seventeenth century. Pomata defines epistemic genres as those kinds of texts that develop in tandem with scientific practices and are in the eyes of their authors linked to the practice of producing knowledge (whichever way it is culturally defined in a particular era). Being a genre means that no single individual case at any point in time can represent the whole spectrum in its dynamics. Let me elaborate further. An epistemic genre evolves and is transformed, especially when prevailing norms are questioned, need to be adjusted to new situations, or when several theoretical/ conceptual schemes are concurrently considered to be paradigmatic and are held and promoted by differing networks. Yet, being a genre also implies a measure of stability, both functional and social, as a shared collective tool for its users, which produces knowledge: observational–cognitive–textual. Therefore, its transformation would mean a transformation of modes of acting and of thinking, both collectively and individually. Andrew Mendelsohn (2017) argues that such cases enable the transfer of knowledge among physicians, as they are added to an invisible reference library of the accumulated known as well as of the changing unknown; a library that has been constantly used by physicians from around 1800 onward. Such cases usually reported extraordinary or novel features, combinations of symptoms, and phenomena that did not easily fit in with known categorizations (Ankeny, 2011).

To consider the medical case as an epistemic genre makes it possible to look into what was regarded at Jackson's time as belonging to 'practice,' referring to the empirical, experiential, or material practices, and what was considered scientific/theoretical knowledge, including *medical* habits-of-the-mind—rules, conventions, norms. The latter components were expressed by the very act of publishing, which made cases useful. This enabled a seemingly smooth transition from the particular case to a generalizable claim and even a universal one.

Certain significant features become discernible by looking at Jackson's theoretical and institutional context in Britain (and Ribot's in France) through the lens of the medical cases as an epistemic genre:

1. *Plurality of competing conceptual/theoretical frames*; especially relevant were those regarding mind–body relations, relations between mind–mental activities, consciousness, the unconscious, and the nervous system. Other relevant frames pertained to the meaning and mechanisms of *evolution* and hence the relations between heredity and environment as well as the features and components of the social.
2. *Plurality of what was considered 'empirical'* and thus indispensable for the production of scientific knowledge in the realm between psychology and neurology, yet was thought to form part of or be related to medicine. Disagreements about 'the empirical' abounded: clinical work within institutional frameworks, animal experiments, laboratory measurement-experiments with humans, microscopic work, autopsy analyses, hypnosis-induced states, suggestion cases, automatic writing, questionnaires, and surveys, and some elaborate forms of introspection.
3. *Plurality of means of collective scientific-knowledge production through sharing*: scientific journals and hence scientific articles, including medical-case articles (Bartrip, 1992; Bynum & Wilson, 1992; Csiszar, 2019); national and international conferences; national and international specialized associations; and institutional networks. The constitution of disciplinary fields and the agreement on basic case components and measurements, on norm-standardization and processes of specialization were mutually enhancing. Such were also emerging agreed-upon procedures for case analysis, contextualization, and narrative sequence and structure. That allowed for the transfer and reproduction of cases among specializing communities and professional journals.[3] Cases were usually

[3] On nineteenth-century journals in Britain, see Bynum & Wilson, 1992; Rylance, 2006.

published in venues called 'journal,' 'bulletin,' 'gazette,' 'circular,' and 'reports' as well as in conference transactions. If they appeared in other frames, it was often under the rubric 'letter to the editor.'

Since the medical case functioned as an epistemic genre during the second half of the nineteenth century,[4] the aggregation of cases could be posited as constituting a gradually and collectively constructed normalcy or as a new categorization, especially when the aggregation was in a format of a 'review' of similar cases or cases posited as close, relevant, or related. Such presentation could even be expounded as a possible challenge to the accepted perception of physiological laws. This was also perceived as enabling an insight into yet un-clustered collective phenomena, even through an exceptional individual medical case (e.g., Geroulanos & Meyers, 2018). At the turn of the century, cases became not only an informal mode of introducing medical novices into their profession, but also an officially recognized didactic tool of the medical education curriculum, such as with Walter B. Cannon (1900).

For both Jackson and Ribot, the biological-medical was predominant in the case narrative and structure and provided objectivity. Even when the 'voice of the patient' was directly or more often indirectly represented, it was perceived as reporting on a subjectivity objectified through biological-medical categories. Fissel (1991) argues that already by the mid-nineteenth century the voice of the patient disappeared from institutional medical cases in Britain. Before the turn of the century, patient files were diversely organized within institutional frameworks: by diagnosis, alphabetically, by bookcases arranged by year, by principal physician. Some historians dealing with medical records from the late eighteenth and the nineteenth centuries, particularly institutional ones, argue that both records and published cases conveyed *a process-narrative* rather than a static-state description. That narrative stretched until the end of the treatment, or until hospital discharge or death (Hurwitz, 2006, 2017). The majority of cases related to lower-class often very poor people institutionally hospitalized or treated in a hospital's outpatient clinics. Neither they nor their families were asked for permission to publish the case, whether bearing a name or anonymously, or to perform an autopsy. The results of autopsy, when done, were included in it. Their very inclusion, especially in neurological cases, projected a retroactive causal directionality. As such, a process-narrative medical case implicitly delineated also the boundaries of its possible interpretation.

Jackson's very early articles, published in *Medical Times and Gazette* and later in the *Reports* of West Riding Asylum, consisted of a comprehensive and detailed presentation and elaboration of a medical case, where the concrete details were very tangible. This concreteness was evident even in summaries of professional reporting on his presentations within diverse frameworks. In somewhat later publications, but also in articles presenting a more generalized argument, there were often

[4] Especially: Ackerknecht, 1967; Anderson, 2013; Berg & Bowker, 1997; Berlant, 2007; Carroy, 2005; Charon, 1992; Class, 2014; Forrester, 1996; Geroulanos & Meyers, 2018; Gerring, 2017; Hurwitz, 2006, 2017; Lawrence, 1985; Hunter, 2021; Nowell-Smith, 1995; Risse & Warner, 1992; Rylance, 2006; Smith, 1999; Wieviorka, 1992.

'collections of cases,' three to four, that were presented in a much more succinct reviewing manner. Many of his articles, particularly the later ones, had also long theoretically and methodologically loaden footnotes, which at times added additional shortened 'cases.' I would argue that each case presentation was in a sense singular, combining elements whose combination was irreproducible, thus grounding its empirical status, and at the same time it was also assumed to elucidate other cases, or to develop into a tool for this or other articles relating to a similar *problématique*. Thus, each case was both unique and a component in the endeavor to constitute the 'meta-case,' groping toward a novel generalization, usually diverging from the currently accepted one. Jackson deployed the cases in his articles either to offer an analysis pointing to such a possible general interpretation or to indicate that though he had not yet developed a good enough generalized view, he could not abide by the prevalent ones. From rather early on, these cases were placed almost always within a Spencerian-Lamarckian matrix projected onto the nervous system. They played a role in a wider clinical/theoretical endeavor, which can be grasped more readily when one follows Jackson's accumulated articles. This was a gradually evolving and not always straightforward, linear process of constituting 'medical facts' in Fleck's sense ([1935] 1979), i.e., establishing and classifying the 'medical-biological fact of epilepsy' and the 'medical-biological fact of aphasia.'

4.2.2 Medical Context: Medical Specialization, Caring Frameworks for 'Nervous Diseases,' Recording Medical Data

Jackson was a practicing and theorizing physician who looked upon himself as specializing in the nervous system, thus as a neurologist (see also, Bynum, 1990; Bynum & Wilson, 1992; Reiser, 1978; Stacey, 1992). I have argued that he strove to generalize from his clinical work, based on a scientifically viable methodology of individual patients as case studies that made generalizing possible, knowledge of the human nervous system (of the individual, *any* individual), and criteria for classification, practical and theoretical, of neurological diseases/disorders.

There is a historiographical inclination to posit him as the founder of the discipline of neurology in Great Britain. This attribution should be qualified by present-day discussions on the role and pace of specializations and the formation of new sub-disciplines within medicine, which took place in Britain, where for a variety of factors the dominant tenor was that of 'general medical practice,' only toward the latter part of the nineteenth century (Casper, 2014, 2016; Shorvon, 2014; Shorthorn & Compston, 2018). The line dividing the production of theoretical knowledge from that of acquiring practical, i.e., clinical, skills was almost non-existent within mid-nineteenth–early-twentieth-century medicine. Yet, meaningful changes took place between the early 1860s and the early twentieth century, when Jackson fully retired, and nervous-system medicine acquired distinguishing features.

Specializing institutions went earlier through more profound medicalization and often preceded the establishment of a sub-field as an autonomous specialization within medicine. Most of the medical textbooks in the second half of the century were written by physicians serving in specialized hospitals.

During the second half of the nineteenth century, there were a number of institutional frameworks, public ones—state, town, and county, supported through charity, under the Poor Law—as well as private ones, which hosted, treated, and catered for an amorphously defined group of 'nervous diseases' sufferers. The four principal types were asylum, general hospital, specialized hospital, and country colony (and the somewhat earlier 'retreat'). All asylums, as a non-punitive mode of distancing poor people, were presumably regulated—also medically—in the wake of the Lunacy Act 1845. The main treatment was moral/'civilizing,' in a framework of regulated 'domesticity,' and 'organic' degeneration-discourse etiology, with long-term hospitalized clinical follow-up and autopsy of those who died there (Hurren, 2012). The research, in which Jackson took part, in West Riding Pauper Lunatic Asylum in the 1860s and 1870s was exceptional, yet it was based on earlier lower-level common practices.

The country colony, such as Chalfont (1894), related to the national hospital, is controversial in historiography: a vision to rehabilitate or to exclude and isolate? It was located in rural areas, populated mostly with epileptics of various degrees of severity who, because of the assumed therapeutic quality of such life, received lower dosages of bromide (Scott, 1992; Turner, 1897). Such 'farm'-colonies (versus urban-industrial projects) were being established all over Europe and the USA during those years (Hamlett, 2015; Star, 1989; Sterling, 2010; Scott, 1992). Jackson first encountered 'mental disease' cases when trained as a very young man at the 'York retreat' of the Tuke family. The difference in the patient populations among these four types was the stage of disorder and social status. Later, hospital reform changed the organization of all hospitals and closed down state-funded asylums and workhouses. Following Florence Nightingale's public involvement, the very structure and internal organization of hospital buildings underwent changes.[5]

The specialized hospitals, such as the one Jackson worked in, were small, 40–90 beds at the most, and often had an outpatient clinic, separating its patients completely from the hospital ones. These clinics enabled a selection of patients to be hospitalized for short periods, combined with a long-term follow-up. The field of neurology was based on some general hospitals, University College Hospital, King's College Hospital, and London Hospital (where Jackson worked partially until the mid-1890s), as well as on three specialized ones, which formed the foundation for systematic neurological research: the National Hospital at Queen Square, the one in Maida Vale founded in the 1860s, and West End Hospital founded in 1878.

[5] E.g., Bristowe and Holmes report, 1863, on the hospitals of the United Kingdom; Lancet report on workhouses, infirmaries, 1865; Metropolitan Poor Act 1867 (Granshaw, 1992; Taylor, 1915, 1925).

The National Hospital for the Relief and Cure of the Paralysed and Epileptic at Queen Square (National Hospital) where Jackson worked from 1862—at age 27—until his retirement, was founded in 1859 and opened in 1860 as a private charitable institution, but it gradually came to be mostly run by the physicians. It was intended for the 'deserving poor,' considered as 'laboring, non-alcoholic, non-debauched,' etc. (recall Henry Mayhew's work on *London Poor* in mid-century, and Charles Booth's division of the London poor into six categories in the 1890s). Since its foundation, the National Hospital was a specialized hospital, only catering to paralyzed and epileptic patients. It did not accept chronic, i.e., long-term patients, and had an outpatient clinic. Thus, Jackson treated short-term hospitalized patients and attended to their follow-up in that clinic. This hospital was soon acknowledged as necessary since its potential patients would not be accepted into general hospitals, and its founding was received with approval, e.g., by the *Lancet* (1859 74:624–25), *British Medical Journal*, and London newspapers (e.g., *Observer* February 1863).

The Society of Neurology was founded in 1886, with Jackson as one of its founders and its first president, while Spencer was elected as an honorary member. In 1907, the Royal Society of Medicine was established, and all the specialized societies became sections of it. Some of the principal members of the Society of Neurology became members of the Royal Academy, including Jackson. There were two main journals: *Mind* (1876), philosophically-psychologically inclined; and *Brain* (1878), clinically-physiologically inclined, with Jackson as one of the founders, since in Britain physiology of the brain was not coupled with psychology of the mind (as for example, in Germany). Between 1871 and 1876, Crichton-Browne, who was involved with the founding of both journals, edited a journal *West Riding Pauper Lunatic Asylum Medical Reports*, reporting on the advanced physiological and pathological research done in the laboratory and the asylum of West Riding, including by Jackson and David Ferrier (Finn, 2012). The Society of Neurology and *Brain* encouraged an inclusive view of neurology – as combined with psychology, physiology, anatomy, pathology, as well as ophthalmology and neural surgery, while the separation of kinds of patients, of diseases, of those who catered for them was gradual. During the last third of the nineteenth century, there were still some who were both psychiatrists and neurologists, e.g., Crichton-Browne. Yet, the field of neurology carved a new space for itself, both medical and institutional.

Jackson's work was carried out through the hospital-cum-outpatient-clinic format, which formed an enabling condition for the construction of specialization in neurology as an autonomous field. The National Hospital soon turned out to be the best specialized one, a preferred study place for young physicians, and a diagnosing center for patients sent from the general hospitals. It was organized around a number of main figures, and it did not have any direct connection to university medical courses. Salpêtriére Hospital and Charcot presented another model, of one person identified with the field and the 'school,' intimately connected to the university school of medicine. In both the USA and Germany, contrary to Britain and France, most hospital departments included both neurologists and psychiatrists, and thus dealt also with insanity cases (Goetz et al., 1995).

I propose to look into the evolving of medical-data recording at the National Hospital into a significant tool in the development of a 'medical-case experimentality,' its roles in the constitution of neurology, specifically in Jackson's work, in his persona as a clinical scientist, and in the construction of epilepsy as a neurological disorder. Between the mid-1860s and the late 1880s, there was a marked increase in the list of checks, biochemical and microscopic tests and examinations—physical, neural, functional—that any patient was subjected to either daily or on another regular basis, as well as a widening spectrum of the required information on patients' health history, personal and social context, and treatment outside the hospital. These checks, tests, reports, etc., included from the late 1860s former illnesses, detailed descriptions of symptoms, mental state (memory, speech, ability to answer questions), cranial and spinal nerves, duration of illness, physical reactions to treatment—electrical reactions to faradism and galvanism; from the 1870s—family history and social history, measurement of hand-grip strength with dyanometer, examination of tactile localization, changes in reflex and sensory reactions; from the 1880s—continued treatments and medicines, seizures frequency and intensity while in hospital and outside it, body temperature, pulse pace, sweating and weight, urine analysis, results of tests of sensory and perimetry charts; from 1888—also microscopic examination of the tumor; from the early 1890s—living environment, occupation, hygiene, diet; hemoglobin in blood (1893), and women's menstrual cycle. As surgery (and localization theory) became more prevalent, relevant drawings and photographs were added to patient files. Most of the data collection, including autopsies, was performed by junior physicians who were responsible for the files (both hospital and outpatient), and later on also by nurses.

Around the mid-1880s, this evolving recording methodology became uniform and standardized. Concern for objectivity/scientificity was an important consideration, hence the growing emphasis on standardization of the clusters of tests, measurements, etc.[6] There emerged a generally agreed-upon procedure of analysis, contextualization, narrative sequence, and structure, as a component of the scientification of medicine. The patient's body (and its past medical history) was regarded through these procedures, considered objective, *a site of 'nature's experiment.*' Jackson was deeply involved in the data-recording process and devised some of the measuring procedures, as part of his invested efforts to constitute the new field of neurology as both clinical and scientific (e.g., 1958, 1:276 [1879]). In 1885, at a meeting of the ophthalmological society, Jackson dedicated his keynote lecture to the evolution of the nervous system and expounded on neurology as an *integrative* field, a unique yet autonomous specialization, drawing an analogy between social, medical, and neural division of labor (1958, 2:346–58, esp. 346, 352–53, 357–58).

---

[6] For Charcot, standardization was not a central concern; his case files were markedly differed.

4.3 Spencerian-Lamarckian Evolutionary Models and Mechanisms as Applied to the Description and Analysis of the Nervous System

I am going to analyze here how Jackson deployed an evolutionary SL model and its mechanisms to understand the nervous system and explicate some of its disorders. Interpretations of Jackson and his neurological legacy have concentrated on the concept of dissolution; yet, its meaning and use were anchored in Jackson's conceptualizing of evolutionary hierarchy and the relations between complexity and organization within that hierarchy. The clinical and theoretical significance of this way of conceptualizing 'dissolution' depended on the presupposition that there was a continuity between states of health and of disease (e.g., 1958, 1:123, fn. [1875], 324 [1880], 2:3–28 [1881]), that the latter was a reversal of evolution's direction, and that one could infer from one to the other, both ways. An apt expression thereof was the often-repeated trope of the clinical investigator looking upon disease as an occasion of 'experimenting' or of 'Nature's experiment': "there is no part of the nervous system which is not experimented on by disease" (1958, 2:265 [1872]). Discussing 'petit mal' epilepsies: "Scientifically, the slightest degree of post epileptic mental automatisms are more important, for these cases are simpler experiments on the organ of mind" (123 [1875]). Disease was considered an exaggerated form of interruption on the continuous range of the normal course, e.g.: "My reason for continuing to use the term 'discharging lesion' is that it keeps us well in mind that the epileptic process is but an exaggeration, although a vast one, of normal nervous discharges – of normal physiological process" (319n2 [1888]).

This theorizing was crucial to positing clinical practice as scientific. Following his teacher, Thomas Laycock (1860), Jackson looked upon the spinal and cerebral ganglia as forming one continuous nervous system, subject to the same laws of reflex action, indicating that the reflex arc was also his model.[7] Jackson's growing tendency to dispel psychological concepts and categories from the discussion of the nervous system as a whole and to present it primarily as sensory-motoric could also be attributed to Laycock's influence. Jackson mapped the Spencerian evolutionary framework unto this view of the nervous system and discussed the relations of brain, mind, consciousness, and 'self/person' from that perspective.

Since the mid-1860s, Jackson quoted and referred to Spencer, in all forms of his presentations, primarily the *Princ.Psycho.*ii. The first volume of the *Princ.Psycho.*ii was mostly concerned with the nervous system, evolution, functioning, structure, notions of mind and brain—concepts that had formerly been considered as separate faculties: feelings, intelligence, instinct, memory, will, and reason. Spencer referred to Jackson's articles of December 1867 in that volume (1870/72, 1:62n*). It seems that Jackson had already been corresponding with Spencer in 1866–67, as Spencer reacted to an article Jackson published that year in the *Medical Times and Gazette*

[7] Jackson's paper on localization of movements (1873, 177–95) opened with a long quote to that effect by Laycock (Laycock, 1845, 298).

(1867, 669). By then, Jackson was already familiar with the principal assumptions and models that Spencer had elaborated from the mid-1850s on, including those expounded in *Princ.Psycho.*i, "The Social Organism," *First Princ.*, *Princ.Bio.* as well as the Journal installments of *Princ.Psycho.*ii during 1868–1869.[8]

Four components of what I have called 'SL models and mechanisms' as found in Jackson's thinking are particularly relevant here:

1. The nervous system was a product of an evolutionary process, higher levels having evolved from lower ones, resulting in layers, where earlier meant lower in the order of evolutionary development (e.g., 1958, 2:29–30, 42–4 [1882]; 2:46 [1884]; 2:78–9, 80–1 [1887]; 2:395–96 [1889]). Their locations were not distributed, as had been conceived earlier on, along the spinal versus cerebral division. Furthermore, the nervous system could go on evolving, though in a different way and in an individualized–universal mode.
2. The layers were hierarchical and linearly inclusive. They consisted of differing nervous centers: higher, middle, and low, of growing inclusion, with higher layers controlling, constraining, and suppressing activities of those beneath them (see above). The locations of the nervous centers too were not distributed along the spinal versus cerebral division.
3. Pathology in such a system was defined as 'dissolution' (a notion that became central for Ribot). 'Dissolution,' a term adopted from Spencer's *First Princ.*, consisted of a state (sometimes called 'regression' by Ribot) where a lower center, whose activities would be 'more automatic,' took over the functions of a temporarily or permanently dis-functioning higher center. Such pathologies embodied the best adaptable reaction of the lower layers (e.g., 1958, 2:112n4 [1887]). Since each layer also pointed to an evolutionary stage, the process of dissolution revealed the evolutionary history of the nervous system. Note that the idea of 'evolutionary *plastic* adequacy' could also be transferred to dealing with 'the mind' as that which responded to make do with whatever was lacking, was deficient, or was dis-functioning, e.g., in Freud.
4. There were 'constant' (Kantian) elements, such as perception of time, space, and mass, whose evolutionary origins were physiological (Gissis, 2020).

I succinctly explicate the significance of the hierarchical order: Jackson's view of the nervous system developed throughout the mid-1860s until the late 1880. The whole body, in terms of its sensory-motor activities, was looked upon as being represented, partially or wholly, in the nervous centers of each of these layers. However, only in the lowest level was the representation direct, though partial, and it became triply mediated, yet totalizing, in the highest one. Thus, the system of representations was nested and dynamic. Different representations between the levels, among the centers and their components, emphasized differing facets of the represented. This division was not a morphological one but related to the 'representational

[8] In a *much-criticized* book, Raitiere (2012) argues that Spencer hid that he suffered from a particular kind of epilepsy; Jackson became his private physician until the late 1880s, and Spencer served as a valuable *informant* on subjective aspects of the disease.

anatomy' of the centers and their layers. 'Representation' allowed for the whole organism to be (differentially) organized through its sensory-motor activities. Adaptation to the environment was carried out through that ever-changing organization of the representation. The notion of 'representation' was deployed to construct conceptual connections between mind and behavior, mind and brain, evading the underlying issue of whether behavior was part of 'nature' or of whatever was considered as 'nurture' then, and it assumed associationist networks. Jackson professed a seemingly parallelist position. The 'representation assumption' meant that the activity of physiological processes as initiated by the various brain centers was 'translated' in specific ways from one field of parallelism to the other (e.g., mind and brain). This was manifested particularly in movements, as traced in disfunction such as epilepsy; in words, as traced in disfunction such as aphasias. This 'translation' meant that the clinician or investigator could focus on the 'brain part' and leave 'mind issues' to others, the psychologist, the psychiatrist, 'alienists' (e.g., 1958, 2:354–55 [1886], 104–11 [1887]).

For Jackson, the necessary upward direction of evolution ('Spencerian progress') meant that there was a correlation of functioning and structure between the degree of complexity, extent and measure of specialization, coordination/interconnectedness, and independence, i.e., a relation between levels of function and levels of anatomy. Such a notion of hierarchy had been worked out by Spencer in both *Princ.Psycho.* and *Prin.Bio.* In *Princ.Bio.*, it meant that evolutionarily earlier/lower (biological) individuals were simpler, less differentiated internally, with less coordination and integration and less interdependence among parts. Thus, they were more homogeneous, less adaptive, i.e., with less plasticity in relation to environmental changes, and their modes of activity tended to be rigid and repetitive. For the SL, *the higher, more complex, heterogeneous*, and thus *highly organized* biological units would also be endowed with *more plasticity* and would be *more resilient* in the long term because of their ability to regain their equilibrium, to innovate, in the face of changes. Furthermore, there would be a delicate but significant balance between the partial autonomy of parts and their inseparable interdependence. Conversely, for Spencer, the more homogeneous living entities would turn out to be more rigid, less organized, and thus less resilient in the long term.

I claim that Jackson differed. His notion of dissolution indicated where such processes could be expounded as the imprint of external events beyond the then-accepted physiological boundaries of the organism (e.g., 1958, 2:354–55 [1886], 104–11 [1887]), and as expressed in the notion of a specific 'brain locality' of the 'occurrence of disease.' For Jackson, *the lower, less complex* were also *the more highly organized*, more *rigid*, and more *stable*, while *the higher, more complex*, were posited as more interconnected but the *less organized, less interdependent.* As he assumed a continuum of all centers, 'independence' meant that the higher centers evolved to become relatively semi-detached from the lower ones: "our highest sensory and highest motor centres (together the 'organ of the mind') can energise, to a large degree, independently of the lower centers" (1958, 1:375 [1888/89]). This allowed them to continue their activities with relative ('triply detached') detachment from the accustomed and continued sensory-motor reactions—adaptations to the environment.

Yet, the units of these higher centers were supposed to be more interconnected among themselves, which (in Spencerian language) meant that wherever there was complexity, there would be better cooperation among parts of the complex structure. In accordance with these Jacksonian features (more complex and interconnected, less organized), the higher centers would be capable of producing novel modifications, activities, learning, etc. (1958, 2:80–1, 98, 112n3 [1887]). By virtue of that, the higher centers had more plasticity, were more adaptable to change, and open to innovation, i.e., the opening and production of 'new neural paths,' and 'neural connectivity,' the possibility of associated neural networks being reorganized. Yet, they were also more vulnerable and much *less resilient* in the long term. For Jackson, change consisted of a threat to needed long-term stability; thus, lower centers with little change provided that stability. One consideration could be the need to provide such locus of stability to the new science of neurology, to rid it of any 'metaphysical speculation.' K. Wilson, one of Jackson's last students, wrote in his account of his conversations with Jackson that he drew a sharp distinction between being highly organized and being complex. Humans could continue to learn because parts of their brains were not highly organized (cortical and nonmodular centers in humans were complex), while a dog could not, since its brain was so highly organized (J.P. Martin, 1975). *Thus Jackson uncoupled and inverted the Spencerian direct relation between complexity, organization, plasticity, and resilience (stability).*

Jackson did not seem to be aware of the deep difference between the two views of evolutionary organizational hierarchies, and apparently neither did his contemporaries, nor ours, perhaps because for both Spencer and Jackson plasticity and adaptability had to reside with the higher levels of the hierarchical order, whether levels of the nervous system or those of the order of organisms. It seems that this inversion, stemming from his clinical-empirical practice, was highly significant in the reception of his work and its gradually growing resonance within neurology.

I put forward a possible interpretation of this inversion in conceiving the relation between organization and plasticity. The most significant feature of the higher centers, in the higher level, was their suppressive or inhibitory (the latter is a term Jackson rarely used) power vis-à-vis the lower ones. In Jackson's *diagnostic* analysis, it meant that the principal explanatory and systematizing factors of the spectrum of disorders and pathologies of the nervous system were revealed to be the disappearance of that regulatory/controlling/inhibitory power—its dissolution and the ensuing consequences.

In a letter of January 9, 1883, to Edward Youmans, Spencer wrote: "I enclose some pages from the *Medical Times and Gazette* (Jan. 6), sent to me the other day by Dr. Hughlings Jackson. The initiative he made years ago by applying the doctrine of dissolution to the interpretation of nervous disorders – an initiative that is now being followed and in that direction – seems likely to lead to other results … The paper … is to me just as much a revelation as was that which Hughlings Jackson made of the doctrine" (see also, Lightman 2016a, b; Pearce, 2020). In another letter to Youmans, where Spencer discussed the role of the cerebrum, he referred to two papers by Jackson published in *Medical Times and Gazette* on December 14 and 21, 1876 (Spencer, 1908, 23).

In pathology, for Jackson, the middle and low levels and their respective centers 'took over,' became autonomous, and 'more easily dischargeable.' Clinical practice could and perhaps should be looked upon as the epistemological starting point. It meant that the activities produced or 'permitted' (i.e., symptoms of disorders) would exhibit less complexity and specialization, would tend to be more automatic and thus less 'voluntary' (a term that Jackson regarded as non-neural/physiological and suggested substitutes for it in neurological discourse). Certain functioning that would be considered pathological at the highest level would in fact be normal for the lower—e.g., middle level—when 'disinhibited.' Furthermore, when higher centers became deleted or their functioning temporarily canceled out, then the disorders actually embodied the best *adaptable* reaction of the lower centers. One could perhaps view it as a manifestation of the range of organismal plasticity. Such a state exposed an evolutionarily earlier state of functional development of the nervous system. The description could not be merely physiological but also evolutionary developmental.

In a rather early article, while referring to the nervous system in palsy and in convulsion, Jackson wrote: "The word 'destruction' is scarcely the correct word to use. By it is not meant that the nerve fibres are necessarily broken up … but simply that there is a change in them which *destroys their function*" (1958, 1:9n1 [1870]). In a late article from 1892 (Hunterian lecture), he discussed a possible exaggerated activity of the lower centers in particular cases considered as insanity: "we have here loss of function co-existing with over functioning" (1925, 49). Yet, in 1895, at a later stage of his theorizing, he also said: "The foregoing and what is to follow on the hierarchy of nervous centres is, of course, hypothetical" (1925, 120), looking upon this as a transfer of Spencerian principles onto medical language. Recovery meant that the unharmed parts would continue functioning as before, and thus a case of lesion would reveal a hierarchy of what one could term 'functional strategies' of the nervous system at large. James Jackson Putnam described Jackson's lectures at London Hospital in the 1880s: "It was Dr. Jackson's custom to draw a pyramid upon the blackboard which should stand for the hierarchy of the cerebral functions, the more fundamental of them being represented by the basal portions of the pyramid, the more complex and recently acquired by the apex portion … . [in case of derangement] the attempt at a re-establishment of some sort of equilibrium is always such that the new arrangement tends to safeguard itself by accentuating the more fundamental of its powers, while sacrificing, so far as necessary, the more elaborate" (quoted in Swash, 1989/2013, 3–10, 9). This implied that the nervous system as a whole and its various parts evolved as based on the same 'program.' Jackson repeatedly discussed the 'double aspect' of dissolution: a 'positive' aspect of a disorder, the functioning of the middle or low level; and a negative one, the disappearance of a higher one. In some places, he also mentioned a 'super-positive' aspect, overtaking of the functioning by the constancy of automatism, as illustrated by aphasic processes (1958, 2:58–9 [1884]).

L.S. Jacyna (1981, 1982) has suggested contextualizing the arguments on automatisms and voluntary action of various authors writing on brain and mind between the 1850s and the late 1870s through the Victorian view on insanity as a lessening of the constraint of willpower, with or without relating this to notions of degeneration, emphasizing the vested interests of the medical profession then in anchoring the mind and its pathologies in its dependence on physiological organization. Writing on inhibition and elaborating on these issues, R. Smith (1992) has emphasized notions of 'loss of control' in their socio-cultural-political resonances particularly through Maudsley's and Mercier's 'civilizational transfers' and their impact. Significantly, this is where collectivities appeared in Jackson, as he offered analogies of such disorders to Victorian views on bureaucratic or military frameworks and to the collective ruling of a state in general (1958, 2:58–9 [1884])—the need for an elite to rule and direct from above, the fear of anarchy from below.

Following Spencer's efforts to provide alternative evolutionary explication for the philosophical/psychological functions of the Kantian categories and forms of intuition of space and time, Jackson too offered a detailed evolutionary explication of the perception of time and considered space and mass to fall within the same pattern. He posited them as elements within the nervous system, considered '*constants*' in that they were inherited trans-generationally. These 'constants' were based on physiological adaptive activities that, through an evolutionary process of development and then 'deletion' of physiological origins, turned into inherited modes of experiencing (e.g., 1958, 2:98–9, 114–15 [1887]). These were members of a broader grouping of inherited molds or patterns that changed little during a then-present lifetime, in contrast to potentialities that would evolve during humans' lifetime, where Jackson saw the great importance of education. In the Hunterian lecture, he stated: "The two factors in education were next spoken of – the inherited nervous arrangements, organized experiences of Spencer, and the action of external circumstances ... The inherited experiences may be arranged physiologically from the lower and more automatic, to the higher or more voluntary (general to special). The lower – those of respiration for example – are almost 'certainties' at birth" (1958, 2:267 [1872]). This came out emphatically in 1895: "Were not the cardiac, the vaso-constrictor and the respiratory centres of the lowest level more strongly organized inheritedly, than are the centres of the highest level, life would be impossible" (1925, 125). Spencer referred to Jackson's work as "supporting the belief that functionally wrought modifications descend from parents to offspring" (1908, 313, and again 694).

In Jackson's view, there was a continuous evolvement within the nervous system. It took place at the high-level centers, which were even called "those now making" (1958, 2:98n1 [1889]). "There is internal evolution [i.e., an on-going one]" (1958, 2:81, title [1887a]). "They [the highest centers] are ... the centers whereby the organism as a whole is adjusted to the environment ... the ever organizing ... in which evolution is most actively going on" (82); "from those comparatively perfectly evolved at birth – to those ... which are only incipient structures, and which are evolving during the latest new mental states" (1958, 2:396 [1888]). What did Jackson mean here? That humans had a further-development prospect evolutionarily? or just cultural innovation and reformation? Spencer certainly had a biological-utopian perspective, but I tend to think that the second alternative would fit Jackson's position. Jackson used the example of language-acquisition-and-usage to illustrate such an evolvement that depended on repeated highly specialized 'stimuli' being a neural 'possibility.' Within this evolutionary hierarchical view of the nervous system, changes would come about through continuous activity of adaptational formation and growth. This is significant since for Jackson 'consciousness' was tied to the ability to use symbols, both linguistic and visual (Harrington, 1987, chaps 7–8).

The epistemic space of Jackson's clinical practice enabled him to invert the meaning of Spencer's evolutionary hierarchy and thus also disconnected him from tying dissolution to contemporaneous degeneration discourse. Yet, can one discuss 'hierarchical order' without taking into account the deep cultural lure of SL mechanisms, concepts and metaphors avoiding Darwinian selection (in contrast to 'competition')? As I have argued (2018), Jackson rarely used the expression 'selection of the fittest' in the same secondary limited sense as Spencer. When one considers the deep impact in Britain and beyond that Jackson's specific deployment had—his particular evolutionary concept of the hierarchy of the nervous system and what it entailed, and his notion of dissolution—it seems to me that it would not have had that impact within neurology and psychology had it not resonated for his contemporaries with the pattern of human scale regression suggested by the varieties of contemporaneous cultural degeneration-discourse.

4.4 Localization—Aphasias, Epilepsies, and Clinical Practice as Science

The historiography on localization in nineteenth-century Britain/France/Germany has been wide, rich, and diverse. Within this historiography, Jackson's writings on the issue have been interpreted as positing him as either a midway figure with certain inner tensions stemming from the somewhat incompatible conceptual frames he used, or as a figure who contributed to, participated in, and encouraged localizationism. Given that rich historiography, I shall offer a succinct note, working under two assumptions:

(a) Jackson's position on this issue, as on numerous others, such as consciousness/unconscious states, evolved and changed throughout his almost 40 years of intense activity both as a clinician and as a theorizer. Furthermore, even at particular stages of his theorizing, he could hold on to not fully compatible and consistent positions when discussing different issues based on diverging clinical resources. Any attempt to force a theoretical consistency, let alone homogeneity, throughout seems to me unproductive.

(b) The SL evolutionism he adopted as his theoretical stance from the early 1860s on made for what I propose to regard as a *processual hierarchical localizationism* which was much more difficult to represent visually in contemporaneous maps of the brain, than an anatomical-spatial-horizontal one.

In his work on aphasic patients, Jackson commented that speech was interwoven with memory and with emotion (and its expression) and thus could not be limited to/localized in one brain region. Given his SL evolutionism, it may be the case that localization and holism would be better viewed in his work on a continuum than totally dichotomous, e.g., in 1895, "yet, insanity is an abnormal state of a whole person" (1925, 127); "for my part I think the whole body is organ of mind" (1958, 2:85 [1887]). Yet, from a wider perspective, Jackson contributed to the 'localization trend' developed experimentally by Ferrier, within British psychology and neurology, especially in Jackson's large corpus on epilepsies.[9]

Jackson position on localization was crystallized principally in his work on aphasias, though his discussion of epilepsy assumed it, mostly implicitly.[10] The phenomena of aphasia were such that they compelled whoever looked into them at that time to ask questions that were initially considered to be psychological—relating to the mental—rather than physiological or at least as related solely to the nervous system. Jackson's evolving explanation, which was interlaced with his deepening application of his SL evolutionary conception to the nervous system, served to sharpen both his position on the relationship of the mental and the physiological/neurological, and his position on localizing functioning and disfunction in the 'geography' of the nervous system. Thus, I suggest that the adoption of the SL framework could not allow for any (static) anatomical localization of functioning and thus of disfunction. I do not reconstruct nor summarize here the evolving of Jackson's position, rather, I use it to succinctly discuss the relations between Jackson's view of the evolutionized nervous system and the issue of localization. Jackson's decoupling of complexity and organization discussed above also meant delineating the path along which the human brain, in its higher centers, arrived at new forms of behavior, of knowledge. Yet, it did not mean either reduction to localized physiological mechanisms nor resort to any 'other force,' vitalist or spiritual. Jackson pointed to the interconnections of such centers and thus to their integrative action, i.e., to the action of the brain as a whole. I exemplify localized–whole and organization-level issues with a present-day example: the cerebellum is ultra highly organized and yet merely a machine controlling automatisms; the cerebral cortical layers, by contrast, exhibit little more but excessive connectivity without clear patterns of strict organization, certainly not in the highly organized cerebellar way, and precisely this complexity beyond a recognizable level of high organization enables the cortex to execute its integrating action.[11]

[9] See e.g., Finn, 2012, 48n161; Harrington, 1987; Young, 1968, 1970.

[10] My discussion relates to Jackson articles on aphasia and speech gathered by Henry Head in *Brain* 38 (1915): 1–190; a summary of his BAAS-talk in 1868; his articles on epilepsy and movement in *West Riding Medical Reports* 1867–1875; and Jackson, 1958, mostly vol. 2.

[11] I am very grateful to Cornelius Borck for this suggestion.

4.4.1 Aphasias and Localization

From early on, Jackson sought to understand the relations among the range of diverse phenomena in his short articles published in 1864, detailing case descriptions of speech difficulty-or-loss together with right-side hemiplegia. These appeared mostly in *Lancet*, *Medical Times and Gazette*, and *BMJ*, which continued to be his main venues of publication on this topic until the early 1870s. There, as well as in later articles, such as in *his report in the Lancet 1868*, Jackson mentioned cases with right-hand hemiplegia that were generalized to mark left-handed patients. These included speech impairment (mental/behavioral), physiological phenomena that seemed to be concurrent with those of speech impairment, in particular right-side hemiplegia, speech mechanisms considered mental, and disfunction of the nervous system. Jackson pointed to a speech variability in the abilities of each of his observed patients at different times when acting in general and particularly when responding to questions and requests.

In his articles published in 1866–68, Jackson already addressed a number of important distinctions between emotional language, exhibiting feelings (e.g., curses), and intellectual language, composing/uttering propositions (or higher and lower speech), between writing and copying—though one could, as a neurologist, study only the physiological bases, nervous arrangements, of emotions. Other distinctions were between willed and non-willed speech, i.e., between 'voluntary/least automatic' (also self-reflective) and automatic speech (until the late 1880s called at times 'unconscious'). The measure of each on the continuum between the two extremes was looked upon as specific to each patient (1958, 2:68–9 [1884]). In various articles, Jackson addressed 'the will' issue, and thus in 1874 he discussed the trodden mode of psychological description of higher centers' activity: "we are obliged for lack of others to use words which have psychological implications. The words voluntary and automatic are such words, but they are also used physiologically" (1958, 2:132n2); and "It is a physiological investigation, and our method must be physiological" (133). In an 1890 article, Jackson said: "properly from *the most automatic to least automatic in evolution* … 'Most voluntary movement' is objectionable because it is a mixture of psychological and anatomico-physiological language" (1958, 1:426n2). Dixon (2003) argues for a somewhat analogous development in the use of 'emotion,' though a more modern term.

Another distinction was added soon after. The distinction of internal and external speech as preceded by a pre-speech stage of the process, that of conceiving a proposition, of transforming images/signs/symbols. "what I call image is a state of mind (each person's) a 'ghost', standing as a symbol of something not us, of the nature of which something we know nothing" (Jackson, 1958, 1:384n1 [1888/89]); also "Words and other symbols in mentation," 2:205–12 [1893]). Jackson was possibly alerted to this by Browne-Séquard's work, but some authors (Gere, 2013; Greenblatt, 1970, 1999; Lorch, 2004, 2008) discuss this distinction in relation to Jean-Baptiste Bouillaud who was his teacher. In later articles, Jackson stressed in tandem the importance of relations among things, among images, and the ability to symbolize both such relations together, i.e., *to form propositions*. Thus, loss of speech was not seen as necessarily including a loss of the ability to read; loss of the symbolizing ability was. Therefore, Jackson considered aphasia a language-disorder and thus a thought-disorder, or even more broadly an expression-disorder (since he included pantomime movements as symbolizing for the deaf-mute) rather than merely a speech-disorder.

In seeking to elucidate the normal through the abnormal, Jackson 'translated' these distinctions into his self-bounded frame of explanation, i.e., of the neurological, into the sensory-motor, e.g., stating "The anatomical substrata of ideas (of the more important evidently) are sensori-motor processes … the anatomical substrata of words were motor articulatory processes" (1958,1:169ff [1874]); "movement enters into the anatomical substratum of every 'image'" (2:92, 140n2 [1874]).

The contemporaneous explanations of relations between these two kinds of phenomena relied heavily on notions of specific, anatomically localized brain function, e.g., speech, thus direct, localized brain damage (e.g., Bouillaud, Broca, Bateman, Bastian). To Jackson, these seemed to be insufficient in explaining the wide variability of the phenomena encountered in clinical practice and in providing for continuity between the normal and the abnormal. With Jackson's implicit associationist assumptions, which he shared with Laycock, SL evolutionized psychology became a productive resource. Jackson's distinction between 'positive' and 'negative' aspects of the dissolution process allowed for attributing localist causality, e.g., in the case of brain lesion and loss of speech-hemiplegia, the negative aspect was the disappearance of the higher centers' activity in their specificity. Concurrently, it made room for detailed description of the positive aspect, i.e., the continued functioning—though less specific and 'uninhibited'—of the lower centers. Thus, 'clinical signs' were produced—resulting from remaining integrative action of the brain. It did not mean the locality of any 'mental capacity,' but the specific 'brain locality' of the 'occurrence of disease,' as was exemplified in Jackson's criticism of Broca: "While we may localise the damage which makes a man speechless, we do not localise language. It will reside in the whole brain (or the whole body)" (1958, 2:669–70 [1867]).

A certain division of labor could thus be established by Jackson whereby theorizing on speech disfunction—aphasias—was presented as arising from close clinical observation of phenomena combined with medical-notes records. It was explicated through theorizing based on the evolutionarily hierarchical structure and functioning of the nervous system, both non-localizationist. Yet, this work was often presented as if empirically supported by animal experimentation and brain-autopsy work, and somewhat later via brain surgery, all carried out with localizationist assumptions. West Riding Asylum and specialized hospitals were significant institutions for Jackson's work, and for other neurologists, in that they provided almost unlimited (by-law) opportunities for autopsy of patients' brains, especially of the poorer ones. In addition, in the last decades of the nineteenth century, the renovated National Hospital provided an up-to-date and well-equipped autopsy theater as well as an operations theater, where Victor Horsley was prominent. These resources pointed to important contemporaneous medical-scientific figures and institutions, as did the rallying against the anti-vivisection-movement in the Ferrier trial, uniting such medical and evolutionary-biological figures and institutions.

It seems that Jackson could only sustain his edifice theoretically within some version of *parallelism.*

4.4.2 Epilepsy and Localization

The interplay between clinical practice, theorizing, and localization issues also comes up in Jackson's work on epilepsy. His interest in and observation of epilepsy dated to the very beginning of his medical path and became more focused from the mid to late 1860s. Yet, it was really from the early 1870s on that his theorizing and efforts to reconstruct the course of the disease's seizures (then called 'paroxysms') became more diverse and nuanced, with the growing accumulation of complex cases. His reconstruction went from preliminary manifestations, such as a range of 'warning symptoms,' throughout early to late stages, to the after-stage manifestations (esp. the sequence of spasm and temporary paralysis). His accumulated data were based on a long-term follow-up of close observation of patients and their data records combined with attentive listening to their self-descriptions and self-reports on sensations, feelings, 'illusions,' and hallucinations. Jackson adopted some of the self-descriptions as technical terms, such as 'dream state' (instead of the then current 'intellectual aura') and 'double consciousness (mental diplopia).' Presumably Jackson learned some of the terms from his colleague and epileptic patient Dr. Myers, named Z (see, e.g., 1958, 1:308–17 [1880]; 385–405 [1888]; also, his articles published in *Brain* in the late 1890s). Quite a number of these follow-up cases ended in autopsies. From the late 1870s, the articles on epilepsy started to set out a neurological-medical layered classification of the *range of epilepsies* (rather than the mixed medical-social, found elsewhere). This classification gradually (already in the early 1870s but especially from the 1880s) stabilized its definitional boundaries, excluding some forms of convulsive disorders and delineating basic common denominators. In 1873, the following was the common denominator: "*Epilepsy is the name for occasional, sudden, excessive, rapid and local discharges of grey matter*" (1958, 1:100). In his 1890 talk "On convulsive seizures," Jackson suggested to add 'explosive' to the list of adjectives and emphasized the interdependence between damaged versus intact 'levels' and specific features of seizures, 'fits' (esp. 1958, 1:413–16).

When explaining epilepsy, Jackson deployed the distinction between voluntary and automatic ('unconscious' until the mid-to-late 1880s). He deployed the same division of phenomena into negative and positive aspects of the process of dissolution, which could either be distributed throughout the nervous system or be local since 'discharges' could spread far beyond their originary sites. The direction of the dissolution process was from the more specialized (most-consciousness) 'down' to more generalized. Analogously to his discussion of aphasias, Jackson related patients' medical histories of the disease and their observed phenomena to possible damage in various areas of the brain cortex. Jackson discussed etiological localization of pathology, such as lesions, yet did not explicate 'normal.' This localizing relationship was especially conspicuous when he analyzed the march-like progress of early movement-symptoms, which, according to him, signified a coordination of motor functions and localized lesions. In the case of epilepsy, the peculiar mediating role of 'representations-of-movements' in the theorized hierarchy of nervous centers ('levels') played a more conspicuous role, since there was hardly a need to relate to, or 'translate' a 'mental-behavioral typology,' given that it belonged more clearly to the physiological/neurological, even though some of the pre-and-post seizure symptoms were perceived as 'mental.' It seems that there is enough evidence to argue, as Shorter (1995) does, that there was a marked drop in the frequency of 'pseudo-epilepsy,' particularly in the second half of the nineteenth century. This could be interpreted as a sign of willing medical reception of the new conceptualizing of epilepsy, or as differing presentations of symptoms by patients, perhaps filtered through cultural-historical changes, or as a negotiation between the two. Either way, the splitting of the neurological and the psychological might have played a role in it. Even though there was a significant disagreement between Charcot and Jackson, Goetz and his colleagues (1995) argue that Charcot was familiar with at least some of Jackson's work (they mention 1872, 1875, and 1888, and his book on localizations) and quoted it in his teaching lectures; Jackson's portrait hung in Charcot's hospital office.

4.5 Philosophical Concerns of a Clinical Scientist: Parallelism, the Mental—Consciousness, Unconscious States, 'Self'

Several issues arose in Jackson's positions, both changing and persisting, on what could now be called neurology and psychology and their relations: the meaning and implications of a parallelist position; what was consciousness; what was voluntary or automatic in human actions, including language usage; what was unconscious; how were positions on localization of mental states related to these; what was subjectivity/being a subject, self, person, having agency, self-consciousness.

Any investigation of the brain, then and now, implies some philosophical assumptions on brain and mind, on brain and consciousness, on body and mind. In the period discussed, there was a wide spectrum of positions ranging not only from a reductionist version of monism to an extreme version of dualism, but also including other combinations, epistemological, methodological, and ontological, with related implicit hierarchies of the sciences. The principal virtue of versions of explicit or implicit materialism was the presumed success of empirical sciences. Yet, universalizing physiology was considered to have left areas of familiar phenomena unexplained, at times turned inexplainable within the boundaries of empirical (or clinically empirical) science: 'consciousness' as integrated, selective and adaptive, the representation of one's body, the world, and their relationships, an experienced sense of 'self' and of 'agency,' as unified, of 'subjectivity,' the meaning of the behavior of others, especially as conveyed through the use of language.

4.5.1 Parallelism

It was in dealing with the high centers that issues of consciousness and self/person came up more sharply, but Jackson's more general concern was to clearly *delineate the spheres* in which he was working, to distinguish between nervous-system-brain (and spine) and mind, between physiological and mental/psychical, as well as between physiology and morphology. He looked upon the nervous system as 'representing' the two basic modes of activity (which were, I put forward, the equivalent of the 'basic units' in the empiricist sense), the sensory and the motoric. In the 'interpreting quotes debate' on whether Jackson, Spencer (and Freud) had been 'parallelists,' or held another, more complex and perhaps intractable, position on the brain/mind issue, there are interpretive gains related to each position. I suggest that from rather early on Jackson's motivation had been to establish neurology as a scientific discipline. He was not unique in Britain in the effort to separate brain and mind, while in Germany neuro-psychiatrists, such as Meynert and Wernicke, and even later scientists, went the other way, seeking close correlations between neural mechanisms and mental processes (Wassmann, 2005). Note that there were very few sites of institutionalized university research of experimental psychology/physiology in Britain at the time, primarily the physiology laboratory at University College, London (where Ferrier worked for quite a while), and thus 'the clinic' and the related hospital laboratory were highly significant.

In order to critically clear the way for this aim, Jackson wanted to disengage description, analysis, and theory from what he perceived as two philosophical proclivities that would harm such a course, namely 'metaphysics' and 'materialism,' as he referred to them. His position on the latter could be confusing. He used the term in order to denote what he called "the material basis of mental disorders" (1958, 2:9 [1881]), which is a particular attitude to scientific endeavor and its object, but presented himself as continuing the tradition of Spencer, Huxley, and Tyndall of not being materialists philosophically (1958, 2:63–75 [1884]).[12]

Clinically, Jackson developed and practiced a close-observation-path (detailed above), aiming at 'neutrality' and 'objectivity.' I propose that he adopted a position that can be looked upon as 'methodological parallelism'—which J.M. Baldwin and G.F. Stout defined in Baldwin's *Dictionary* as follows: "the affirmation that conscious process varies concomitantly with synchronous process in the nervous system, whether the two processes have a direct relation or not. This is the *most*

[12] On public science in Britain 1880–1919, see Turner, 1993.

generally accepted view of the relation between mind and body *at the present day* … The principle of psychophysical parallelism involves no explanatory theory of the connection between conscious and nervous process" (1901–05, 2:258–59—emphasis added). For the purpose of his work, Jackson assumed that brain states and mind states were intrinsically different, with no (simple, mechanistic) causal connection between them in either direction; and causal chains were internal to each chain. Since they appeared as temporally parallel, he admitted they were 'concomitant.' Given that Wundt also claimed to be a parallelist, and that 'concomitance' was a Leibnizian term for parallelism, one could perhaps view the use of the term 'concomitant' as a negation of interactionist position. In the Croonian lectures in 1884, Jackson composed an *imaginary* mixed-bag group adhering to such positions, with himself as a member: Hamilton, J.S. Mill, Spencer, Max Müller, A. Bain, T.H. Huxley, T. Laycock, David Ferrier, Tyndall, and Du Bois-Raymond (1958, 2:45–53, 63–75).

It seems that Jackson considered this position also as non-materialist (nor even an epiphenomenalist one, such as T.H. Huxley was interpreted to be committed to). Within his evolutionary framework it meant that there would be solely a correlation—'concomitance'—between the system of evolutionary functional brain levels and relevant mental phenomena—whether considered subjective or not. Such a separation allowed Jackson to discuss, including causal analysis/medical etiology, solely the nervous system and to theorize basing himself on his and others' clinical-empirical findings. He sometimes named the purpose of this theorizing 'rational generalization': "Before we can make rational generalizations we must have made empirical observations and generalizations … I submit that on the principle of evolution we have the widest basis for rational generalizations of diseases" (1888, 59—following Spencer, *Data of Ethics* 61, whom he quoted there; see also 1958, 2:265 [1872]). Hutchinson in his obituary characterizing Jackson: "his two greatest characteristics - if his unique modesty be omitted - were an unceasing and laborious attention to the minutest details coupled with a power of widest generalizations" (*BMJ* 1911/2, October 14, 950). Yet, Jackson also claimed to follow both J.S. Mill and Spencer. Andermann (1997) argues that during the early to mid-1860s, Jackson was a member of a group of young doctors who shared workplaces, their findings, professional associations, and weekly journals. Significantly, they also shared a commitment to an ideology concerning the proper method for the pursuit of their clinical work and research, which adopted J.S. Mill's recommended reasoning. Already in 1866, in an article presenting clinical cases, Jackson wrote: "Unless a man can put the particular phenomena he himself sees under general Laws, or unless he tries to do this, he can scarcely be said to know or to be studying a thing in a very valuable sense" (Jackson, 1866, 4; Andermann, 1997, 476). This claim is partially supported by Jacyna (2011) who argues that Jackson's method—described after Paul Forman (2010) as 'procedurism'—is one most suitable for collective endeavors. The founding of the journal *Brain* in 1878 by Crichton-Browne, Ferrier, Jackson, and J.C. Bucknill, a journal with a marked emphasis on practical-medical issues in the wider field of neurology, could be considered a later facet of Jackson's involvement in collective endeavors.

In my view, *methodological* parallelism allowed Jackson to separate 'the mental' (at times called 'psychical') from the neuro-physiological and to claim to deal scientifically only with the latter. This also allowed for self-evidently close relations between physiologically and experimentally based neurology and general medicine. Thus, Jackson often declared that his neurological theorizing should not be applied to asylum residents as long as their diagnosis turned solely on the mental. Yet, he applied that theorizing, putting on a par epilepsy, drunkenness, psychosis, and dreaming: "*A parallelism being assumed*, all that is attempted is to discover the nature and conditions of activity of those nervous arrangements which are assumed to be the substrata of states of mind. In other words, whilst admitting, as of course, that prior psychological analysis is necessary, we are directly concerned not with psychology, but with certain questions in the anatomy and physiology of the nervous system" (1958, 1:41 [1875/1873]—emphasis added); "It is impossible to study cases of disease of the brain *methodically* if we confuse psychical states with nervous ones. We must be thoroughly materialistic *in our method* so far as is practicable ... study without psychological bias the material basis of mental disorders" (1958, 2:9 [1881]—emphasis added). "It seems to me that the doctrine of concomitance, is at any rate *convenient in the study* of nervous diseases … Those who accept the doctrine ... would give, or try to find, materialistic explanations of physical inabilities" (1958, 2:72 [1884]—emphasis added).

Since Jackson often claimed that he did not wish to assume two separate substances, and his criticism of the 'two substances' position was repeated in numerous papers, while reasserting his methodological position, I think that the way he quoted G.H. Lewes gives an insight into his own position: "all the evidence points to the very different fact *that the neural process and the feeling are one and the same process viewed under different aspects*. Viewed from the physical or objective side, it is a neural process; viewed from the psychological or subjective side, it is a sentient process." Jackson continued: "Mr Lewes's view does not conflict with mine *in this inquiry* … it is I think, indeed convenient to make the distinction, even if it be purely artificial" (1958,1:42 [1875/1873]—quoted in Lewes 1879, 2:459; see also Jackson, 1958, 2:84 [1887]). Note that Lewes already articulated his position in *The Physiology of Common Life* (1860) as well as in *The Physical Basis of Mind* (1877). Furthermore, I think that in spite of Jackson mentioning the Leibnizian simile of the two synchronized clocks, his pronouncements on the 'self' and his disquisition on subject-states and object-states (see below) revealed him as a supporter of the position summarized so aptly by Lewes rather than the concomitance position, which as I argue was adopted for methodological/ disciplinary/clinical purposes. I concur with C.U.M. Smith (1976, 307, 1982) on this.

York and Steinberg (2002, 316) strongly oppose the claim of some of Jackson's interpreters and biographers that he constantly combined medicine and philosophy and argue that he had a medical pragmatic position. Such a position supplied the proper framework for his struggle to provide for scientific neurology and yet indicate that the nervous system was the 'anatomical basis' of mental activities, as Jackson wrote, for example: "let it be granted for the sake of argument … that the doctrine (2) ["that activities of the highest centers and mental states are one and the same thing, or are different sides of the same thing" (1958, 2:84)] is ascertained to be the true one. I then ask that the doctrine of concomitance be provisionally accepted *as an artifice*, in order that we may study the most complex diseases of the nervous system more easily ... our concern as medical men is with the body ... Negative and positive mental symptoms are for us only signs of what is not going on, or of what is going on wrong, in the highest sensori-motor centers" (1958, 2:85 [1887]).

Eccles (1989) argues strongly for a diametrically opposed interpretation, which attempts to turn Jackson into a pre-Popperian 'three worlds' parallelist, though not a very consistent one (also Popper & Eccles, 1985). It would have been difficult, if not impossible, to include insanities, to posit an integrated explanation for both simple and complex functioning (and former 'faculties'), and offer an evolutionarily based generalization of/on them without delineating an analogous evolutionary framework for the mental functioning. I consider Jackson's suggested distinction between object states and subject states of consciousness as another gesture in that direction.

4.5.2 Relations of Brain, Mind, Consciousness, Unconscious States, 'Self/Person'

I again refer to Baldwin's own definition in his 1901 dictionary as an attempt to express a broad consensus: "*Mind* – the individual's conscious process, together with the dispositions and pre-dispositions which condition it [i.e., an attempt to bring the unconscious in]. It is thus the individual's consciousness with its capabilities; its capabilities include all faculties, powers, capacities, aptitudes, and dispositions, acquired and innate" (Baldwin, 1901–05, 2:82). The definition of consciousness, written by Baldwin and Stout echoes this inclusion effort: "*Consciousness* – the distinctive character of whatever may be called mental life. It is the point of division between mind and not mind ... wherever there is not total unconsciousness, in the sense which we attribute unconsciousness to a table or a log of wood, the existence of some form of mind we denote by the word 'consciousness'" (Baldwin, 1901–05, 1: 216–17).

It would not be far-fetched to assume that there were certain broad consensual presuppositions concerning features of consciousness among those dealing with brain–mind–body (in the world) on the range of medicine–philosophy–biology–psychology positions that denied a unitary-substance/psyche. They sought a non-theological, preferably naturalist (even when not fully materialist) description/analysis since most of them held either reductionist or parallelist views. Thus, it seemed to them that there was no need to assume another force of nature to explain the mental. There was an ongoing secondary debate on whether 'mind,' 'the mental,' could be considered as completely subsumed under, i.e., co-extensive with, consciousness, or whether unconscious states should be included as well. Jackson did change his position on this issue. These presuppositions included that consciousness was located in some way in or closely related to the nervous system; it was characterized by global, sensory-motor selective activity ('the mental'), whose products were widely accessible, its activity was directed toward or turned about objects (some thought that this included itself), related to the body, in-the-world: organizing, coordinating, differentiating, integrating. Whatever explanations were adduced for the relations between brain, mind, body, it was agreed upon that there was some sense of *binding* as well as of *agency*, even *ownership*, or at least a specific/particular *continued perspective of experiencing*. That sense was termed at times 'self' or 'person.'

A common feature of this range of positions on consciousness was that they were deemed by contemporaries to oblige or even constrain them to produce either a meaningful description of such relations via some bridging mechanism, or explain why such bridging was superfluous, or else posit a gap-less construction. One may argue that with Spencer consciousness was (weakly) emergent, and that he offered, within his framework, biological-evolutionary markers for it. In present-day theorizing, especially among philosophers of mind, this constraint is still brought up under many guises. In recent years, there have been a plethora of publications aspiring to reconnect, reshape, narrow that gap, or else call for a radical reformulation of the issue both through novel (esp. imaging) technologies in the neurosciences, development of novel (computational) modeling, thought experiments, and theorizing on notions of emerging evolutionary subjectivity. In view of the tortuous history of thinking about 'consciousness' and of deploying it as a concept, it seems that Jackson, who contributed significantly to the body of work on relations of the mental, consciousness, the unconscious, and the nervous system, especially the brain, at some point too had to account for the meaning of 'self.'

On a more everyday level of the contemporaneous medical profession, these presuppositions were questioned and tested, in particular whether and how cerebral changes—ordinarily part of medical etiology—impacted on mental states, on the functioning of the mind, on consciousness. It meant noting changes that ordinarily formed part of the 'mental symptoms' description, further correlating the cerebral ones with body changes, also ordinarily forming part of the symptom description, thus marking a difference between normal and abnormal. Jackson's insistence on the importance of ophthalmological examination is an instance thereof. This was expressed in a manifesto style in a 'News section' item in *Mind* 1878, quoting the announcement of the coming publication of the first issue of *Brain* (1878). "The journal … will include in its scope all that relates to the anatomy, physiology, pathology and therapeutics of the nervous system. The functions and diseases of the nervous system will be discussed both in their physiological and psychological aspects; but mental phenomena will be treated only in correlation with their anatomical substrata, and mental disease will be investigated as far as possible by the methods applicable to nervous diseases in general" (*Mind* 1878, 295). It was reflected in the ambivalent relations from the 1870s to the1890s between neurology and psychology, for the physiological basis of the mind was at the center of both. Was this in any way instrumental in the evolving of experimental psychology in Britain? One tends to brush aside that W.H.R. Rivers was mentored by Jackson in the early 1890s (see also Bynum 1985).

For Jackson, in the formulation of his parallelist position and in his daily work as summarized in the numerous case studies, the nervous system was distinct and separate from the mind, and yet formed its foundation, was involved in the formation of mental experience, and thus the 'non-mental'/the nervous system became the foundation of the healthy functioning of the mental. Awareness and reactivity were the generalized features of consciousness, but also the two associationist-empiricist ones of perception and ideation. Since traditional categories of the 'mental' had already been reformulated through the nervous system's activities by Spencer's evolutionizing, Jackson would not accept categorizations of any disfunction based on them, such as 'paralysis of the will.' In "Factors of insanities" (1958, 2:411–21 [1894]), he offered the neurological explication that 'mental disorders' and 'insanities' ought to be posited as "diseases of the highest centers." This relationship together with the notion of 'representation' constituted a further step along the path of 'en-nerving' the mind, i.e., neurologizing it, and yet was haunted by it. This comes out more clearly in Jackson's discussions of epilepsies, where he excluded easy psychologist explanations of 'mental symptoms,' concurrently discussing these symptoms with his patients, as he incorporated them into the construction of his analysis and explication. "There is no physiology of the mind any more than there is psychology of the nervous system. On the basis of concomitance, mental symptoms (synonymously abnormal states of consciousness) are, strictly speaking, only signs to physicians of what is not going on or of what is going on wrongly ... the excessive discharge, begins in some part of the 'organ of mind', or equivalently, highest centres of the cerebral system; the physical process in these and all other kinds of fits is our proper concern as medical men" (1958, 1:385–405 [1888]; also Forrester 1980).

Three interrelated questions on *the mental and consciousness* seem to be relevant:

1. *If a nervous correlate was assumed for every mental act, would the converse necessarily follow?* It seems that after a short wavering, Jackson—in line with both his methodological parallelism and the evolutionary hierarchy—did not allow for the converse, all the more sharply from the 1880s on, and it therefore led to a change in his view on the possibility of unconscious states *of mind.*
2. *Was it assumed that since higher centers' activities were accompanied by more complex mental ones, consciousness could only be a correlate of the higher, in fact highest, centers, and only of them?* Indeed, such a position crystallized already in the 1870s and even somewhat earlier. "The highest centers are those which form the anatomical strata of consciousness, and they differ from the lower centers in compound degree only" (1958, 1:171–72 [1874]). This resulted in a further complexification of the high-level centers. In a very late article discussing insanities and thus much more meticulous on the correlation of nervous high centers and the mental/psychical, one finds detailed/particularized descriptions of the higher-level centers as subdivided into four sub-levels, with a description (diagrammatic as well) of a state when all the sub-levels were 'rendered functionless,' deployed in a discussion of addictions, insanities, and subjectivity (1958, 2:411–21 [1894]). This amounted to a complete deletion of the higher centers (see fourth depth illustration) (Fig. 4.2).

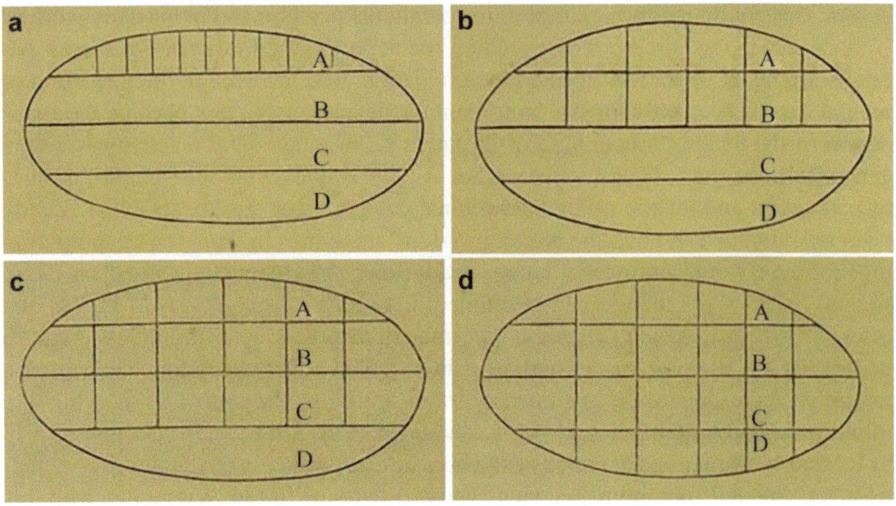

Fig. 4.2 John Hughlings Jackson : illustrations of the four depths of dissolution, *Selected writings of John Hughlings* Jackson, Hodder & Stouhgton, 1931, "Factors of Insanities", pp. 414, 416, 417 (**a**) First depth :"…we may say topmost layer of the highest cerebral centers is rendered functionless by some pathological process" (**b**) Second depth: "Now, the negative mental element, defect of consciousness, is greater and the positive less; less consciousness is retained" (**c**) Third depth: "Now, the negative mental element is very great, and the positive very little. Correspondingly the dissolution is very deep; the range of evolution remaining, only D, is very shallow" (**d**) Fourth depth: "Here the negative mental affection is greatest, is indeed total; …there is no mind or consciousness. There is complete dissolution, and thus no lower range of evolution remains…In the fourth depth there is no *person*, but only a living creature". Source: Wellcome collection, London

3. *Could one assume unconscious* mental *functioning?* The entry on the unconscious in Baldwin's *Dictionary* might help in understanding Jackson's change of positions: "unconscious … not mental, not possessed of mind or consciousness … sometimes used … to denote psychophysical processes which, for various reasons, lack their normal conscious correlates" (Baldwin, 1901–05, 2:724). Given the use of hierarchy, Jackson did not deploy the supposition of unconscious, when he held that there were such states, as enabling a regression to 'the primitive'—feelings, impulses, instincts. It stands in line with negating the socio-criminal aspects of epilepsy. The dictionary's definition (written by Baldwin) equated conscious and mental and denied both epistemological and ontological understanding of the unconscious. Danziger (1982) argues that for Laycock consciousness was an attribute of the nervous system, a coincidental one to its causal processes, but he did not equate mind and consciousness, since he considered mind to be objective. Quick (2014) relates this attitude to Laycock's impact on Jackson as well as on Crichton-Browne. In the earlier articles, Jackson gave room for such a possibility, which arose mostly in the case-by-case descriptions and analyses of epilepsies, where he discussed states

of sub- rather than un-consciousness (in particular 1958, 2: 242–44 [1874]) and could conceive of mental activity that was not fully conscious. Such activity would be supposed as concomitant of lower centers, given that consciousness was considered the concomitant of the higher ones. In the third Croonian lecture in 1884, Jackson already expressed his doubts about whether there could be any mental activities correlated to the lower centers' functioning: "But whether the activities of the lower nervous arrangements gave attendant states of mind, however faint or not, is disputable. Less consciousness attends activities of nervous arrangements, the more organized or automatic they are or have become" (1958, 2:73). Later on, more emphatically between 1884 and 1887, his position on the first two questions above became firmer, and Jackson totally equated mind and consciousness. Consequently, he negated any possibility of mental activity that was not also conscious. In an article on Fechner, Romand (2012) compares the positions of German psychologists and some evolutionary biologists and British physiologists of that period on the possibility of unconscious. He argues that on the German part, reflex was considered a purely physiological mechanism, while Carpenter, Maudsley, and Laycock related reflex activity to 'unconscious cerebrations'—mental activity taking place automatically—that contributed to the appearance of conscious phenomena (also Romand, 2010).

This move by Jackson made his theorizing seemingly more coherent, for within its proclaimed framework, the unconscious could only be posited as a level of mental activity that became latent—thus begging the question: how could it be empirically exemplified through 'nature's experiment'?! Indeed, York and Steinberg (2013) explain Jackson's decision as an act of consistent loyalty to his own criteria of empirical work. More importantly, it enabled a full conceptual neurologizing explication of the mental, since 'unconscious' could not be (by definition) a state of consciousness, i.e., of mind; it was non-mind: "A distinction is made by many between mind and consciousness … I take consciousness and mind to be synonymous terms; if all consciousness is lost, all mind is lost. Unconscious states of mind are sometimes spoken of, which seems to me to involve a contradiction" (1958, 2:85 [1887]). It could be compared to Wundt's grounds for changing his position on unconscious.

These positions, resulting from Jackson's particular evolutionary conception of the nervous system, together with his adherence to a rigorous methodological parallelism, had a deep impact on both Ribot and Freud, who did acknowledge unconscious states.

There was another aspect of that combination that found resonance in later developments of psychology systems. Jackson's continued struggles with the brain–mind issue seemed more pronounced once he broadened the scope of his work to include the wide range of both aphasias, epilepsies, and insanities and constructed a generalizing scheme for them. He drew a complicated distinction between what he called (after Lewes) *object-states and subject-states*, a distinction already found in the early 1870s (e.g., 1958, 1:161–265 [1874]).[13] The difference between the two was not a mirror of the distinction between the neurological/physiological and the mental/psychical since both states were characterized as falling under the latter, also related to the dual-hemisphere brain. Rather, it dealt with the possibility of unmediated introspection as a source of reliable knowledge on the constitution and functioning of both. Jackson's position was complicated but firm. On the one hand, there was no way to provide 'an explanation' within his evolutionary framework for the traditionally named faculties (will, memory, reason, emotion) except as modes of functioning of consciousness perceived as 'attendant' on or correlate—i.e., concomitant with certain activities/configurations of the higher nervous centers (1958, 2:63–75 [1884]). The latter were, indeed, a legitimate topic of research. Significantly, Jackson already wrote in a footnote to his 1868 article on language and thought: "Although we 'will' in the sense that we do what we like, provided there are no external hindrances, we do not desire as we like (see Spencer on the 'will', 'Principles of Psychology' ...) I must use some such word and I use the words 'will', etc., pretty much as 'effort', 'reflection', etc. are used" (1958, 2:66n1).

From Jackson's point of view, medicine as a scientific field, particularly neurology, could only deal with objective, embodied, and thus 'third person' states, provide explanations and generalizations, and thus constitute robust concepts of illness. Yet, the pathological phenomenology dealt with clinically, included 'subjective,' 'mental,' first-person accounts, though only posited as possible, not necessary correlates/concomitants of the first-group states. Canguilhem (1978, chap. 4) holds that Jackson was laying the foundation to consider the pathological as a novel dimension, for within his view states of disturbance called for specification of the disturbance itself, rather than listing what was lost or missing (e.g., Jackson, 1925, 17:144 [1896]); already sketched in 1864, "On the Study of diseases of the nervous system," reprinted Jackson 1903). Jackson argued: "Although there are different *kinds* of insanity (different insanities) implying disease of different areas of the highest level, yet any insanity is an abnormal state *of a whole person*" (1925, 127 [1895]—emphasis added). The activities of consciousness bore the mark of *belonging to 'somebody.'* Yet, that somebody could only be gotten acquainted with, even to him–/herself, through 'his/her' activities—his/her reactions to his/her environment, their product, including 'himself/herself,' etc. These were 'object states.' Already in 1874 and then more elaborately in 1876, 'subject states' were considered to be generally unconscious and prior to 'object states' and in a sense served as their enabling condition. The issue then became how to guarantee the continuity of that 'belonging' within methodological parallelism. Jackson's solution was to do so by anchoring it in 'anatomical substrata of subject consciousness,' which were described within the higher centers in terms that echoed the descriptions of the

[13] For differing interpretations, see, e.g., Bassiri, 2015; Harrington, 1987; Riese, 1956; Smith, 1982b.

robustness/resilience, rigidity, and un-changeability of lower levels in the evolutionary hierarchy of the nervous system. They were described as "more perfectly organized ... in comparatively fixed relation to one another ... comparatively un-modifiable in new ways" (1958, 2:96–7 [1887]) by whatever activities/neural configurations, since "All states of object consciousness 'come out' of subject consciousness" (95), i.e., belong to that particular system (and the effects of the environment on it).

I suggest that in the above explication that Jackson characterized as 'speculative,' he pointed to the necessary, in fact indispensable condition for his science—i.e., the individual, *the 'person' or 'self'*—which nevertheless was indescribable in the terms and tools of methodological parallelism but could be assumed if one adopted Lewes's position. This doubling was also supposed to explain the keen sense of a subject's continuity of consciousness as against its discrete empirical reality (1958, 2:199 [1879/80]). Another aspect was Jackson's emphasis in the third Croonian lecture, where he considered the degree of dissolution as determined by its expression, e.g., the measure of occurring automatism, to be individualized, emphasizing his *clinical scientist* point of view: "in using such an expression as *the* 'automatic', we must bear in mind that what is very automatic to one man is not automatic to another" (1958, 2:63–75 [1884], also 2:416 [1894]).

I refer the reader to the first chapter on Lamarck, while reading Jackson's description of the phenomenology of being 'a self'/'a person,' 'the sense of existence,' since Lamarck undertook one of the earlier attempts to define that phenomenology in terms that aspired to offer a scientific-materialistic report on the brain rather than on 'the mind.' I would argue that both Jackson and Lamarck started with 'a third-person description'—to use Damasio's expression (1999, 199, also 2006)—but found out that these (Jackson's 'object states') did conceptually need a first-person foundation, or experience, or explicated meaning. Thus, Jackson: "Subject consciousness is something deeper than knowledge; it is that by which knowledge is possible. Perhaps we may say that it is an awareness of our own existence as individuals, as persons having object states making up for each the (his) Universe; it *is* us in an emphatic sense. Subject consciousness is the comparatively unchanging, the most unchanging. It is thus a 'constant' [in the sense explicated above] to object consciousness which is continually changing. Most unchanging, that to which all more changing is relative, is the same as unchanging" (1958, 2:96 [1887]). In some sense, Jackson adopted methodological parallelism in order to deal solely with the organism's nervous system, its vicissitudes and disintegration, to deal with the patient as the scientific object of a newly constituted medical specialty of a neurologist, rather than with the individual's, the person's dis/continuous consciousness, damaged mind, disturbed psychological states, and destroyed agency. Agency could be found (or lost) in the evolutionary hierarchical structures of the nervous system and its functioning and adaptations. Yet, Jackson could not quite ignore the phenomena, the lived experience, of the 'self.'

4.6 Concluding Comment

I close the discussion of Jackson's evolutionary conceptualization by going back to the relations between organization and plasticity, which underlay most of the issues discussed in most of the sections. Though it could be argued that both Spencer and Jackson used some model of evolutionary development of organisms, I think that their 'model-organism' was different. In Spencer's *Principles of Psychology* (both editions), the passage from most elementary, unicellular organisms to more complex ones served as an ever-present model, though in the background. It seems that for Jackson, having had medical education with its focus on the human body, the model was much more highly developed organisms. Clarke and Jacyna (1987) argue that most of the nervous-system researchers from the late 1840s on did not take into consideration possible differences between non-human and human systems, though earlier comparative questions on function–structure relations and on the evolvement of brain and spine across the whole range were instrumental in bringing about a new model.

For Spencer, as for many of the Lamarckian (and neo-Lamarckian) authors, interactions with the environment and their effects were of paramount importance. The degree of organization indicated the extent of *constantly changing* entwining interdependencies between the organism, parts of the organism, and the environment, thus delineating the range of possible adaptational, innovative change. Consequently, stability and continuity were inherent in *the pattern of change and its plasticity* rather than in less-or-non changing constellations. That meant that for these Lamarckian authors the relation between degree of complexity and degree of organization was direct rather than inverse. For Spencer, there was theoretically and methodically a legitimate inter-generational back-and-forth passage between social/psychological inheritance and biological heredity that was from the mental/psychological/social to the physiological/neurological, resulting in 'hybrid categories.' The sharp divide that, as I have argued, Jackson exercised for methodological/ clinical-scientific purposes could not allow for such 'hybrid categories.' When Jackson discussed 'constants' or trans-generationally inherited modes of experience, he could give room/allow for long-term gradual physiological evolutionary change. The results of such change, once again, would be analyzed on the neural level (with 'attendant'/concomitant mental consequences), as he argued for a processual-evolution view of brain processes.

From Jackson's *clinical* angle, it was important to explain why certain neural sub-systems subsisted in bodily crisis while others ceased. He interpreted a high degree of organization as less receptivity and/or vulnerability to change, in fact resistance to it, and this seemed to be a key to endurance and duration. It seems that within contemporaneous conceptual frameworks, Jackson could not conceive of a complex sub-system whereby many small diffused and local changes would result in a novel overall effect, which would therefore imply plasticity together with a high degree of organization and resilience.

Materials from the following articles have been used by permission of :

Elsevier

2020 – Transfer of Lamarckisms and Emerging 'Scientific' psychologies: 19th – early 20th Centuries Britain and France, *Studies in History and Philosophy of Biological and Biomedical Sciences, 83*.

References

Ackerknecht, E. H. (1967). A plea for a "Behaviorist" approach in writing the history of medicine. *Journal of the History of Medicine and Allied Sciences, 22*(3), 211–214.

Algazi, G. (2016). Exemplum and Wundertier: Three concepts of scholarly persona. *Low Countries Historical Review, 31*(4), 8–32.

Andermann, A. A. J. (1997). Hughlings-Jackson's deductive science of the nervous system: A product of his thought collective and formative years. *Neurology, 48*(2), 471–481.

Anderson, W. (2013). The case of the archive. *Critical Inquiry, 39*(3), 532–547.

Ankeny, R. (2011). Using cases to establish novel diagnoses: Creating generic facts by making facts travel together. In M. Morgan & P. Howlett (Eds.), *How well do facts travel? The dissemination of reliable knowledge* (pp. 252–272). Cambridge University Press.

Baldwin, J. M., Ed. (1901–1905). *Dictionary of philosophy and psychology*, 3 vols. McMillan.

Bartrip, P. W. (1992). The British medical journal: A retrospect. In W. F. Bynum, S. Lock, & R. Porter (Eds.), *Medical journals and medical knowledge: Historical essays*. Routledge.

Bassiri, N. (2015). Epileptic insanity and personal identity – John Hughlings-Jackson and the formation of the neuropathic self. In N. Bassiri & D. Bates (Eds.), *Plasticity and pathology: On the formation of the neural subject* (pp. 65–111). Fordham University Press.

Berg, M., & Bowker, G. (1997). The multiple bodies of the medical record: Toward a sociology of an artifact. *The Sociological Quarterly, 38*(3), 513–537.

Berlant, J. L. (2007). On the case- introduction. *Critical Inquiry, 33*(4), 363–372.

Boring, E. G. (1929). *A history of experimental psychology*. The Century Co.

Bynum, W. F. (1990). The nervous patient in 18th and 19th century Britain: The psychiatric origins of British neurology. In R. M. Murray & T. H. Turner (Eds.), *Lectures on the history of psychiatry: The Squibb series* (pp. 115–127). Gaskell.

Bynum, W. F., & Wilson, J. C. (1992). Periodical knowledge: Medical journals and their editors in nineteenth century Britain. In W. F. Bynum, S. Lock, & R. Porter (Eds.), *Medical journals and medical knowledge: Historical essays*. Routledge.

Canguilhem, G. (1978). *On the normal and the pathological* (C. Fawcett, Trans.). D. Reidel.

Cannon, W. B. (1900). The case method of teaching systematic medicine. *The Boston Medical and Surgical Journal, 142*(2), 31–36.

Carroy, J. (2005). L'etude de cas psychologique et psychanalytique (XIX' siecle-debut du xx' siecle). In J.-C. Passeron & J. Revel (Eds.), *Penser par cas* (pp. 201–228). Éditions de l'EHESS.

Casper, S. T. (2014). *The neurologists: A history of a medical specialty in modern Britain c.1789–2000*. Manchester University Press.

Casper, S. T. (2016). The currency of consciousness: Neurology, specialization and the global practices of medicine. *Canadian Bulletin of Medical History, 33*(2), 321–363.

Chamberlain, J. E., & Gilman, S. L. (Eds.). (1985). *Degeneration: The dark side of Progress*. Columbia University Press.

Charon, R. (1992). To build a case: Medical histories as traditions in conflict. *Literature and Medicine, 11*(1), 115–132.

Chirimuuta, M. (2019). Synthesis of contraries: Hughlings Jackson on sensory motor representation in the brain. *Studies in History and Philosophy of Biological and Biomedical Sciences, 75*, 34–44.
Clarke, E., & Jacyna, L. S. (1987). *Nineteenth-century origins of neuroscientific concepts*. University of California Press.
Class, M. (2014). Introduction: Medical case histories as genre: New approaches. *Literature and Medicine, 32*(1), vii–xvi.
Critchley, M., & Critchley, E. A. (1998). *John Hughlings Jackson: Father of English neurology*. Oxford University Press.
Csiszar, A. (2019). *The scientific journal*. University of Chicago Press.
Danziger, K. (1982). Mid-19th century British psycho-physiology: A neglected chapter in the history of psychology. In W. R. Woodward & M. G. Ash (Eds.), *The problematic science: Psychology in 19th century thought* (pp. 119–146). Praeger.
Danziger, K. (1990). *Constructing the subject: Historical origins of psychological research*. Cambridge University Press.
Daston, L. (1978). British responses to psychophysiology 1860-1900. *Isis, 69*(2), 192–208.
Daston, L., & Sibum, O. (2003). Introduction: Scientific personae and their histories. *Science in Context, 16*(1–2), 1–8.
Dewhurst, K. (1982). *Hughlings Jackson on psychiatry*. Sandford Publications.
Dixon, T. M. (2003). *From passions to emotions: The creation of a secular psychological category*. Cambridge University Press.
Eccles, J. C. (1989). Hughlings Jackson's views on consciousness. In C. Kennard & M. Swash (Eds.), *Hierarchies in neurology: A reappraisal of a Jacksonian concept* (pp. 39–48). Springer.
Finn, Michael A. (2012). *The West riding lunatic asylum and the making of the modern brain sciences in the nineteenth century* (PhD dissertation). University of Leeds.
Fissel, M. E. (1991). The disappearance of the Patient's narrative and the invention of hospital medicine. In R. French & A. Wear (Eds.), *British medicine in an age of reform* (pp. 92–109). Routledge.
Fleck, L. ([1935] 1979). *Genesis and development of a scientific fact*. The University of Chicago Press.
Forrester, J. (1996). If p, then what? Thinking in cases. *History of the Human Sciences, 9*(3), 1–25.
Gere, C. (2011). 'Nature's experiment': Epilepsy, localization of brain function and the emergence of the cerebral subject. In F. Ortega & F. Vidal (Eds.), *Neurocultures: Glimpse into an expanding Uni-verse* (pp. 235–247). P. Lang.
Gere, C. (2013). Curating aphasia: Pierre Paul Broca's Museological science. *Interdisciplinary Science Reviews, 38*(3), 200–209.
Geroulanos, S., & Meyers, T. (2018). *The human body in the age of catastrophe: Brittleness, integration, science, and the great war*. Chicago University Press.
Gerring, J. (2017). *Case study research: Principles and practices* (2nd ed.). Cambridge University Press.
Goetz, C. G., Bonduelle, M., & Gelfand, T. (1995). *Charcot: Constructingn neurology*. Oxford University Press.
Gissis, S. B. (2020). Transfer of Lamarckisms and emerging 'scientific' psychologies: 19th – early 20th centuries Britain and France. *Studies in History and Philosophy of Biological and Biomedical Sciences, 83*(October), 1–13.
Gowers, W. R. (1880). The Gulstonian lectures, on epilepsy. *British Medical Journal, 1*(3), 353.
Granshaw, L. (1992). The rise of the modern hospital in Britain. In A. Wear (Ed.), *Medicine in society: Historical essays*. Cambridge University Press.
Greenblatt, S. H. (1970). Hughlings Jackson's first encounter with the work of Paul Broca: The physiological and philosophical background. *Bulletin of the History of Medicine, 44*(6), 555–570.
Greenblatt, S. H. (1999). John Hughlings Jackson and the conceptual foundations of the neurosciences. *Physis, 36*(2), 367–386.

Hamlett, J. (2015). *At home in the institution: Material life in asylums, lodging houses and schools in Victorian and Edwardian England*. Palgrave Macmillan.

Harrington, A. (1987). *Medicine, mind, and the double brain: A study in 19th century thought*. Princeton University Press.

Harrington, A. (1991). Beyond phrenology: Localization theory in the era. In P. Corsi (Ed.), *The enchanted loom: Chapters in the history of neuroscience*. Oxford University Press.

Harris, M. (1992). Social diseases? – Crime and medicine in the Victorian press. In W. F. Bynum, S. Lock, & R. Porter (Eds.), *Medical journals and medical knowledge: Historical essays*. Routledge.

Head, H. (1926). *Aphasia and kindred disorders of speech* (Vol. 1). Cambridge University Press.

Hunter, K. M. (2021). Remaking the case. *Literature and Medicine, 11*(1), 163–179.

Hurren, E. T. (2012). *Dying for Victorian medicine: English anatomy and its trade in the dead poor, c.1834–1929*. Palgrave Macmillan.

Hurwitz, B. (2006). Form and representation in clinical case report. *Literature and Medicine, 25*(2), 216–240.

Hurwitz, B. (2017). Narrative constructs in clinical case reporting. *Studies in History and Philosophy of Science, 62*, 65–73.

Jackson, J. H. (1864). On the study of diseases of the nervous system. *London Hospital Reports*, vol. 1 (Reprinted 1903, *Brain, 26*(3), 367–82).

Jackson, J. H. (1866). A Physician's note on ophthalmology. *The Royal London Ophthalmic Hospital Reports and Journal of Ophthalmic Medicine and Surgery, 5*, 279.

Jackson, J. H. (1867). Remarks on the disorderly movements of chorea and convulsion, and on localisation. *Medical Times and Gazette*, (December 14), 642–43, (December 21), 669–70.

Jackson, J. H. (1868). Notes on the physiology and pathology of the nervous system. *Medical Times and Gazette*, (2), 177–179.

Jackson, J. H. (1873). Observations on the localisation of movements in the cerebral hemispheres, as revealed by cases of convulsion, chorea and 'aphasia.' *West Riding Lunatic Asylum Medical Reports, 3*, 177–195.

Jackson, J. H. (1925). *Neurological fragments by J. Hughlings-Jackson* (James Taylor, Ed.). Oxford University Press.

Jackson, J. H. ([1931–1932] 1958). *Selected writings of John Hughlings-Jackson* (J. Taylor, G. Holmes, & F. M. R. Walshe, Eds.), 2 vols. Staples Press.

Jackson, J. H. (n.d.-a). *Untitled commentary on the factors of evolution*. Unpublished Papers ref. QSA/870. Queen Square Archives.

Jackson, J. H. (n.d.-b). *Untitled comments on Ferrier's and Soury's ideas on cortical centres*. Unpublished Papers, ref.QSA/867. Queen Square Archives.

Jacyna, L. S. (1981). The physiology of mind, the unity of nature and the moral order in Victorian thought. *British Journal for the History of Science, 14*(2), 109–132.

Jacyna, L. S. (1982). Somatic theories of mind and the interests of medicine in Britain, 1850-1879. *Medical History, 26*(3), 233–258.

Jacyna, L. S. (2000). *Lost words: Narratives of language and the brain, 1825–1926*. Princeton University Press.

Jacyna, L. S. (2008). *Medicine and modernism: A biography of Sir Henry Head*. Pickering & Chatto.

Jacyna, L. S. (2011). Process and progress: John Hughlings Jackson's philosophy of science. *Brain, 134*(10), 3121–3126.

Lawrence, C. (1985). Incommunicable knowledge: Science, technology and the clinical art in Britain 1850-1914. *Journal of Contemporary History, 20*, 503–420.

Laycock, T. (1845). On the function of the brain. *British and Foreign Medical Review, 19*, 298–311.

Laycock, T. (1860). *Mind and brain; or, the correlations of consciousness and organisation; with their applications to philosophy, zoology, physiology, mental pathology, and the practice of medicine*. Sutherland and Knox.

Lekka, V. (2015). *The neurological emergence of epilepsy: The National Hospital for the paralysed and epileptic (1870–1895)*. Springer.
Lightman, B. (2016a). Spencer's American disciples: Fiske, Youmans, and the appropriation of the system. In B. Lightman (Ed.), *Global Spencerism: The communication and appropriation of a British evolutionist*. Brill.
Lightman, B. (2016b). Popularizers, from 1860s to the participation and the transformations of nineteenth century publishing: From the 1860s to the 1880s. *Notes and Records of the Royal Society of London, 70*(4), 343–359.
López Piñero, J. M. (2010). The work of John Hughlings Jackson, edited and translated by GE Berrios. *History of Psychiatry, 21*, 85–99; *21*, 224–36.
Lorch Marjorie Perelman. (2004). The unknown source of John Hughlings Jackson's early interest in aphasia and epilepsy. *Cognitive Neurology, 17*(3), 124–132.
Lorch Marjorie Perelman. (2008). The merest logomachy: The 1868 Norwich discussion of aphasia by Hughlings Jackson and Broca. *Brain, 131*(6), 1658–1670.
Martin, J. P. (1975). Kinnier Wilson's notes of conversations with Hughlings Jackson. *Journal of Neurology, Neurosurgery, and Psychiatry, 38*, 313–316.
Mendelsohn, A. (2017). Empiricism in the library: Medicine's case histories. In L. Daston (Ed.), *Science in the archives: Pasts, presents, futures* (pp. 85–110). Chicago University Press.
Nowell-Smith, H. (1995). Nineteenth-century narrative case histories: An inquiry into stylistics and history. *Canadian Bulletin of Medical History, 2*(1), 47–67.
Nye, R. A. (1984). *Crime, madness, and politics in modern France*. Princeton University Press.
Oppenheim, J. (1991). *"Shattered nerves": Doctors, patients, and depression in Victorian England*. Oxford University Press.
Pearce, T. (2020). *Pragmatism's evolution: Organism and environment in American philosophy*. University of Chicago Press.
Pick, D. (1989). *Faces of degeneration: A European disorder, c.1848-c.1918*. Cambridge University Press.
Pomata, G. (2010). Sharing cases: The *Observationes* in early modern medicine. *Early Science and Medicine, 15*(3), 193–236.
Pomata, G. (2013). The recipe and the case: Epistemic genres and the dynamics of cognitive practices. In K. Greyerz, S. Flubacher, & P. Senn (Eds.), *Connecting science and knowledge: Scenes of research* (pp. 131–154). V&R Unipress.
Pomata, G. (2014). The medical case narrative: Distant reading of an epistemic genre. *Literature and Medicine, 32*(1), 1–23.
Popper, K. R., & Eccles, J. C. (1985). *The self and its brain*. Springer.
Quick, T. (2014). From phrenology to the laboratory: Physiological psychology and the institution of science in Britain (1830-1880). *History of the Human Science, 27*(5), 54–73.
Radick, G. (2007). *The Simian Tongue: The long debate about animal language*. Chicago University Press.
Raitiere, M. N. (2012). *The complicity of friends: How George Eliot, G. H. Lewes, and John Hughlings-Jackson Encoded Herbert Spencer's Secret*. Bucknell University Press.
Reiser, S. J. (1978). *Medicine and the reign of technology*. Cambridge University Press.
Riese, W. (1950). *Principles of neurology in the light of history and their use*. Nervous and Mental Diseases Monographs no. 80.
Riese, W. (1954). Hughlings Jackson's doctrine of consciousness: Sources, versions and elaborations. *Journal of Nervous and Mental Disease, 120*(5), 330–337.
Riese, W. (1956). The sources of Jacksonian neurology. *Journal of Nervous and Mental Disease, 124*(2), 125–134.
Riese, W. (1965). The sources of Hughlings Jackson's view on aphasia. *Brain, 88*(4), 811–822.
Risse, G. B., & Warner, J. H. (1992). Reconstructing clinical activities: Patient records in medical history. *Social History of Medicine, 5*(2), 183–205.
Romand, D. (2010). Das Körper-Seele problem. *Revue de Synthèse, 131*(1), 35–51.
Romand, D. (2012). Fechner as a pioneering theorist of unconscious cognition. *Consciousness and Cognition, 21*(1), 562–572.

Rylance, R. (2000). *Victorian psychology and British culture, 1850–1880*. Oxford University Press.

Rylance, R. (2006). The theatre and the granary: Observations on nineteenth-century medical narratives. *Literature and Medicine, 25*(2), 255–276.

Scott, D. F. (1992). The discovery of anti-epileptic drugs. *Journal of the History of the Neurosciences, 1*(2), 11–118.

Shorter, E. (1995). The borderland between neurology and history. *Neurologic Clinics, 13*(2), 229–239.

Shorvon, S. (2014). The evolution of epilepsy theory and practice at the National Hospital for the relief and cure of epilepsy, Queen Square between 1860 and 1910. *Epilepsy and Behavior, 31*, 228–242.

Shorvon, S., & Compston, A. (2018). *Queen square: A history of the national hospital and its institute of neurology*. Cambridge University Press.

Smith, C. U. M. (1982a). Evolution and the problem of mind. Part I: Herbert Spencer. *Journal of the History of Biology, 15*(1), 55–88.

Smith, C. U. M. (1982b). Evolution and the problem of mind: Part II. John Hughlings-Jackson. *Journal of the History of Biology, 15*(2), 241–262.

Smith, C. U. M. (1993). The use and abuse of metaphors in the history of brain science. *Journal of the History of the Neurosciences, 2*(4), 283–301.

Smith, C. U. M. (2012). Philosophy's loss, neurology's gain: The endeavor of John Hughlings-Jackson. *Perspectives in Biology and Medicine, 55*(1), 81–91.

Smith, D. L. (1999). *Freud's philosophy of the unconscious*. Kluver Academic publishers.

Smith, R. (1992). *Inhibition: History and meaning in the sciences of mind and brain*. University of California Press.

Smith, R. (2016). History of psychology, what for? In K. S. Hroar & R. Smith (Eds.), *Centrality of history for theory construction in psychology* (pp. 3–28). Springer Nature.

Smith, R. (2020). Inhibition and top-down metaphor of organization. *Studies in History and Philosophy of the Biological and Biomedical Sciences, 80*(October), 1–9.

Spencer, H. (1908). *Life and letters of Herbert Spencer* (David Duncan, Ed.). Methuen.

Stacey, M. (1992). *Regulating British medicine: The general medical council*. Wiley.

Star, S. L. (1989). *Regions of the mind: Brain research and the quest for scientific certainty*. Stanford University Press.

Star, S. L. (1992). The skin, the skull, and the self: Towards a sociology of the brain. In A. Harrington (Ed.), *So human a brain: Knowledge and values in the neurosciences* (pp. 204–228). BirkHäuser.

Sterling, J. (2010). *Representing epilepsy: Myth and matter*. Liverpool University Press.

Swash, M. (1989). John Hughlings Jackson: A historical introduction. In C. Kennard & M. Swash (Eds.), *Hierarchies in neurology: A reappraisal of a Jacksonian concept* (pp. 3–10). Springer.

Taylor, J. (1915). The ophthalmological observations of Hughlings Jackson and their bearing on nervous and other diseases. *Brain, 38*, 391–417.

Taylor, J. (1925). A biography. In J. Taylor (Ed.), *Neurological fragments of John Hughlings Jackson* (pp. 1–26). Oxford University Press.

Temkin, O. (1994). *The falling sickness: A history of epilepsy from the Greeks to the beginnings of modern neurology*. Johns Hopkins University Press.

Turner, W. A. (1897). The care and Management of Epileptics in colonies. *Lancet, 149*, 1733–1735.

Turner, F. M. (1993). *Contesting cultural authority: Essays in Victorian intellectual life*. Cambridge University Press.

Wallesch, C. W. (1989). Hughlings Jackson and European neurology. In C. Kennard & M. Swash (Eds.), *Hierarchies in neurology: A reappraisal of a Jacksonian concept* (pp. 16–24). Springer.

Wassmann, C. (2005). The science of emotion: Studying emotions in Germany, France and the US, 1860–1920, vol. 1 (PhD dissertation). University of Chicago.

Wieviorka, M. (1992). Case studies, history or sociology? In C. Ragin & H. S. Becker (Eds.), *What is a case? Exploring the foundations of social inquiry* (pp. 159–172). Cambridge University Press.

York, G. K., & Steinberg, D. A. (1993). Hughlings Jackson rejection of the unconscious. *Journal of the History of the Neurosciences, 2*(1), 65–78.

York, G. K., & Steinberg, D. A. (1994). Hughlings Jackson rejection of the unconscious. *Journal of the History of the Neurosciences, 3*(3), 153–168.

York, G. K., & Steinberg, D. A. (2002). The philosophy of John Hughlings Jackson. *Journal of the Royal Society of Medicine, 95*, 314–318.

York, G. K., & Steinberg, D. A. (2006). *An introduction to the life and work of John Hughlings Jackson*. Welcome Trust Centre for the history of medicine at UCL.

York, G. K., & Steinberg, D. A. (2013). Jackson's neurological ideas. *Brain, 134*(10), 3106–3113.

Young, R. M. (1968). Association of Ideas. In P. P. Wiener (Ed.), *Dictionary of the history of ideas*. Charles Scribner's Sons.

Young, R. M. (1970). *Mind, brain and adaptation in the nineteenth century: Cerebral localization and its biological context from gall to Ferrier*. Oxford University Press.

Chapter 5
Théodule Armand Ribot: 'Scientific Psychology' in France

5.1 Introduction

Théodule Armand Ribot is the fourth person in my framework of founders who can be looked upon as a representative of a community aspiring to establish a new discipline, 'scientific psychology,' in contemporaneous France of the Third Republic. In the last third of the nineteenth century, he fulfilled a controversial, seminal, and dual role as an important cultural-scientific agent of change. The threads connecting him to both Herbert Spencer and John Hughlings Jackson were articulated in his works early on, through SL, later neo-Lamarckism (NL) views on evolution and heredity. Both Jackson and Ribot used medical models to investigate the pathological. Yet, Jackson was a physician turned clinical scientist, who deployed philosophical-evolutionary sources to establish a new clinical-scientific discipline—neurology; whereas Ribot was a thinker, who had both the training and the ability to deploy certain medical-scientific as well as philosophical evolutionary materials in his cultural–philosophical–scientific endeavor to establish psychology as a scientific discipline. This distinction pertains to the broader fields in which their endeavors were carried out (Fig. 5.1).

In recent decades, there has been an abundance of publications, primarily in French, on French psychology and psychiatry during the nineteenth and early twentieth centuries and within that on Ribot. Here, I attempt to illuminate a circumscribed perspective—Ribot's work as a component within the broader argument on SL/NL transfer and the emergence of new 'scientific' social sciences in Britain and in France. My focus will be on those works, until the early twentieth century, relevant primarily to the transfer of SL/NL assumptions and mechanisms, including heredity, and to the resulting depictions of mind–brain relations, the notion of self (individual), the scope of the mental, role and scope of conscious and unconscious phenomena, memory, constructing individuals and collectivities. I am going to discuss Ribot's enterprise as an endeavor to constitute a new field, scientific-experimental psychology, forging a

S. B. Gissis, *Lamarckism and the Emergence of 'Scientific' Social Sciences in Nineteenth-Century Britain and France*, History, Philosophy and Theory of the Life Sciences 36, https://doi.org/10.1007/978-3-031-52756-2_5

Fig. 5.1 Théodule Armand Ribot portrait (1839–1916). Photogravure by Synnberg Photo-gravure Co., 1898. Source: Wellcome collection, London

tool to scientifically classify emotions ('sentiments'), pathologies, and human groups, and to establish it as an institutionalized discipline, thereby also participating in the endeavor of the Third-Republic regime to forge new non-religious citizenry in the cultural field.

I single out the special role of positivism-strands in the tangled causal and correlational web of the emergence of 'scientific psychology' and 'scientific sociology' in France. Almost all the participants in the political/social/cultural polemics of the early decades of the Third Republic were former students of philosophy, going through the spiritualist heritage of Victor Cousin and his 'eclectic' heirs. Comtean positivism too had changed since the 1840s. Between the 1870s and the 1890s, its legacy was marked by inner divisions, turning the system into a general *Weltanschauung*, open to individualized transfer of discrete components. The principal commonality was a faith and trust that philosophy, methodology, moral attitudes, ethics, and social reform should be derived from science as the locus of knowledge residing in a single/unitary 'Scientific Method.' It allowed for a singularity of each field of science, being empirical or inductive but always rational, guaranteeing objectivity, universality, and accumulative progress. Biology was seen as a principal component, its use of 'structures' and 'functions' as appropriate for organic life generally, including social life. The centrality of evolution for the *new strands* of positivism (not for Comte!) was conceptualized in the 1870s within the NL framework, introducing also 'organization' and 'milieu of the organism.'

This diffuse 'positivism,' resonating in Ribot's work, was a '*cultural sieve*' among the new republican elites resulting in a belief in the possibility of (progressive) remolding of human nature in its social aspects; the distinctions between the normal and the pathological, illness, and health were transferred from the medical to the social/political/colonial discourse of the 1880s and 1890s.

Ribot was one of the pioneering evolutionists among the nucleus of people who attempted to radically change the field of psychology. Yet, there is disagreement on the type of evolutionism he represented, and on the extent his position underwent modification, recalling that Darwin's *Origin* was translated into French somewhat later than into other European languages (by Royer, 1862; by Barbie, 1876), as the long-lasting influence of 'Lamarckian' evolutionary views has usually been pointed to as a possible explanation of this slow introduction (Conry, 1974; Bowler, 1991; Delage & Mary, 1909; Loison, 2010). Three of Ribot's enterprises are relevant here:

1. His and Alfred Espinas' translation of Spencer's *Princ.Psycho.*ii, and his active encouragement of translations of German works in psychology in the 1870s and 1880s, carried out in spite of the deeply ambivalent French attitude toward German culture and science in the wake of the 1870 defeat (Digeon, 1959).
2. His own writings on new British and German works in psychology; his book, based on his doctoral dissertation on psychological heredity; the three short monographs on pathologies of memory, volition, and self; and the book on sentiments from the mid-1890s.
3. The wide variety of his psychological writings, many of which with evolutionary presuppositions, published in the 1870s and early1880s in the *Revue philosophique*, as well as articles and studies published elsewhere until the mid-to-late 1890s defining the field and its boundaries and constraints.

Ribot was the long-time single-handed powerful, courageous, yet balancing, open, and tolerant editor of the *Revue philosophique* (*RP*), which quickly stood out and became one of the most influential journals among the Third-Republic intellectual-professional elites of the human, social, and medical sciences. Fabiani (1988) rightly stresses that in the second edition of *Psychological Heredity* (*L'hérédité psychologique*, 1882), Ribot did not mention his title or degree beside: "the director of the RP"—an assertion of his power position, independently of academia. Ribot was a significant, at times original, synthesizing contributor to the establishment of a 'new,' 'scientific' ('empirical,' 'experimental,' 'physiological,' 'concrete,' 'modern,' 'pathological') French psychology, prolific author, high-brow popularizer, important translator. His ability to disseminate perspectives beyond the mainstream and his rise to a position within academic institutions, without being a regular academic, depended on his being an innovative *cultural agent.*

Ribot occupied a controversial place in the history of modern and French psychology. He fulfilled a seminal role in the transition from philosophical perspectives on the self/person and his/(her) mental functioning to perspectives aspiring to be non-metaphysical, scientific, and experimental (in whichever way these attributes were deployed then). Boring's assessment of him: "Beside the physiologists there were a number of psychologists who may be described as belonging more or less on the periphery of the new experimental psychology … And then there were French psychologists like Ribot and Binet, who belong properly in the French tradition but who nevertheless were on the outskirts of the German movement" (Boring, 1929, 426). "For France plays its major part in the history of psychology within the more strictly physiological psychology and in psychopathology where she assumes a leading role. France had *Théodule Armand Ribot* (1839–1916), who undertook to interpret the new psychologies of England (1870) and Germany (1879) to his country … He was not an experimentalist" (430). Indeed, Ribot began to constitute the new field's proper objects of investigation, its boundaries, and its methods through importing two modes of psychology—the German and the English, the latter forcefully presenting the SL evolutionary approach and associationism both of which he adopted, together with the notion of a blurred boundary-line between the physiological and the psychological.

I put forward four aspects for investigation in order to assess Ribot's efforts to establish 'scientific psychology' as a discipline:

(a) To what extent did the field become regarded as *autonomous*—with specific *problématiques*, conceptual tools, laboratories, and professional journals?
(b) Were there any professional associations? With regulatory functions? Publicity?
(c) What sort of quasi-disciplinary activities, such as public activities, semi-professional ones, and national and international conferences, were undertaken?
(d) Was there any *academization*, such as courses, chairs, curricula, and degrees?

Ribot invested a lot in organizing the first international congress of psychology, chaired by Charcot, in the summer of 1889, in which some of the predominant figures in France in psychology, psychiatry, and neurology participated, e.g., Espinas, Taine, Richet, Pierre Janet, and Valentin Magnan, attracting the most outstanding international figures of the emerging field, e.g., Wundt, James, Bain, Lombroso, Morselli, young Durkheim, and Freud as well, and much publicity. There was a chasm between universities and the diverse contemporaneous teaching/research frames that emerged during the last part of the nineteenth century and until 1914. Nevertheless, Ribot's contributions to all aspects, especially autonomy and progress toward academization, were remarkable, though the psychology departments later established did not follow his vision; he established a teaching position under a close enough title and a couple of laboratories. Ribot was aware of two pressing and not quite compatible needs: The first one is to strike a balance between the need to introduce and make room within the cultural-intellectual field at large for 'scientific psychology,' its point of view, analyses, and the necessity to draw distinctions between these and other discourses, close enough to be perceived as belonging together, while concurrently supporting 'scientific psychology's' claim to academic institutionalization. And the other is to make the field accessible enough to larger audiences as a way of turning it into a valuable tool for the political regime on the long road to academization.

By the end of the century, there reigned a diversity of models, internal division of the general field, and professional societies, some in alliance with various strands in philosophy, some with medicine. Yet, the difficulty of bounding was also related to the entwining of contemporaneous literature with these psychologies—in a mode not dissimilar to Lepenies' (1988) theses on the emergence of sociology between science and literature.

5.2 Methodological Issues

In my view, it is worthwhile to succinctly look into the complex role adopted by Ribot as both a critic of the contemporaneous philosophy/psychology discourse and its transformer through the transfer of implicit and explicit evolutionary notions into it, thus formulating the conceptual frames of 'scientific psychology.' His importations offered new-to-that-place models, interpretive codes, and repertoires of metaphors and analogies, which might change the accepted ordering of signification.

5.2.1 *Early Positions*

Ribot's publications during the early to mid-1870s encapsulated the core of his endeavor, as he looked upon it then, serving as a guideline in interpreting his work until the late 1880s. The list consists of (1) *English Psychology* (*Eng.Psy – La Psychologie anglaise contemporaine*, 1870), especially its introduction; (2) a number of articles on aspects of German psychology, in particular on Wundt, in the *Revue scientifique* (1874–1875); (3) the statement on the inauguration of *RP*, his article on duration, and the very table of contents of the first issues of *RP*; (4) the article on French philosophy in *Mind* (1877), the enterprise of the unpublished translation of Spencer's *Princ.Psycho*.i, and a couple of years later the published translation of *Princ.Psycho*.ii with Espinas (1874). In May 1871, Ribot wrote to Espinas, his old-time Normalien-colleague, translation-partner, life long close friend, and correspondent, when discussing the translation work, that Spencer complimented him on his *Eng.Psy* in the introduction to the book (Ribot, 1957 [May, 1871]).

There has been a certain controversy about whether the introduction to *Eng.Psy.* in itself could be regarded as a Manifesto, especially given the reactions of eminent figures to it.[1] Based on his private correspondence, Ribot looked upon it as an effort to both clearly delineate the boundary separating psychology from philosophy and to introduce the French reading public to a barely-known English psychology. To my mind, the summary and critical comments, forming a past–present–future analysis of the field in the ensemble of chapters—David Hartley (from second edition, 1873 only), J.S. Mill, Herbert Spencer, Alexander Bain, G.H. Lewes, Samuel

[1] E.g., ambivalent Vacherot in *Séances et Travaux* (1870, 467–72); especially supportive Taine in *Journal des Debats*, 13.3 (1870, 2).

Bailey—constituted an early positioning and foundation narrative. Paraphrasing Durkheim's assertion almost two decades later, one could say: "(scientific) Psychology exists, it has an object, it has a method."

The relevant programmatic issues in *Eng.Psy.* were: a historical quasi-Comtean view of the separation of scientific fields from philosophy/metaphysics; the place of psychology posited as a field with a specific range of objects, well-defined questions, '*facts*' as the object of investigation of the specific phenomena of psychology, and its causes and laws. "There are facts of a special nature, difficult to observe, still more difficult to classify, but which do not the less constitute the most solid and the most indisputable portion of the science. It is the pure and simple study of these facts which can constitute an independent science" (Ribot, 1873a, 15 [1870]). Reiterated with a difference: "the facts of consciousness, their laws, their immediate conditions and causes" (17). Conceptual acrobatics were called for to negate the foundational assumptions of Victor Cousin's heirs, yet preserve an awareness of the complexity of the phenomena. Thus, there were side-by-side novel and traditional terms, such as 'memory,' 'will,' etc., and a carefully bounded role for self-observation/introspection, presented as a report (presumably from 'the inside') on states of consciousness, following both Spencer's wording and attribution of function (Spencer, 1870/1872, 1:97–128). This equivocal attitude concerning introspection haunted all of Ribot's writings and was particularly conspicuous in his later criticism of Wundt, especially in his 1900 essay.

Arguing that if the field were to be scientific, then its *method* should be experimental, physiological, and inductive (but following J.S. Mill, also 'deductive'); Ribot followed Lewes in positing physiology and psychology on a continuum, yet with an 'autonomy' caveat: "as an independent branch of inquiry, its [psychology's] results cannot be amenable to physiological canons, their validity cannot be decided by agreement or disagreement with physiological laws" (Ribot, 1873a, 29 [1870]). Ribot pointed to Lewes' significance in emphasizing that *the method of psychology was bifocal—both objective and subjective* (vis-à-vis Spencer and Bain). *Objectivity* was to be enhanced by a comparative study of animals/other organisms. Using pathology meant an occasion of 'nature's experiment,' carried out observationally from the outside: "the morbid derangement of the organism which produce intellectual disorders … are to us as experiences prepared by nature" (25). The application of *an evolutionary view* to the nervous system, introducing a dynamic view of the object of psychology, and a progressivist view of its development were foundational. Following Spencer and relying heavily on Jackson, Ribot assumed an *evolutionary continuum* between the simplest, most elementary organisms and humans. It meant that such a continuum also existed between the simplest phenomena of human nervous physiology and the most complex and sophisticated mental states. Heredity already played a role in *Eng.Psy.*, primarily in reflecting on the role of inheritance in both Spencer and Lewes (e.g., Ribot, 1873a, 312 [1870]). His principal criticism was directed to the poverty of discussion of emotions/sentiments within the English tradition, in particular Spencer.

I put forward here to consider as a group the series of articles on contemporaneous experimental psychology in Germany, introducing German psychology. They appeared in the *Revue scientifique de la France et de l'etranger* during 1874–1876 and in *RP* in 1876, later worked into the chapter on Wundt in *German Psychology* (1879—translated to English by J.M. Baldwin, 1886). Together with *Eng.Psy*, they pointed to a projected synthetic-methodological direction in the endeavor of 'scientific psychology.' These proposed holding on to SL evolutionary notions and their implications to a somewhat curtailed associationist framework as well as to the ideal of 'experimental' to investigate the psychological *problématique*. The articles stressed that it was an ideal that could be realized through measurable, quantifiable precision work. Each of these components underwent changes in Ribot's work from the late 1880s onward.

Two articles (Ribot, 1874, 1876c) discussed the method and results of duration-measurement experiments, two others (1875a, b) elaborated on varied facets of Wundt's books of the 1860s, especially *Grundzüge der physiologischen Psychologie* (Ribot, 1874), laying the ground for stretching the experimental-physiological into the realm of the traditional objects of philosophical psychology: memory, will, person(ality), and sentiments. Ribot carefully introduced the theorizing and experimental practice that were both new and unfamiliar to his readers, using them to undermine well-entrenched views on theoretical and methodological matters, thus performing cultural tasks through seemingly straightforward summarizing reports. Wundt—as reported on by Ribot—turned the order of importance of phenomena upside down, since he claimed that the 'unconscious' was the basis and condition for 'consciousness,' the traditional object of philosophy/psychology, especially in the spiritualist-eclectic-Cousinian tradition. By the 'unconscious' Ribot meant the biological underpinning, the physiological foundation (1875a, 721–32). The phenomena of the psychological order were regarded as corresponding to phenomena of the physiological order, as well as being under the impact of the inorganic world. Such a view had far-reaching consequences in its repercussions on the then-prevailing mainstream position on the object and method of psychology. It impacted methodology and undermined the status of the isolated individual mind/individual consciousness, and its findings, claimed to refute the assumption of their unity and continuity since consciousness was posited as dependent on the nervous system for both. It denied the conception of 'mental faculties' and their classifications and the significant role attributed to 'innateness.' Part of this critique was also relevant to the associationism framework elaborated upon in *Eng.Psy*.

Ribot explicated 'the experimental attitude' of contemporaneous German psychology at large, praising the effort to produce what he considered a measurable, precise, calculable, and therefore 'rigorous science.' He specified that not only sensations and movements were found to be causally dependent on the nervous system, but also ideas, the more complex sentiments, 'will,' relations between internal and external, pointing to the manner in which Wundt wiggled out of the 'free will' controversy (yet, Ribot did not adopt it in *Will).* All would be de-composed into the more elementary, discrete isolatable components—sensations etc.—assuming there was an identical mechanism and the same 'logic' all the way through (Ribot, 1875a,751–60). In Ribot's report/analysis, psychology was done by posing simple focused questions regarding the functioning of the nervous system in order to experiment on these components, assuming that it would yield objective results through precise measurement of duration of acts. Implicitly assuming also that progress of instrumentation would bring deeper and wider understanding.

In his own work, Ribot acknowledged the de-facto *limited results* of the experimental measurement method of German physiological psychology, pointing to their significance as a common scientific practice, though it could not provide an explication of 'higher functions,' discerning the relative duration of the psychic act, rather than the direct (inapproachable) causes of these effects—the nervous system in the case of psychology (1876c, 267–88). He considered it as progress in purging and scientification in both philosophy and psychology. Consequently, the traditional tool of introspection/self-observation could play for him merely a minor role, while external observation, experimentation, and induction replaced it as the high road in search for general laws (1875a, 721–32, 751–60) (Fig. 5.2).

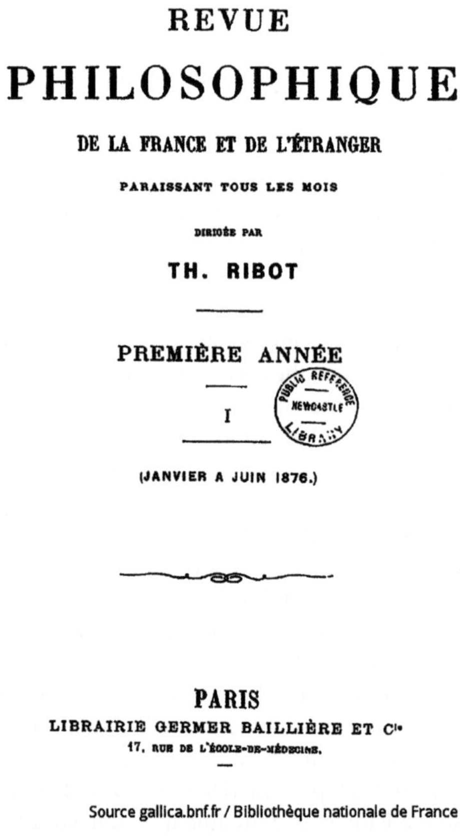
REVUE

PHILOSOPHIQUE

DE LA FRANCE ET DE L'ÉTRANGER

PARAISSANT TOUS LES MOIS

DIRIGÉE PAR

TH. RIBOT

PREMIÈRE ANNÉE

I

(JANVIER A JUIN 1876.)

PARIS

LIBRAIRIE GERMER BAILLIÈRE ET Cie

17, RUE DE L'ÉCOLE-DE-MÉDECINE.

Source gallica.bnf.fr / Bibliothèque nationale de France

Fig. 5.2 Title page of the first issue of the *Revue philosophique de la France et de l'étranger* 1876. Source: gallica.bnf.fr

The first issue of *RP* and the 1877 article in *Mind* did not address his endeavor directly, yet contributed to its delineation. The fact that *RP* started simultaneously with *Mind* in Britain was noted in the new section of the latter (Mind, 1876, 156). "Simultaneously with *Mind*, a French philosophical journal, very similar in its scope, begins to appear. The *Revue Philosophique de la France et de L'Etranger* (Germer Bailliere) will be issued henceforth on the first of every month, under the direction of M. Th. Ribot, well-known in this country by his *Psychologie Anglaise* and other works." Csiszar (2018) traces the shift whereby scientists and scholars in the later nineteenth century started to attribute growing importance to scientific papers in scientific periodicals and in specialized serials. One result of this was the changing features of seemingly generalist periodicals, which led to greater instability in the boundary distinction between the specialist and generalist periodicals and their readerships. In France and Britain, forming specialized learned societies was part and parcel of this. A parallel effort to *RP* but in 'the sciences' was the *Revue scientifique*. The 1870 defeat was in the background of the founding of The French Society for the Advancement of Science, and of a shift in the editing direction by Emile Alglave, the editor of the *Revue de cours scientifique*, renamed *Revue scientifique*. When Charles Richet took over in 1880, he reimagined it as unifying the diverse scientific sub-communities. A first attempt at a specialized psychology journal was the *Archives de physiologie normale et pathologique* founded by Charcot, Brown-Séquard, and Vulpian in 1868. There was also a short-lived attempt to attach a journal to the newly founded Association of Psychology (1885) by Charcot, Ribot, and Richet: *Bulletin de la société de la psychologie physiologique.* The impact of *RP* started to wane mostly in the mid-to-late 1890s after a number of new specialized journals in French came out, such as the mentioned *Bulletin* (1886), *L'Année psychologique* (1894), *Revue de psychologie* (1902, Swiss), *Journal de psychologie normale et pathologique* (1904), and *L'Année sociologique* (1898), a number of journals in medicine, and the philosophical neo-Spiritualism inclined *Revue de metaphysique et de morale* (*RMM*–1893). An important factor in *RP*'s economy was the high-school subscriptions, which went down markedly after the founding of *RMM*. Both *RP* and *RMM* had a number of common topics and had an interest in presenting novel fields and new authors. Around half of the authors published in both journals, though *RMM* posited some as against *RP*; more articles on medicine and psychology were published in *RP*, and more on mathematics and physics in *RMM*. A similar process occurred a bit earlier in Germany and almost concurrently in Britain.

A number of detailed analyses of Ribot's *RP* have appeared recently (Carroy, 2006; Fabiani, 1988; Mucchielli, 1998/2006; Nicolas, 2005). The *RP* was Ribot's vehicle for weaving a wide-reaching heterogeneous network deploying his schooling communities, establishing a loosely bound group of people committed to instituting the field of psychology with differing theorizing and practice. Comparing the origins of psychology/*RP*-people and those of the Durkheimian group, the majority of the psychology people came from outside Paris (all except Janet) and were sons of either physicians or pharmacists, while the majority of the sociology people also came from outside Paris, mostly sons of people in education and mid-level provincial governmental positions (Benrubi, 1933, Fabiani, 1988; Karady, 1979). The wider *RP* network consisted of academic and non-academic figures in a variety of fields, both well-founded, and novel and emerging. It included the high-school-teacher communities at a time of far-reaching educational reforms as well as practicing medical and evolutionary life-sciences people, some of whom were members of the Republican political ruling elite, international supporters, authors, reviewers, and participants in the endeavor to constitute novel frameworks—congresses and associations, local and international—of experimental and physiological psychology. The *RP* became renowned for its tolerance, as Ribot wrote in 1875 to Espinas: "it will be characterized as being open" (Ribot, 1957, 14), though not for mere literary flourish. It was equally renowned for its effort to bring forth ever-new perspectives on matters methodological, *biological, psychological, and social.* These were controversial, evolutionary, SL (between 1876 and 1881, 20 articles by Spencer and numerous reviews on him), materialist, non-religious, and positivistically inclined. About a third of each issue was devoted to book reviews, and a large part to the new psychology, both publicizing and legitimizing. *RP* delineated the changing boundaries of the scientific field, the meaning of social-scientific knowledge, and its proper methods, producing and accumulating symbolic capital outside academia for them. The close cooperation between Ribot and the publisher Felix Alcan, who became the journal's publisher a couple of years after its start, was significant, not only in publishing the *RP*, but also in initiating and maintaining a book series on contemporary issues in philosophy. *RP* militated against the claims of the mainstream, consisting mostly of the Spiritualist/eclectic philosophical elite that ruled over the philosophy departments and the philosophy education on all levels of the system, having the prerogative both in emerging social sciences and in philosophy. Marcel Mauss wrote for Ribot's anniversary book in 1939 and summarized his contribution at large as well as to the constitution of both psychology—directly, and sociology—indirectly, ending thus: "We had all worked with him and for him. There we had forged our first weapons" (Mauss, 1939, 138).

In the first issue of *RP*, an anonymous short opening article announced a policy of inclusion, cooperation, and self-definition for each and every philosophical school, with the following specifications: positivism, German experimentalism, French Kantianism, and spiritualism. However, the critical comment, which was much elaborated on in the *Mind* article a year later, was mentioned here solely as a warning against mere rhetoric. Of the five groups of issues mentioned, four were directly related to the endeavor: Psychology (which would need anatomy, physiology, mental pathology, history, and anthropology), naturalized human/social science, natural sciences of the organic/organized (life sciences) as well as evolution and chemistry, critical evolutionary history of philosophy, while the fifth group, Metaphysics had certain caveats attached to it. The journal aimed at supplying what was sorely missing there—'*facts and documents*.' Upon perusing the table of contents, the order of priorities was conspicuous above and beyond the need to follow a policy of 'openness and inclusion' in order to survive. Throughout the first-year volume, seven out of 20 original articles were on psychology, including Taine, Spencer, Howe (on Laura Bridgman); 10 out of 26 reviews were on psychology/

social-science topics, including Brentano, Jackson, Luys, and Wundt; among the 11 journals reported on were *Mind*, *Journal of Mental Science*, and *Revue scientifique*.

Ribot's article in *Mind* (1877) was seemingly a more explicit presentation of the same policy, as earlier that year Ribot had visited London, following the success of his book on heredity (1873b), and reported to Espinas in 1877 that he met there beside Spencer and Tyndall, also Bain and Robertson, the founding editors of *Mind* (Ribot, 1964). About a third of his article was devoted to delineating an alternative to contemporaneous predominant philosophy, named '*experientialism*,' intentionally lumping together English and German psychology. 'English contemporary philosophy' represented a path between purist positivism and positivism as a methodological attitude, represented by J.S. Mill, Spencer, Bain, Lewes, Huxley, and Tyndall. German physiological psychology was represented by Helmholtz, Fechner, and Wundt. Taine was posited as a representative of a gradual change in that direction within French philosophy. Scientists were included in the list provided for English philosophy; and two in the list representing change in France: Berthelot and Bernard. I would argue that the main thrust of the article was political rather than a mapping of French philosophy. Ribot offered a historical narrative of the development of nineteenth-century French philosophy. The division into schools and figures was less important than the following three factors, which were much more detailed. The dominant philosophy, whether eclecticism founded by Cousin, or contemporary varieties of spiritualism led by his heirs, had been intertwined with all the political regimes from the 1820s on. There was a process of mutual adjustment between the two, in which religion and its institutional power also played a role. Ribot's sharp definition of spiritualism there: "[it] consists in a collection of opinions founded on common sense and adapted to the religious beliefs of the majority. If we extract from the different religions subsisting in Europe the common basis that is called deism or natural religion, and deduce from this deism the theology, the morals, and the psychology which it involves, we shall have Spiritualism" (1877, 384–85). Recently, Vicenti (2020) has argued that spiritualism, especially the second generation one, e.g., Ravaisson, Lachelier, Lemoine, and Paul Janet, had a more composite doctrine than usually presented, that the relations between these and Ribot were complex. Ribot rejected their principal tenets, yet on a deeper level he could be looked upon as adopting/continuing some of their positions. Lachelier and Lemoine had been his teachers, and for many years, he had ongoing intellectual relations with Paul Janet, who was often mentioned in Ribot's work. Janet became a member of the Société de psychologie physiologique, which Ribot founded, and was instrumental in bringing about Ribot's chair at the College de France (Janet, 1888).

The structure of the public education system was such that the spiritualist philosophy (which was until the early 1870s in alliance with the political regime) had its ideas and worldview inculcated throughout the school system, in lessons and examinations specifically devoted to 'philosophy and morals,' supported by a multilevel mechanism of strict supervision of contents. The higher-grade philosophical education of those who became teachers of the teachers/instructors going to lycées and colleges was part and parcel of the dominance of philosophers who represented the ruling philosophy in a centralized country within a centralized educational system. From their seats at the École Normale, they controlled and supervised present and future generations' philosophical worldviews and position-nominations all over the country and were members of the administrative and political machinery, at times also of the intellectual elite.

Ribot pointed to possibilities of resistance and transformation through potential allies, especially men of science and medicine, and by forging independent paths outside that system. Implicit was a hope, a not-so-muted call to the new political regime of the Republic to undo that long-standing predominance. The *RP* soon became financially dependent on subscriptions by heads of lycées and colleges, since, due to its policy of 'openness and inclusion,' innovation was tempered and qualified through numerous articles by prominent members of that very spiritualist variety (Goldstein, 2005, 2013). Such sharp political exposition appeared only in English. Note also that Ribot did not openly participate or express a clear position in the heated controversies around secularization during the educational-reform years, nor concerning the Dreyfus affair.

The formula provided in *RP* for scientific psychology was: adopting *scientific method,* dealing with generalization—laws and regularities—and '*facts*,' found and verified through observation and experiment—both considered empirical. Within the French context, it implied emphasis on physiology and embodiment, but also on (medical) pathology, assuming a continuum between the normal and the pathological. Besides the 'Wundtian states of consciousness,' it thus included 'unconscious states,' at times called, with certain distinctions 'subconscious'; 'somnabulistic' states; both hypnotism and suggestion-inducing states, such as presented by Paul Janet in his four articles in the *Revue politique et littérraire* 1884 (Carroy & Plas, 2000; Vicenti, 2019); automatic writing, as already discussed by Taine (1878 [1869]), but especially by Pierre Janet and early Binet (Bacopoulos-Viau, 2019). Concurrently, it excluded 'psychic phenomena' and their research, in stark contrast to Richet, with whom Ribot often cooperated, who looked upon conscious and unconscious phenomena, as well as reflex and intellectual cognition as end-terms of continua, where all could be reduced to brain-physiology explanations (Harrington, 1985, vol. 3; Hacking et al., 1995; Plas, 2012). Ribot's formula was repeated with variations over the next 20 years in the writings of the newly emerging social sciences.

5.2.2 *Late Positions*

Until the late 1890s, Ribot published, besides his books, a number of theoretical-programmatic texts on the field of psychology. For a couple of years in the late 1880s, he taught an 'open course,' officially titled "cours de Psychologie expérimentale et comparée" at the College de France. Such courses were open to the public without preliminary knowledge requirements. The opening lesson of 1887, in which Ribot attempted to provide a list of activities in diverse subfields in France, was published in *Revue de l'hypnotisme expérimental et thérapeutique* (1888, 344–47), on whose board were Charcot, Bernheim, Pierre Janet, biologists Grasset and Luys, and anthropologist Lacassgane. Ribot ended his brief exposition with an emphatic mention of the founding and continued flourishing of the Society for Physiological Psychology (Paris, 1885a). Here I do not go beyond 1900, but one could find such statements and critical comments also during the first decade of the twentieth century. Ribot's exchange of letters with Henri Pieron during 1907–1915, especially around 1912–14 (Ribot, 2005), sharpened, primarily on Pieron's part, the difference between physiological psychology and other concurrent and competing conceptions of experimental psychology (e.g., Binet's). Parot (2012) considers Pieron as the single person responsible for the status of psychology as a field, as a discipline, before World War I and in the Interwar period. She relates this primarily to what she considers his narrow and inadequate view of psychology, as he posited sensation as the core and focus of a reductive physiological view of the field and held all the significant positions of power within relevant institutional frameworks, though Pierre Janet ruled over the educational aspect. Around the turn of the century, Pieron was interested in 'zoological psychology' (Pieron on the Sorbonne lab, Mengal, 1994; Thomas, 2003) and cooperated with prominent neo-Lamarckists such as Giard to promote it. Yet, Ribot, who encouraged Pieron and helped to advance him to positions of power after the turn of the century, wrote to Pieron in 1907 that Psychology was "*at once a natural and a moral science, forming the bridge*" (Ribot, 2005, 345).

Moreover, in 1909, Ribot published a detailed and argumentative article in a book on methodology in the sciences, which included programmatic articles by prominent figures in the natural sciences as well as two newly established social sciences: Durkheim on sociology and Ribot on psychology. In that article, the scientific method consisted of defining and bounding the object of investigation, searching for specific forms of law-like regularities rather than laws, and using objective methods of research, with a caveat on necessary but limited legitimacy for introspection. The 'objective methods' pointed to the presumed inherent connection of the field with both biological sciences and (pathological) medicine, thus to an evolutionary-developmental account as a necessary component of methodology. There were two significant qualifying comments: none of the observational methods—whether direct or indirect (including tests and questionnaires)—could exhaust their object given the uniqueness of individuals (1928, 284–85 [1909]); introspection was not totally exiled from the legitimate repertoire of methods, though its scope was strictly delimited. The lacuna of how to deal with collectivities was dealt with through evolutionary mechanisms.

5.2.3 'Introspection'

'Introspection' was the self-evident method of the Cousinian school and its exclusively mentalist approach. Goldstein (2013) argues that even Pierre Janet, who revised 1902 the high-school textbook and program (based on his 1894 book) for psychology studies within philosophy, still endorsed 'introspection' though epitomizing the 'non-scientific'—being internal, not given to replication, and by its very private character, closed to rules of public procedure and reporting. 'Introspection' was severely criticized by Comte, who eliminated psychology from his list of scientific disciplines. So did his principal followers, especially Emile Littré. Spencer, in his *Princ.Psycho.*i, well within the associationist tradition, pointed out its inadequacies, yet without giving it up entirely. In his first books, Ribot (heroically, ambivalently) attempted to strike a middle way, so as to avoid a strong reductionist position on brain–mind matters. He stated that it could, perhaps, serve as a starting point for the investigation, being an internal observation, though on a limited field of phenomena. In his opening address, "On the status of contemporary psychology" (1889), presented in the summary of the First Congress of Physiological Psychology (held in Paris in 1889) summarized by William James in *Mind*, Ribot stated: "showing in simple but impressive words how it advances by combining physiological and pathological observation and experiment with the older introspective method" (1889, 614); yet, in an article published in 1900, he claimed "nothing begins without it, nothing is achieved solely with it" (1900, 282–84).

Ribot propagated first the Wundtian physiological-experimental model and then the British (Galtonian) questionnaire method as conducive to psychological-scientific *measurement*. In 1904, critical comments on the questionnaire method could already be found in an article (1904, esp. 8–9) and followed his own experimenting with such forms of investigation in his work (1891–1892). In a later book (1896) on 'General Ideas,' Ribot assumed that there was some parallel between physiological processes of the brain and those of association and verbalization. His criticism helped to further enhance the 'case-by-case (i.e., individual-focused) methodology' whose model was 'nature's experiment' of medical pathology, rather than inducing and intervening/'provoking' experimentation, or aggregated quantitative work. In the 1909 article, Ribot introduced an elitist political undertone into his criticism of the quantitative-questionnaire work, thereby countering Binet (Ribot, 1909, 289). Historians of psychology discussing the contemporaneous complex positions on 'introspection' (Carson, 2007, chap. 4; Danziger, 1980, 1990; Carroy & Schmidgen, 2002) point out that even the experimentalist method practiced at Wundt's laboratory and their instrumentation, measuring the duration of perception, stimulation, etc., relied on the subject's introspective report within laboratory settings.

Furthermore, the questionnaire method was considered to induce an introspective process in those who replied. Both methods were criticized by various parties within contemporaneous psychology-medicine communities for missing out on the qualitative aspects, on the individualized ones, and in general on issues considered by French contemporaries as those most pressing and fundamental in psychological research. Thus, the question was: Could 'introspection' become 'scientific,' without endorsing the tenets of the spiritualist-eclectic school, in any of its extant variations, including the much transformed approach of Pierre Janet? (Beaunis, 1909). *Or else:* How could one avoid radical reductionism *and* preserve that 'uniqueness of individuals' (e.g., Ribot, 1909, 282–84), which came up in fields other than psychology as well? As Libby Schweber aptly put it in the introduction of *Disciplining Statistics*: "In England, political liberalism went along with an individualist, probabilist understanding of population rates. In France, in contrast, politicians, administrators and legislators resisted this interpretation. When statisticians reported to them on the mortality of the nation, they responded by asking who or what those figures referred to and whether one could reasonably *aggregate across radically different individuals* ... Eventually, French administrators and scholars resolved their problem by positing a radical difference between collective, statistical entities and the individuals who made them up" (2006, 1, 5—emphasis added). In late nineteenth-century France, as Carroy and Plas (1996) rightly argue, the more prevalent model practiced did include elements of 'induced introspection/self-observation' during a long period of vacillating over research methods. In his report from 1893 on his stay at four German laboratories (Wundt, Müller, Martius, Ebbinghaus), Victor Henry specified that his attempt to get them interested in the French method of 'self-observation' and its value totally failed, commenting critically on the interpretive subordination of individual differences to broad conclusions. This was also true of some of the groups of neurological and psychiatric professionals.

Ribot was thus criticized by positivists (e.g., Littré) for endorsing a limited acceptability of introspection and negated by spiritualists for sharply circumscribing it. In the introduction to his 1896 book *The Psychology of Sentiments (Sentiments)*, Ribot discussed the difficulties confronting any empirical investigation in this field, warning that introspection could not be deployed since this psychology dealt with the physiological, the unconscious (1896, esp. vii–viii, 93, 438ff). In his 1909 essay on psychology, Ribot tried to strike a balance between the status of the individual, being interested in 'differences,' and the methodological need for 'scientific objectivity' (1909, 282–84). Binet's work/experimentation on his daughters (1890–1903) was a contemporaneous case of combining introspection and German measuring experimentalist methods. Implicit throughout the controversies on introspection was the assumption on the uniqueness of the 'mental object'—the individual, beyond neural functioning and its possible laws.

5.2.4 The Use of Medical Cases and the 'Pathology Model'

In the late nineteenth century, there was a noticeable move in case presentations in current psychological discourses toward greater subjective-affective phenomena, as well as descriptions of states of consciousness versus its absence, and even 'literary flair.' Concurrently, literary authors wrote more on what would now be considered 'psychological cases.' The most famous was Stevenson's *The strange case of Dr Jekyll and Mr Hyde* (1886—Finn, 2017; Sotirios, 2018). Ribot's use of medical cases is conspicuous, especially in his three 'diseases' books: *Diseases of Memory* 1881, translated 1882 (*Memory*); *Diseases of the Will* 1883, translated 1894 (*Will*); and *Diseases of Personality* 1885, translated 1895 (*Personality*), differed from Jackson's, since Ribot presented already crystallized and carefully formulated theoretical positions, having gone over numerous medical cases. As part of his study of medicine, he was present in numerous sessions of Charcot 'inducing' hysteria and actually appeared in the well-known 1887 painting by P.A.A. Brouillet of such a session (Harris, 2005). Ribot used contemporary relevant theories, SL and French NL (FNL) assumptions, and had a well-founded familiarity with the work, both experimental in any sense and theoretical, done in the wide field of contemporaneous psychology in France and beyond. The detailed medical cases he presented were *deployed as examples* of topics discussed, at best elucidating, often merely illustrative. Occasionally single cases were used as counter-examples whose very existence should undermine a contrarian theoretical position (anachronistically 'a Popper-like fashion'). It seems that he used these cases as a form of medical-psychological literacy in which the cognitive practice was implicitly inscribed.

The central role of the '*patholo*gy model' in French psychology has been controversial in historiography. The contemporaneous psychology discourse toward the last third of the nineteenth century was carried out in a variety of modes, with sharp disagreements on the proper sites of 'scientific' investigation, whether named 'empirical,' 'experimental,' and 'physiological':

1. Physiological/experimental mode, principally following the one carried out especially in two differing ways in Germany, on duration in perception and motion (Henri, 1893).
2. Physical-anthropology discourse investigating the relationships of skulls and intelligence.
3. Work on 'psychological pathology,' particularly on topics such as pathological states of consciousness, unconscious states, automatic action, 'somnambulism,' contrasted with the 'non-scientific' philosophical/metaphysical investigation with 'introspection' in any extant variant as the *sole* method.

Issues of embodiment-corporality, environment, and the nature of subjectivity could not be avoided even within this latter philosophical investigative mode. Thus, the controversy was also relevant to the relations between budding 'scientific psychologies' and emerging sociologies in France. New fields of sociology, pedagogy, and psychology were included in Paris curricula much later than elsewhere, only when special chairs were instituted for them. The new fields, which had 'appropriated' topics/objects from the field of philosophy, were firmly blocked by the strong figures of philosophy, especially at the Sorbonne. Having medical education could open a parallel path in psychology for some, e.g., Pierre Janet and Georges Dumas. The change after these strong philosophical figures disappeared was gradual but conspicuous.

Thus: Was Ribot in fact a thinker who did not separate psychology from philosophy, in spite of biological SL/FNL presuppositions, since both in his theories and within the actual communities to which he belonged and viewed himself as affiliated with he still belonged primarily to the philosophers? He used philosophical argumentation against a particular kind of philosophy and not against the psychology discourse within philosophy in general. Furthermore, it is argued that it was solely through philosophy, by means of a strategy of co-existing insertion and differentiation, that Ribot's psychology found public and institutional space within which it could flourish. If a new discipline did emerge, it was only after compromise and with limited institutionalization. Recent proponents of this position are Brooks (1998), Carroy and Plas (2006), and Ohayon (2008). In a survey carried out by Binet in 1907, it seemed that young people studying philosophy as part of their high-school curriculum were interested in 'psychopathology' (Binet, 1908). One could infer from the survey that Pierre Janet's model, and specifically his textbook, won the day (Goldstein, 2013). It combined elements from Paul Janet, Charcot, and Ribot. Recall Ribot's claim in 1907 that psychology was simultaneously a natural and a moral science.

Or else: Did Ribot strive consistently to disconnect the new psychology from philosophy, not only from the Cousinian-'spiritualist'-eclectic current, but from philosophy in general, though with varying degrees of success? It is argued that he strove to establish a separate distinct field that adopted both regulative ideals and methodology as scientific as possible within the then-existing positions. Moreover, he tried to affiliate it to the medical sciences, particularly physiology and neurology. The main proponents of this view are Mucchielli (2006), Nicolas (1998), and Nicolas et al. (2016).

Nevertheless, both parties, though with different emphases, have promoted the view that the (medical) pathological model was predominant in French psychology at large, including Ribot. Both parties refer to the editorial *Nécrologie* for Ribot in *RMM*, which emphatically stated: "But what distinguishes the conception of Th. Ribot and that which made his work forceful … the study of pathological cases" (1917, 129–32).

Mental pathology, assuming that insanity revealed the basic forms of normalcy, already appeared in the title of the alienists-psychiatrists' journal, *Annales médico-psychologiques* (founded 1843), whose editors/founders, including Baillarger, were related to hospital/prison system/the academy of medicine. Taine, especially in numerous editions of *De l'intelligence* (1869–1914), elaborated on the importance of learning through the pathological about the normal. Ribot followed by constructing 'pathology' as 'experimental,' relying on Bernard. In Canguilhem's interpretation ([1978] 1991), it meant that once identity of object and method was established, the method of the clinic and that of the laboratory could be conceived as identical, as argued especially in Bernard's widely read *Introduction* … (1865) and *Leçons* (1878/79), stressing that from Bernard's point of view, there was an irreducible difference between disciplinary perspective on those states of the organism and their 'biological value.'

Ribot's 'diseases' books exhibited intertwining of 'pathology' themes from the French psychiatric tradition (e.g., degeneration, forms of mental alienation), with foundational notions of Jackson's SL, reworked into an integrated pathology frame, also adopting components of Charcot's practices. It was conspicuous in *Memory*, yet earlier in *Psychological Heredity 1873 (Psy.Her)*, allowing Ribot a smooth move between psychology as physiological and as medical, and a backdoor path to reconnect psychology to philosophy. Among contemporaries, Binet reverted to experimenting with normal subjects, but for Pierre Janet the model of pathology remained central, founding in 1904 a rival journal to Binet's *l'Année psychologique*, named *Journal de psychologie normale et pathologique,* and shortly thereafter an institute bearing the same title.

From yet another point of view the adoption of the pathological model meant to offer causes for any pathology investigated, usually related to the nervous system and its physiological functioning. Toward the end of the nineteenth century, many biologists hoped that empirical criteria of 'the normal' versus 'the pathological' would be discovered, founded on quantification of the intensity of certain 'vital functions' of the organism. 'Normal' would then be a statistical average, while 'pathological' would form both extremes of such measured function. Yet, issues such as states of hysteria and vicissitudes of memory raised controversies on whether 'objective physiologically dependent states' of consciousness were sufficient, or perhaps were irrelevant to the phenomenology with predominant subjectivity of the individual, examining behavior under verbal suggestion, the unconscious as a space of reflex activity and impulses. Such positions could potentially challenge the status of 'rationality' and 'free will,' and their positioning in the evolutionary hierarchy of functions, while still deploying the pathology model.

Ribot was among the principal agents of diffusion of the new conception of 'the normal' and 'the pathological.' Yet, his conceptualizing of 'scientific psychology' and his methodological position changed within a context which he greatly contributed to and helped to shape, through his continuous publications, teaching, and significantly in the establishment and support of psychological research laboratories and the weaving of networks of medical and evolutionary physiological professionals around the newly emerging field.[2] The methodological emphasis on individuals, which went hand in hand with the 'individual-centered' view of the field, could at least partially explain the need to deploy SL/FNL evolution, heredity, and memory to universalize and collectivize when self-presentation as 'scientific' was of paramount importance.[3]

[2] On laboratories, networks in France, see Carroy and Schmidgen (2002), Carson (2007), and Mucchielli (1998).

[3] Ribot's *self-presentation:* published letters to Taine, Espinas, Pieron; Feuerhahn and Trochu (2016) on further correspondence.

5.3 Psychological Heredity: *L'hérédité, étude Psychologique* (1873)

> Mr. Ribot is proud of science, and science leads to morality in seeking nothing but the truth. (Taine, 1873)

I suggest distinguishing between certain facets of Ribot's discussion of evolution, evolutionary mechanisms, and heredity during this period. Ribot's overall view of evolution was emphatically Spencerian and thus Lamarckian to the extent that Spencer's was. In his later writings, especially since the mid-1890s, Ribot continued to emphatically support SL/FNL on the inheritance of acquired modifications. Yet, the discussion of certain aspects of heredity in his doctoral dissertation in French—later *L'hérédité, psychologique* (1873b)—drew from a variety of additional sources, mostly Darwin's *Origin*, Galton's publications of that period, Lucas, even Morel and Magnan between late 1850s to the early 1870s. The dissertation was approved in 1873, though blatantly anti-spiritualist. It has been argued that it could not have happened before the establishment of the Third Republic, and 'behind the scenes' firm support of both Taine and Renan. Ribot often used there the rhetoric of survival, selection, degeneration, and atavisms. For example, "All progress, or, to speak more precisely, all development, pre-supposes evolution and heredity. Without the former there is no change; without the latter there is no fixity. But the action of heredity has its limits. As we have seen in the physiological introduction, deviations tend to disappear, and after a few generations the reversion to the primitive type is complete. In the moral order there are facts of the same nature – as, reversion to the savage life and to nomadic instincts" (1873b, 379). Strikingly, a very subsidiary role was allotted to environment, and heredity was conceived, besides its role in conservation, as a central regulating factor in evolution, significant for both progress and degeneration. Staum (2011) holds that there is room for nothing but heredity in Ribot's work, and Guillin (2004, 2016) supports this position, adding that since the environment has hardly any role, the plasticity discussed is not a real one. The last part of Ribot's book expounded on the social meanings and consequences of 'heredity,' in particular the much-discussed contemporaneous issues of race, degeneration, psychological-cognitive make-up of the individual, social morality, and social institutions.

The book had eleven French editions and was translated into multiple languages. It is considered by Ribot interpreters an important reservoir of concepts developed in his later works, adopting the perspective of individual heredity. Ribot kept gradually updating the book's biological repertoires in later editions; thus, in the 1894 edition, he openly adopted FNL positions, more sharply delineated in the controversies with August Weismann. Its thrust was read as reformist, claiming to begin laying the foundations for a new *scientific* field. Contemporaries in the generalized field of philosophy-psychology regarded it as an explicit effort to replace the Cousinian-spiritualist-eclectic cluster of positions with a physiologically inclined view on the mental-psychological, within an evolutionary framework. Being nominated in 1882 for the chair of Clinical Diseases of the Nervous System, at the School of Medicine in Paris, whose first nominee was Charcot, was thus regarded as a direct blow to Cousinian views. Ribot claimed to reveal 'laws of heredity' expressed not only physiologically but also psychologically. This claim served as a foundational assumption for three moves: a blurring of the boundary between the physiological and the mental which emerged from it (e.g., 1875c, 21); evolutionary gradual/graduated continuity, which was echoed in the graduated continuous rise of the level of organization within the mental; adoption of the Jacksonian view on the evolutionary hierarchy of neural systems, especially human ones, and the inferences drawn from that on how to conceive of normalcy and pathology of mental functioning. Dowbiggin (1991) suggests the adoption of heredity (with degeneration) as a principal causative/explanatory mechanism of insanity by psychiatrists in the second half of the nineteenth century, providing a physiologizing scientific garb, an instrument of professionalization and a means of becoming party to the medical-political discourse, and Ribot's *Psy.Her.* as instrumental in that long-term endeavor.

These three moves accounted for the movement from unconscious states to conscious ones. Memory was thus conceived as a particular mental functioning, whose products were inherited. Heredity was explicated as spreading over all forms of life and levels of organization, though its particular formations might vary, and the means for tracing it might differ: "In conclusion, heredity always governs those broadly general characteristics which determine the species, always those less general characteristics which constitute the variety, and often individual characteristics. Hence the evident conclusion that heredity is the law, non-heredity the exception" (1875c, 144–45 [1873b]).

Ribot often wrote as if the psychological and the physiological orders were distinct and separate (e.g., 1873b, 254–55); in the second edition (1882), the third part was heavily edited and much of the discussion dealing with these issues was taken out. Yet, I would argue that there was a 'surface parallelism' with a foundational-level assumption that the two orders were one. A parallelist position seemed a useful one for positioning psychology as a science. Considering parallelism a surface phenomenon whose deep structure revealed monism had benefits, though laden with inner tensions, such as allowing a unique perspective on 'the mental,' a broader meaning for 'the psychological': "Thus the parallelism is complete between these two orders of fact [of phenomena – the physiological order and the psychological order], *which at bottom are only one*" (1875c, 280—emphasis added). In contrast to Jackson, the 'two heredities' were construed so that the physiological was the cause of the psychological. This 'surface parallelism' supported his argument for a potential analogical pattern of reversion in both orders of phenomena (1873b, 318–19). The 'two heredities' came up when Ribot deployed material of French anthropologists, in particular members of the Societé d'anthropologie de Paris, the more biologically and racially-oriented society, which branched from Broca's group of disciples and was closely related to the faculty of medicine. They and Le Bon, who is considered by some historians as a proper member of the group, subscribed to notions of collective heredity, including forms of 'moral atavism,' attributed primarily to non-European races, but also to certain groups among European ethnicities. Taine's laudatory article too used examples of various indigenous individuals, who were brought up as Western, but 'lost it all,' when they returned to their country of origin and reversed to the supposedly inherited psychological patterns, which, though, could only be defined by physiological conditions; Darwin's Fuegians spring to mind too.

Ribot updated the book further with following editions. He wrote to Espinas on May 22, 1882: "As for 'the heredity': you can regard it as *completely redone*, without bothering anymore about the first edition" (1970, 344—emphasis added). Yet, his emphasis on a more rigid hereditary pattern, with clear laws, was still conspicuous in 1882. Numerous pronouncements resonated with those in the relatively compact yet historically oriented item on heredity, written by Letourneau, in the *Dechambre Dictionary* (1859–1865). One should also add the contemporaneous ascendancy of the hereditarian *medical* model in French culture as well as in French political life, with ample testimony in innumerable essays in the daily press of the period and in many general cultural journals such as *Revue de Paris*, *Revue de deux Mondes*, and in much of the contemporaneous political rhetoric.[4] In the third edition of 1887 of *Psychological Heredity* (1887), Ribot further omitted the part that discussed the relationship between mind/*esprit* and matter. Relevant additions in 1897 (the fifth edition, further editions followed it) were the discussion of unconscious psychological phenomena that became important then, participating thus in the ongoing controversy; and differing emphases on heredity and environment as concisely expressed in his article from 1893, where he firmly supported the inheritance of acquired modifications (characteristics) to be located within the context of the controversy in France and elsewhere on August Weismann's argument on the total separation between hereditary material units (germ-plasma) and somatic ones. At the end of his 1893 *Encyclopédie* article, Ribot said, relating to the controversy and Weismann's re-stated position (apropos his debate with Spencer): "He [Weismann] admits that predispositions are transmissible, this seems to me to reintroduce heredity [of acquired modifications]" (1169). In that article, Ribot emphatically rejected the role of parents as mere receptacles harboring the germplasm materials to be passed on trans-generationally. He argued for their role both in the upbringing and in the passing on of modifications acquired via the environment in its broad sense throughout their own lives, which played an important 'moral role' in generational responsibility (1169). Stressing environment became all-important in *Sentiments* (1896), while in the *Encyclopédie* article Ribot still argued for direct heredity of sentiments and passions. Yet, already in the first edition of *Psy.Her.*, Ribot acknowledged accumulated changes/modifications, both physiological and psychological, that were fixated in the individual through habit. These were the basis of 'progress' and of a growing constitutive chasm intellect/civilization-wise between 'primitive' and modern. Benichou (1989, 88–9, 94n56) argues that Ribot's adherence to SL/NL was intentionally obliterated from the discussion of his work in the 1939 anniversary celebration (on the celebration, Voutsinas, 1988), and that it

[4] Carroy and Plas (2006), Dowbiggin (1985), Ellis (1990), Goldstein (1987), Leonard (1992), and Nye (1984).

was a decisive factor in not including Ribot in a special celebration of 100 years of psychology in France in 1989 (Benichou, 1989, 88–89n56, 94).

To sum up, the specifically innovative aspects of *Psy.Her. for French audiences* were the claims that there existed psychological heredity, that in order to trace, investigate, and analyze it, one had to look at a wide array of phenomena, some of which belonged to the realm of the pathological in the field of the mental; and that following Spencer and the assumption of evolution, instinct, habit, memory, and intelligence formed a continuum. Consequently, the ensemble of these positions called for a re-evaluation of the Cousinian/quasi-Cousinian philosophical positions on the division of the mind into faculties and on the unity of the self. Thus, *Psy.Her.* was read as a contribution at a deeper level to the controversy on the possibility of scientific, i.e., empirical, experimental rather than introspective, investigation of the constitution and functioning of 'the cognitive' and 'the self.'

5.4 Foundational Notions

I shall primarily discuss the combination of evolutionary SL/FNL assumptions and elaboration of principal—possibly foundational—notions as developed in Ribot's books until the end of the nineteenth century, noting also that Ribot, following Spencer (and British psychology), was under the spell of associationism.

5.4.1 Hierarchical Order

Almost concurrently with the translations of Spencer's *Princ.Psycho.*ii, Ribot published *Eng.Psy.*, where he aptly summarized Spencer's views, including the assumption that going down the hierarchical order of mental phenomena meant also decreasing the complexity of the nervous system and its organization, and vice versa (1874, 159, 187 [1870]). In spite of the important role of Spencer in his work then, Ribot actually adopted Jackson's view of the evolutionary hierarchical order, while studying courses in medicine from 1872 on (especially on the nervous system and mental illness). In the 1880s, when he was already teaching psychology, Ribot still recommended the medical path to his students as the most suitable one. He studied with Bernard and Charcot, as well as histology with Charles Robin, Jules Baillarger on heredity, and befriended Jules Luys, Robin's student working then on hereditary diseases and the nervous system (Benichou, 1989). Concurrently, Ribot served as a teaching assistant at the Sorbonne in the courses of Pouchet (1875) and Dastre (1876–1885) on the nervous system. Robin was particularly significant because of his radical biological materialism, especially concerning cell theory, the organism's biological individuality and organization, the normal and the pathological in its physiological and chemical organization. During the Second Empire,

Robin was castigated as an anticlerical positivist, later on related to Littré (Thomas, 2017).

Already in February 1876, a translation of an article by Jackson (1873a) from the *West Riding Asylum Medical Reports* was published in *RP*, and Jackson mentioned Ribot in his work during the 1880s. The adoption of Jackson's view on evolutionary hierarchy, complexity, and organization was particularly conspicuous in Ribot's work on memory and heredity, *The Diseases of Memory* (*Memory*—1881), which ran nine editions, but also in his discussion of Lewes in *Eng.Psy.* Ribot followed Jackson's reversal and his inversion of Spencer's direct relation between complexity, organization, and stability and applied it to heredity. It seems that Pierre Janet too, in *L'automatisme psychologique* (1889), adopted Jackson's notion of hierarchy, perhaps familiar to him through Ribot, as he regarded the exposition of the pathological—of automatic action—with the help of hypnotism, as revealing the low, elementary, actions ordinarily under inhibition of the upper functioning of the normal person. In Janet's later work, the psychological functions were posited in a descending order, though it was neither developmental nor explicitly evolutionary.

Yet, adopting the Jacksonian notion of evolutionary hierarchy undermined the accepted division of the brain, as well as that of the mind's capacities, of the 'self,' relativizing the mind's order of importance. The hierarchical order, as evolutionary and developmental, was interpreted as graduated, manifesting rising complexity in the human nervous system. In the 'diseases' books, the hierarchy was investigated in the variations of differences among human brains exemplified in society between the child and the adult, 'the primitive'/non-civilized and the civilized, and the normal and mentally ill/the pathological. Ribot was the first in the general field of the new psychology to use as his models not only *the child*, *'the primitive'*—exemplifying multifaced infancy of humanity—but also *the mentally affected* (Spencer used only the first two), indicating thus his three principal source-fields: studies of human development and Enlightenment's philosophizing on child development; contemporaneous anthropology and medicine; and 'degeneration discourse.' The notion of evolutionary hierarchical order was crucial in Ribot's evolving thinking on memory and on consciousness.

Delay (1951) has characterized Ribot as an authentic Jacksonian, who preceded the wave of 'neo-Jacksonism,' attributing it to a common source of influence—Spencer. Delay considers three components as fundamental to Jackson: hierarchy and complexity; distinction between the general and the local (especially in 'dissolution'); and distinction between positive and negative symptoms. Neo-Jacksonism assumed, so Delay argues, all three but went beyond that by ascribing a primary power to instinctive dynamism/to the unconscious, i.e., by positing a neurological-mentalist perspective, which was foreign to Jackson, for whom both phenomena and etiology were basically anchored in the biological-organic realm. These issues were prominent in Ribot's work in the late 1880s, as Ribot's use of the Jacksonian hierarchy evolved from neurology to psychology proper, became dynamized, and was deployed as such in the 'diseases' books, especially *Memory* and *Will*.

While Goetz and his colleagues (1995) argue that Charcot biologized/neurologized states that were considered mental, the domain of psychology, whereas Ribot tried to deploy Jackson to explicate psychological *problématique*, without losing the physiological-neurological anchor, which was to guarantee the status of 'scientific psychology.'

5.4.2 The Law of Reversion/Dissolution

> The law which I have formulated is nothing but the psychological expression of a law of life and in its turn pathology in memory then shows us a biological fact. (Ribot, 1881, 99)

I explicate the 'law of reversion/dissolution,' which for a while was named after Ribot, by looking into the endeavor to offer laws in the field of psychology. There were accepted regularities in physiology, often posited as laws, both in the 'Galilean' and the Comtean senses, which supplied knowledge supposed to form the basis for further experimental and observational practice, relying on their universal applicability and validity. This did not apply to medicine, nor to contemporaneous psychiatry. Thus, the intention to lay down a law applicable to psychological functioning within a pathology model indicated the primacy of 'scientificity.' Discussing neurasthenia, Rabinbach (2018) claims that Ribot, discussing diseases of the will, suggested a special psychological-sociological interpretation to 'the law of least effort,' which means, from my perspective, another attempt to posit a psychological law. Even though the 'reversion law' was mentioned already in *Psy.Her.*, it was explicated and widely applied in *Memory*. Hacking (1998, 208) defined it thus: "they offered knowledge. It was new knowledge, scientific knowledge about memory. Real knowledge, scientific laws about memory, even what is still called 'Ribot's law'. It purports to be an objective truth. It follows from facts. The facts in question are from pathological psychiatry. It is a law about loss of memory, about forgetting. Finally, the law covers, in a uniform way, both forgetting caused by physical lesions, and forgetting caused by psychic shock."

The SL inheritance mechanism of back-and-forth movement between the social and the biological was transposed to a movement between the neural-physiological and the psychological to produce a general law of pathological psychology. In a letter to Espinas in October 1882, reacting to Paul Janet's scathing critique of *Memory*, Ribot wrote, referring to the positivist 'ladder of the sciences': "My aim on the contrary is to deploy physiology for the benefit of psychology, to use it as tool. Every science should depend on a lower one: does not physiology do the same with regard to chemistry?" (quoted in Nicolas & Gounden, 2017, 192). 'Hierarchical order' served as an underpinning on several levels: as a view of the physiological as always 'attendant' on the mental, particularly on the psychological, but not vice versa, establishing the foundational role of the physiological in discussing mind and psychology; all psychological functions were perceived as 'embodied,' implicitly dependent on their physiological description/analysis; pathology was conceived as 'nature's experiment.'

Ribot picked Jackson's term, 'dissolution,' adopted from Spencer, and expounded on it further, as a biological law transferred to the psychological arena, conspicuously in his explication of amnesia in *Memory*, which included enumeration of 'cases' typical of medical rather than physiological argumentation, perhaps partially adopting Jackson's model. Contrary to Spencer, whose name could be found in numerous footnotes and quotations, but *not* in the index, Jackson was one of the most frequently mentioned there. The phenomena of amnesia were thought to appear in a particular order, whose direction was the reverse of the evolutionary one. It stretched all the way from the least organized, most complex and 'voluntary' (Jackson preferred 'less automatic'), down to the most organized, simplest, and automatic. Carroy and Plas (2000) interestingly suggest distinguishing between activities considered as *unconscious automatic* within British contemporaneous neurology—which referred mostly to those related to the spine, e.g., Laycock and Carpenter—and those within the French Spiritualist tradition, which regarded sleep, dream, and somnambulism as examples of such activities. Hence, by enlarging the 'unconscious automatic' through pathological physiology, Ribot deployed them primarily *to undermine* the Spiritualist concept of the 'unitary self/personality.'

Furthermore, according to Ribot, two cases followed these stages 'from bottom to top' of the hierarchical order and thus revealed the early stages of the evolutionary development: the evolutionary growth of memory in general as exemplified in 'the child' and in 'the primitive'; and those cases of spontaneous return of lost memory (discussed in the chapter on partial amnesia and following Jackson's work on some forms of epilepsy, neural structures, and functioning). The presumed underpinning and causes of psychological phenomena were deemed 'biological' (e.g., Ribot, 1881, 127), and their spatial locations were described according to the hypothetical functioning of various parts of the neural system (e.g., 129–31).

In *Will*, the same assumptions and a similar methodological approach were adopted. The 'law of regression/reversion' served as a presupposition in the description of pathologies as the traditional faculties of memory and of will were 'de-constructed' into mental functioning, into 'states of x.' In both, there was a physiological hierarchical order whose pathologies followed the 'opposite direction from the evolutionary' as indicated by 'Ribot's law' (Ribot, 1881, e.g., 121–31, esp. 127, 202–04, and significantly in the chapter on "hypermnesia and the law," 183–84).

The chapters on partial amnesia and on hypermnesia exemplified Ribot's deep immersion in the medical, neurological, and psychological literature of the period, though these were very infrequently quoted directly. They underlined his determination to establish the field as scientific, thus having fundamental laws. A short comparison with Bergson could be helpful: Bergson too adopted an evolutionary perspective, Jacksonian to a large extent, concerning the 'law of regression.' He held to both 'habit' as a result of repetition, and 'memory' recalled as personal. Yet, for Bergson, regression was not morbid, did not bring about any decrease; it could have a freeing effect that allowed choice, which could bring about '*a souvenir*.'

Ribot classified and ordered medical/psychological cases to supply further proof for the law of regression, discarding doubtful instances as irrelevant. The analysis presented the pathology of dissolution/disorganization as 'nature's experiment,' enabling the inference of normal organization.

Memory was assumed to work under fixed and stable laws. The direction of the pathological process would be exposed through memories that were less stable because newer, less anchored in webs of repeated associations, more complex and thus more vulnerable to disintegration. The last memories to disintegrate would be older, those often repeated and registered, simpler and therefore stabler. Contrary to Jackson, Ribot never talked of 'local dissolution,' though he did discuss the possibility of a partial one, which was supposed to emphasize the dynamic aspect of the process (1881, 111–12). Ribot did not distinguish between sensory and motor aspects, a significant distinction in Jackson's discussion of aphasia. The same description of the pathological process would apply to the dissolution of the will, whose direction would be from the 'voluntary' to the increasingly automatic. The same applied to the pathologies of personality (in 1885b). Even in *Sentiments* (1896), one finds the same directed mechanism of disintegration of affective life, consisting of four stages. When memory-ability was recovered, the direction would be opposite to that of dissolution; stable memories, which disappeared last, would be the first to return.

There was a significant difference between Jackson and Ribot: For Jackson, the broad array of phenomena of dissolution, from the normal to the pathological, signified damage, hurt, and illness of the nervous system; for Ribot, they also signified degeneration, accompanied by a strong moral undertone. In the 'diseases' books, the varieties were explicated as various forms of 'loss of control' in a moral sense, of social deviance. Le Bon elaborated further on this as a loss of mental control, resulting in regression to infantilism, labile moods, impulsiveness, 'weakness of the will,' attributed to 'primitives,' women, and children in the book on crowds (1895), and on the psychology of peoples (1894). Pierre Janet did not! He discussed in that vein mainly change of personality through mechanisms of automatism. Recall that Charcot correlated hypnotism and hysteria to women in contrast to Bernheim's position that hypnotism could work on both men and women.

Concurrently, Ribot interwove into that 'narrative of law' the functions assigned to conscious and unconscious states, issues of language acquisition and loss of the gamut of linguistic abilities, mechanisms of volition, inhibition and habit formation. For example, "To describe the various diseases of memory, and to seek to determine the law which governs their evolution, has been our endeavor" (1882, 191 [1881]; see also, 202–04; 1884, 116, 122–23 [1883]). The direct result of the wide application of this 'law' was an undermining of the unitary view of the individual/the self, both as substantial and as facultative, without however, giving up the notion of psychology as the science whose object is the individual.

5.4.3 Memory–Heredity Continuum

Ribot's efforts bore the marks of contemporaneous conflicting and entangled national contexts, and his 'scientific psychology' was a complicated mixture of concurrent novel and existing concepts, at times incompatible. His work on memory should be looked upon with this in mind. He struggled to explicate processes of encoding, retention, and retrieval by formulating a biological-physiological activity of production-registration, conservation-fixation, and reproduction, as an alternative to facultative explanations of memory, involving the unified introspective 'soul.' His two principal choices were to do so through investigating amnesia and thus tap the resources of the clinic and to turn the experiential-subjective personalized aspects of memory into contingent and secondary by isolating them from those he posited as the biologically foundational entities and mechanisms. He hoped thus to constitute a thoroughly observation-based, objective, mechanist, scientific, rather than metaphysical scheme of the 'self.'

During the first stages of his endeavor, already in *Psy.Her.*, Ribot constituted both memory and heredity on an evolutionary SL/FNL continuum. He devoted a long discussion to 'the diseases,' with a general pronouncement on their specific etiologies—physiological/psychological. Toward the turn of the century, issues such as the passing on of behavioral and cognitive patterns trans-generationally individually and/or collectively became an enabling condition to discuss heredity in terms taken from the contemporaneous discourse on memory.

I propose to look at two dictionaries: Dechambre's *Dictionnaire encyclopédique*, specifically the very detailed entry on "*Amnesie*" (Falret 1865–70), which had been written somewhat before Ribot's work discussed here; and J.M. Baldwin's *Dictionary*, in particular the entry on "Memory (defects of)" by Jastrow (1902), written around the turn of the century, to give a sense of the 'conceptual space' within which Ribot operated and his changing place within it.

In *Dechambre's item* (Falret 1865–70), there was a strong emphasis on how medicine would and should read amnesia: a symptom related to memory that accompanied other problems of the brain or of other parts of the organism. Memory activity could be exaggerated (*exaltée*), weakened, or perverted. The item went into classifying in great detail, primarily through prevalence and presumed etiology, many examples taken not only from medicine, mostly from the early nineteenth century on, but also from somewhat later phrenological writings (Combe), anthropological reports (e.g., Prichard, 1824). It included concise reports on language disorders, including aphasia, 'nervous diseases' (e.g., somnambulism, 'ecstasy'), and 'mental diseases' (e.g., mania, melancholy, dementia, idiotism, as well as general paralysis, quoting contemporaries like Baillarger). There were also mixed categories. For example, epilepsy was considered there as either physical or mental, while consumption of toxics (stressing alcohol in diverse forms) and old age mixed the social and the medical. Three pages were devoted to descriptive enumeration of the diversified phenomenology of amnesia and then to its progress, how to diagnose, prognosticate, and treat.

In *Baldwin's Dictionary*, the principal relevant item was "defects of memory" by Jastrow (1902). The very initial classification was different: "abnormalities of memory fall readily into the two classes of excess and defect." An equally fundamental distinction was that between affections of general retentiveness and affections of special-acquisition groups, i.e., 'classes of memories.' Amnesia became a much more limited class of 'failure in retentiveness.' The analysis stressed normalcy—which included forgetting—as the departure point for the analysis. It offered a general definition of memory as the functional activity of nervous centers whereby concomitants of experience-residua were formed "whether carried out subconsciously and automatically or with conscious intent." Thus, the item suggested that classification should be according to mechanisms of memory formation, retention, reproduction, and recognition, rather than according to presumed phenomenology or etiology. Yet, in the very last sentences, all the categories in the Dechambre item came toppling in with the timely addition of hysteria and hypnotic suggestions. Ribot was mentioned in the item and his book was referred to in the bibliographic list.

Ribot's work on memory and amnesia at large stood midway between the two views. It was not an extremely detailed combination of description and classifying analysis of the manifestations of amnesia. Yet it was not guided solely by the concept of mechanisms that underlay the entry in Baldwin's *Dictionary*. Ribot sorted and clustered according to types (3) and then subdivided in a crossed mode according to temporal origin and extension.

Spencer's explication and distinction between conscious and unconscious memory, which belonged to the organic, was significant in fashioning Ribot's position. In *Princ.Psycho.*ii, the evolutionizing of psychology enabled Spencer to present 'instinct' as an unconscious organic memory that became more complex and was 'swallowed up' in the evolutionary development of more advanced psychic activity to become a conscious memory (1870–1872, 1:427–31). For Ribot, everyday memory, analyzed psychologically, was indeed a case of the evolutionary process. Organically, it would always be 'true,' which was not the case psychologically, as it acquired its appearance of stability, even unity, due to repetition. The thrust of Ribot's argument on psychology was founded on the assumed biological foundation; thus, development and evolution were applied, with heredity a highly significant component. In *Psy.Her.*, heredity was conceived as a specific case in the mutual dependence between the biological and the psychological-moral, thus psychological modifications becoming organic and organic modifications affecting the psychological makeup of individuals.

Moreover, in Ribot's work, memory was firstly described and analyzed in physiological terms taken in part from the contemporaneous discourse on heredity, conspicuously in discussing states of amnesia. Toward the turn of the century, the passing-on trans-generationally of collective states of mind, behavioral and cognitive patterns, became an enabling condition to discuss heredity in terms taken from the contemporaneous discourse on memory. Such a two-way transfer could be found among FNL as well as the US neo-Lamarckists, e.g., Cope, and was problematized in differing ways by Hering, Butler, and Semon (Gould, 1977; Roger, 1979a, b; Sommer, 2016), while Ribot's own work became a fertile ground for a two-way transfer.

In *Memory*, Ribot deployed 'instinct' as a mediation between reflex action and memory, partaking in certain attributes of both, which could be looked upon as an inherited habit, originally formed through repetition, or else as mediation to well-organized and thus more rigidly-organized memory. The initial significant step taken by Ribot was to assume that one could investigate 'objective, merely biological' states of memory, divorced from 'subjective states' where a memory was somebody's memory. Subjective memories were always in a network of such 'owned memories,' becoming subjective by virtue of that belonging/'being owned' by somebody, and were thus ordered in (his/her) earlier–later temporal relationship. Accordingly, it enabled Ribot to disconnect memories from conscious mental functioning, rendering them (merely) biological memories, organically preserved, operating under evolution, including inheritance of acquired modifications—thus investigated through a variety of automatisms, conceived as products of heredity rather than of learning. Then Ribot resorted to Ewald Hering (Ribot, 1881, 13) and then to Hartley (1881, 14), and was familiar with Sully's work, based on Jacksonian notions (Otis, 1994; Hum, 2014). Memory became a collection of modes of functioning: "that it [memory] is the last term in a long evolutionary series, the product of an extended, but connected, development, having its origin in organic life; in short, that memory is, per se, a biological fact – by accident, a psychological fact" (Ribot, 1882, 9–10, 17ff [1881]). Ribot's analysis of the temporalization of memories emphasized this since organic unconscious memories were considered as being in a spatial, non-temporal, order (esp. 1882, 47ff.)—a feature also found in Freud's description of the contents of the Id.

Ribot considered the investigation of memory through its pathological disorder/disorganization as experimental since it was based on analysis of medical cases, such as *dédoublement* (dual personality) and loss of sense of reality, taken from contemporaneous medical, psychiatric, and psychological, also alienist, publications, and from his firsthand acquaintance with Charcot's sessions at Salpêtrière. Moreover, since Ribot assumed that memory functioned under constant and stable laws, layers exposed as de-composed revealed their original/normal organization; the upper ones held recent/new materials and were less stable and less anchored in previously established nets of associations and interconnections. Consciousness was not a constitutive factor in memory, though crucial for discussing culture, and any memory 'accompanied' by consciousness, thus temporally ordered, was also more vulnerable. Roth (2011) suggests locating the significance of Ribot's and others' dealing with memory and amnesia within contemporaneous French culture of coping with the French collective traumas of that 'long nineteenth century.' In his 'diseases' books, focused on three traditional objects of philosophical psychology—memory, will, personality/self—Ribot adopted the view of layered structures: the foundational layer—the lowest—was assumed to be physiological/organic, unconscious, and inherited. Based on his 'law of reversion,' he argued that one could unearth in the individual, in the ontogenetic, the phylogenetic processes of memory and amnesia—sitting well with 'recapitulation' notions that had become a familiar evolutionary trope that did not call for special elaboration, all the more so among followers of either SL or FNL.

Criticizing Spencer, Ribot included emotions within the field of phenomena to be accounted for, arguing that they formed a suitable object for the methods of scientific psychology, including pathologies, just as the cognitive, the habitual, and the reflexive. When his *Sentiments*, where he tried to explicate 'collective emotions,' appeared (mid-1890s), emotions had already become a topic of great interest to many aspiring psychologists, sociologists, and anthropologists alike.

In sum, Ribot posited memory as primarily biological, calling it organic, histological, and cerebral, depending on which facet was proposed as scientifically foundational, in order to argue for a continuity between memory explicated as inhering in particular organic associations and memory in its psychological status and connotation. Two modes of its functioning were important: for one, the initial framing of a mental state as an object of conservation (functional motor disposition), regarding such a framing as dealing with mental states as 'representations'; second, the repeated production of that state as a pattern of movement, of behavior, of feeling etc., on the biological level. Its repeated production did not entail being considered a 'past event,' i.e., habit, since habit, a central mechanism in Ribot's work, did not necessarily carry with it any 'past' connotation, any 'time tag.' For Ribot, conscious memory, i.e., a memory belonging to someone, was contingent upon the necessary evolutionary-biological one. Such a conscious memory, located within a network of memories with temporal ordering since they belonged to someone, was continuously produced (and reproduced) from the point of the relativized presence of the network's owner. For Ribot, the distance between a reference point in the past and such a relativized present shrank by deleting the in-between points (e.g., 1882, 47 ff, 9–10). This view assumed a certain gradualism. When any memory was first presented to an individual's consciousness, it was as a 'living past,' yet repetitions would blunt and blur some points of contextual reference, and it could become a matter of habit or even disappear. Similar to Nietzsche, Ribot thought that the health of 'memory in general' depended also on a measure of forgetting. There was an unresolved tension between this view and the notion of plasticity and multiplicity of levels that was just as essential for him.

'The unconscious' was considered by some in France at the time as a non-object since thought was supposedly solely conscious; others looked upon it as a locus of the automatic, of reflex action, others yet as the site of interaction of impressions and instinct. Historians of psychology disagree on the dating of the emergence of 'the unconscious' as a significant topic in France, e.g., Carroy (1991) dates it to the 1870s, as Hartmann's book (1869) on the philosophy of the unconscious was translated in 1877; Finn (2017) dates the turning point around 1900 when doubts arose concerning 'reason' and 'will,' but sees the deep interest developing around hypnosis and '*dédoublement*' already in Taine's 1869 *De l'intelligence* and later on in Charcot's hysteria; Harrington (1987) argues that between the 1860s and 1900, the interest in 'the double mind' in France included controversies on the conscious–unconscious, much more so than in Britain and Germany; Parot (2012) relates it closely to the political history of the Third Republic in the 1880s and 1890s. Ribot discussed 'the unconscious' in his articles in *RP*, later turned into *Personality*, but perhaps its main non-professional diffusion was through the work of Eugene Azam: *Hypnotisme, double conscience et altérations de la personalité* (1887).

Discussing various forms of dementia, Ribot enumerated the serial order of dissolution of memory functions: "we find that amnesia, limited at first to recent events, extends to ideas, then to sentiments and affections, and finally to actions. Here we have the data for the *formulation of a law*" (1882, 128–29)—elaborating on the diverse forms of memory and their relations to parts of the nervous system. The havoc wreaked by reversion was described in *Personality* also as 'destruction,' and as '*disorganization*' in *Will*. Charcot adopted this concept of law, especially in his work on language, going back to the early stages of human development: ontogenetically the infant-child, phylogenetically 'the primitive'—a contemporaneous racial-colonial anthropological equivalent.

Just as Spencer (and Lamarck) used organizational notions to define 'life,' so did Ribot, and in doing so enabled, within his SL/FNL evolutionary framework, the passage from memory, serving an adaptive function, to heredity. One mode especially relevant here is 'collective.' Following Laycock, Spencer, and others, Ribot included memory states in this class, which became innate through repeated experience stabilized hereditarily. In contrast to his rather simplistic hereditary view of intellectual attributes in *Psy.Her.*, relying on Galton, Ribot's position became more nuanced in the 1890s, and his involvement with FNL deeper. In his *Grande Encyclopedie* article (1893), Ribot argued that acquired modification "*can be fixed in the species through heredity*." There, he looked upon memory in its widest reference as habit, and for him a physiological-organic functioning would be located on the continuum from intelligence through instinct to that which is automatic-unconscious. Thus, patterns acquired through activities of memorization, such as fixation, repetition, etc., would be considered not only as individualized but also as biologically inherited in the species. From a present-day point of view, his argument would seem to be a description relating to the evolutionary process of individuals, given on the species level. I propose that Ribot considered it a description specific to the human species, which could then be explicated in terms of collectivity notions relevant only to it (see below).

5.4.4 The 'Self'/Personality

In *Personality*, Ribot declared: "Memory is the Me statical, the group of tendencies is the Me dynamical, that consensus of the organism of which the conscious me is but a psychological expression" (1887, 28 [1885b]). For Ribot, individuality was physiologically based. Consequently, the 'self/personality/me,' which Ribot considered the most advanced psychological form of such individuality, constituted its conscious, complex, least stable, not to say *flickering or intermittent*, layer or level, wholly contained within the boundaries of that physiological individuality. In *Personality*, states of consciousness were defined as observational data allowing one to infer cerebral neural activity. Thus: "[dreamless sleep] is sufficient to establish the intermittent character of consciousness" (1887, 4). This position became more predominant after the mid-1890s when emotions ('sentiments') became more significant. Unconscious states and their structure, content, and role in behavior formation, in explaining memory and its relation to heredity, were discussed throughout the 'diseases' books almost solely from a physiological perspective. Such a psychology-endeavor, which made use of evolutionary-biological, i.e., primarily neurological resources, seemed to Ribot then as the proper framework, topically and strategically. Recall his position on Taine in *Mind* 1877. An evolutionary perspective undermined any assumption of an entity, psyche/soul with separate capacities/powers, yet traces of the criticized Cousinian tripartite division into senses, intelligence, and will survived until at least the end of the nineteenth century.

Descriptions of 'self' in *Personality* (1885b, 1887), which ran twenty editions in France, and in *Will* (1883, 1884), which ran 37 editions, manifested similarities to the descriptions of simple to complex biological individuality in Spencer's biological work, with connecting links to Espinas' *Des Sociétés animales; étude de psychologie comparée* (1878), as well as the FNL *Les Colonies animales et la formation des organismes* (Perrier, 1881), who had a deep impact on Durkheim as well. Ribot also quoted from Haeckel and Genegbauer. These descriptions resulted in a new note in both the opening paragraphs and the conclusion—some kind of 'evolutionary collectivity' as a constitutive component of the evolutionary-biological individual. "As we descend in the animal series, we see the psychic individual formed by more or less perfect fusion of less complex individuals – *a colony consciousness* being produced by the co-operation of local consciousnesses. ... and one is led to ask whether the human personality itself is not '*a coalition whole*' whose extreme complexity makes its origin difficult to discover or even inscrutable… *Human personality…* is a concrete whole, *a complex*" (1887, 1—emphasis added).

Ribot did not use this in *Personality* as a leverage for insight into an intimate or constitutive relationship between such evolutionary collectivity, 'the social,' and human individuality. Rather, he further elaborated on his already well-trodden path of the physiological foundation of a self—a 'Me'/*Moi* whose individuality was organically determined. "The physical personality, or in more precise language, its ultimate representation thus appears to us… as a wonderfully complex *inextricable* [missing term in the translation] network where histology, anatomy and physiology are baffled every moment" (1887, 50).[5] Thus, the body was conceived as the carrier of the (embodied) individualized person, of memory and the diverse forms of its disappearance and deletion. The body produced habits and harbored the possibilities for their transformation, e.g., "The organism and the brain, its supreme repre-

[5] "*Comme un lacis prodigieusement enchevêtré et inextricable, ou l'histologie, l'anatomie et la physiologie s'égarent a chaque instant*" (Ribot, 1885a, b, 164).

sentation, is the real personality, containing in itself the reminiscence of what we have been and the possibilities of what we shall be" (1887, 51).

In Ribot's 1896 book, analogously elaborating also on *Will* themes, the argument was developed in two ways relevant to 'self': For one, Ribot argued that in states of pathology, according to the law of reversion/regression, the range of the continuum established was between 'interest-less sentiments'—aesthetic and intellectual—through moral-social to those considered most elementary and indeed 'ancient' evolutionarily, automatic, the most basic nutritional needs. It was put analogously to Richet's arguments and summarized in his 1887 *Essai de psychologie générale*.

In addition, Ribot posited a cerebral, neurological unconscious to subvert the contemporaneous philosophical notion of the autonomy of the self and its identification with the conscious personality. Already in *Personality*, he argued for the wide indispensable foundation of the labor performed by the unconscious (1885b, 5) and then amplified it in discussing the psychology of emotions, while concurrently consciousness in general was looked upon as a mode of adaptation (in Le Bon also as an instrument in the battle for survival). Thus, contrary to Pierre Janet, Ribot did not present conscious activity as humanly superior, but rather emphasized the connections between conscious and unconscious states and their role in normal and in pathological life histories. Though unconscious states were not conceived as merely a residue of earlier evolutionary-developmental states, they did not have the cohesive and coordinated power of conscious ones. Both terms 'Personality' and 'Me'('self'/'*Moi*') were used, not quite interchangeably, since individuality was posited as organism-dependent and all sensitive, perceptive, emotive, cognitive, etc., apparatuses became personalized through inhering in a specific organism. Each conscious state was a state belonging to a particular individual and could be observed and investigated by virtue of that belonging/ownership: "Every one of my states of consciousness possesses the twofold character of being such and such a state and of being *mine … The personal character is not superseded, but inherent*; it is an integral part of the fact, and results from its physiological conditions" (1887, 51—emphasis added), resonating with a long tradition.

The ensemble of conscious states, sometimes called 'the conscious personality,' was posited as limited and smaller than the totality of the neural activities, totally dependent on its organism, specifically on a particular organ, the brain (e.g., Ribot, 1873b, 57, 1882, 38–9 [1881], 1887, 1, 51 [1885b], 1884, 51–2 [1883], 1886, 267–88; also 1888). Ribot's telling analogy was between a topographic drawing and the actual piece of land thereby represented (1887, 50). 'The will' too, an elusive and problematic concept, which Ribot de-constructed into 'states of willing,' appeared as a manifestation of these states, a dependent unstable consequence. Ribot discussed 'abulia' (Janet's term), 'idiocy of the will,' as a mentally caused paralysis, where the will was greatly weakened or disappeared, as if the 'self' was hidden from itself: "the 'I will' has no efficacy without the complex psycho-physiological mechanism, in which alone resides the power to act or restrain" (Ribot, 1887, 44). His analysis related such extreme states to religiosity and/or neurasthenia, both publicly considered socially harmful/dangerous (e.g., Nye, 1984).

Ribot's major examples of the regression of will were (psychological) paralysis, aphasia, and drunkenness. Following his books on Schopenhauer and *German psychology*, 'willing' was posited as a consequence of an infrequent unstable coordination of complex states, some conscious, others unconscious, fully physiological (1885b, 178–79). In Ribot's view, 'will' was not a dominant feature of human situations, whose more frequent feature belonged to automatic behavior and habit, and it was not looked upon as a causal force either. This was succinctly expressed by Ribot saying: "'I want' takes note of a state of things but does not constitute it" (1885b, 79). He regarded it as the latest stage in the evolutionary process of modes of adapting, whose principal and most plastic one was consciousness itself. Since consciousness was the latest evolutionarily, it was also considered the most vulnerable, the first to 'dissolve.' Thus, consciousness, as discussed above, was not regarded by Ribot as an indispensable component of the memory mechanism.

Already in the mid-1870s, through his translations of Wundt, his articles (1874, 1875a, b), and *German Psychology* (where also Herbart loomed large), Ribot argued that consciousness was constituted of non-continuous short-lasting states (1876c, 267–288). Wassmann (2005, 2017) presents Wundt as the one who introduced the investigation of emotion through science and whose work had an inestimably deep impact on Ribot, Dumas, and Binet. The phenomenon of forgetting conveyed the importance of habit, rather than consciousness, as the foundation for attributing to the 'self,' to oneself, modes of feeling and thinking. Analogously to Jackson's 'dissolution,' in Ribot's forgetting or dementia, only the 'lowest forms,' basic habits first-to-be-formed remained of the rich normal mental life, and similarly in the disappearance order of emotions in *Sentiments*. Ribot excluded all phenomena of 'psychic research': parapsychological, 'medium' phenomena, eliciting mental states through hypnotism and suggestion, automatic writing, etc. Yet, he related to work on the dual-personality phenomenon investigated through these techniques (Plas, 2012). Concurrently, Beaunis, who ran the Sorbonne lab then, considered hypnosis as 'moral vivisection,' which exemplified the mechanism of the unconscious (vivisection was then at the center of controversy mostly in Britain; Shmuely, 2018).

Most of the activities ordinarily attributed to the 'psyche' were produced according to Ribot under two forms, the conscious and the unconscious, and the two investigations—of the unconscious/physiological, including habit and automatic acts, and of the conscious/psychological—were along one continuum. This is yet another component of the more general framework of 'laddered continua.' In *Eng.Psy*, Ribot criticized Lewes's conception of consciousness as an attribute of all ganglionic cells (e.g., Ribot, 1870, 301) and suggested that both Spencer's and Bain's accounts had been more accurate and cogent. In *Psy.Her.* (1875c, 57 [1873b]), the dynamic conception encompassed all layers: "Mental life is a constant transformation, the unconscious becoming conscious, and vice versa; but this transformation does not take place by chance: though the laws are unknown, it is not without laws." The network of memory states, once 'personalized,' recalled them as conscious, since 'personalizing' turned them into an individualized psychological matter, rather than unconscious physiological ('organic'). Since conscious states were more complex and plastic and less stable, such a 'self' could not be conceptualized *à la*

Paul or Pierre Janet as either a-priori or as a continuously constructed superior unified consciousness substrate. 'Dissolution' impacted cognitive and affective activities of the 'self' downward, along those laddered continua. Neither conscious states nor volitional ones would accompany any of them. It was clear that 'free will' or 'will' generally was a 'misplaced concreteness,' e.g.: "the 'I will' which indicates a mental situation but which of itself possesses no efficiency" (1884, 2).

Pierre Janet's automatisms were not identical to the unconscious states, whether automatic or not, postulated by Ribot. Janet used only 'subconscious' to indicate non-surface phenomena, holding a particular view on the unity of the self and its consciousness, which made it difficult to accommodate unconscious cerebration, and even subconscious phenomena were accompanied by a low-level consciousness. Indeed, Bacopoulos-Viau (2019) argues that in his work on automatic writing Janet pathologized it within the context of discussion on the 'will.' Ribot used the term 'subconscious' very rarely (e.g., Munsterberg, 1910), but also there he clearly meant 'unconscious.'

The 'self,' personalized/temporalized memory, was seen as but a small fraction of 'physiological experiencing,' in analogy to conscious states versus unconscious ones: "The personal and possessive character of our states of consciousness … *is not superadded but inherent*: it is an integral part of the fact and results from its physiological condition" (Ribot, 1887, 51). "Unconscious cerebration not being subject to restrictions of time, operating, so to speak, only in space, may act in several directions at the same moment. Consciousness is the narrow gate through which a very small part of all this work is able to reach us" (1882, 38–9). Consciousness was thus posited as a relativized point at one end of a continuum, at the other end of which lay unconscious states, habituated patterns, the robust and efficacious foundation of the *narrow 'self,'* looked upon as a sum of such a medley of states at a particular spatial arrangement, that could be independently investigated as psychological act. Such a notion of 'self' as a continuum and the site and role of consciousness in it could call into question the roles of both 'reason' and 'will' in guiding human action, and at some stage around 1890, it did. 'Regression of will' was depicted as a move from the voluntary to the automatic (e.g., Ribot, 1883, 188).

Parot (2012) puts it in a political context as questioning the very notion of rational citizen's political participation. Nye (1975) presents a broader argument on the establishment of the 'Paris School' and its impact on the discussion of bounding consciousness, the unconscious, automatism, and affective states. He argues that it resulted in depicting how the individual became unconscious of his own deeds, as according to the Paris School—its core members included Binet, Pierre Janet, and Paulhan, with followers Dumas, Pieron, and Bourdon—the mind was subjected to nonvoluntary/automatic movements.

Around the turn of the century, the question of where to locate creative/imaginative capacities produced another controversy around the unconscious. In 1906, Ribot published *Essai sur l'imagination creatrice*, and so did Beaunis. In 1893, Brunschvicg severely criticized another book on a close topic. By 1909, more philosophers came out with scathing critiques of the 'unconscious'—e.g., Binet, Fouillée, Parodi (who was skeptical), and Bergson talked about 'pre-conscious.'

The focus of the critique was the assumption that affectivity and creativity were not related to reason/intelligence. Le Bon's position was different—a unique mixture of the components of automatic will, the unconscious as creative and dynamic, a drastic narrowing of the role of reason in individual and collective life.

Ribot's work until around that time could be looked upon as forming a continuous though evolving corpus, where 'evolutionary hierarchy' played a changing but always a central role in discussing memory, will, and self/personality. Even in *Sentiments* (as operative in psychology and not only in physiology, 1896, 427), 'evolution' and 'dissolution' constructed the perspective through which emotion and needs were discussed; 'hierarchy of levels of consciousness' was also used in *La vie inconsciente et les mouvements* (1914). Following the then-current mode of introducing novel fields through presenting their object, methods, principal questions, and use, Ribot claimed that 'facts of consciousness' constituted the object of psychology (first sentence of 1900 article), thus reconstituting the role of introspection. Yet, in his 1906 essay on creative imagination, he rejected current views on the unconscious, which, according to him, either looked upon it as a mere cerebration-physiological activity or else a shrunken activity of consciousness, unrelated to the self (e.g., in the contemporaneous discussions as '*dédoublement*'). In *Imagination*, Ribot insisted on treating unconscious states as 'faits,' mental states calling for more than one mode of explication, arguing that the roles of the unconscious in the processes of inspiration, dream, and states of chemical intoxications were different (1905, 45–6, 53 [1900]).

To sum up, whatever extant explanations were adduced for the relations between brain, mind, and body, there was an almost general agreement among both contemporary psychologists and philosophers on the existence of some *sense of binding* as well as of *agency*, even *ownership*, or at the very least a specific/particular *continued perspective of experiencing*, at times termed 'self'/'person.' They felt constrained to provide a meaningful description of the relations between 'consciousness' and such 'self,' to offer either a bridging mechanism or explanation why such bridging was superfluous, or else an assumption of a gap-less construction of self and consciousness. In present-day theorizing, especially among philosophers of mind, this very constraint is still brought up for discussion under many guises. In view of the tortuous history of thinking about and deploying 'consciousness,' Ribot, was aware of the *problématique* inherent in deploying it. He was a meaningful contributor to the accumulating body of work on the relations between 'the mental,' consciousness, unconscious states, and the nervous system, deploying the concept of evolutionary hierarchy, navigating among experimental and non-experimental technologies.

I have argued that Ribot was a methodological parallelist. Yet, he adopted parallelism primarily as a starting point in order to open the field for scientific psychology since under this assumption each internal event would necessarily be related to a 'concomitant' physical–physiological–neurological one. By investigating the 'concomitant' event, one would indirectly study the conscious act, the psychological event. During the 1870s and 1880s, Ribot expressed the hope that at some further stage psychology would turn completely physiological, yet at other times he granted 'the mental' a special role. There was a semi-intentional opacity on mind–body relations and particularly on the mind–brain actual intimate entanglement, concurrently adhering to the nervous system as its boundary. In his work on non/un-conscious states in pathological cases, Ribot strove to identify such states with physiological states, to which consciousness was once attendant but not anymore. Presumably, he hoped and intended to eliminate any non-materialist/transcendent explanations. It seems that Ribot was aware that the various contemporaneous technologies and procedures used to explore minds and brains impacted their object of investigation, eliciting states and playing a significant role in defining the object within the competing theories.

Peirce's sharp critical comments on Ribot's position turned on what he saw as 'Ribot's monism'! pretending to be 'a psycho-physical double aspect theory,' "a happy compromise between materialism and spiritualism" (2010, 8:14 [1890]). According to Peirce, who supported a dualist position, Ribot's position was actually materialist through and through. In his "Review of Théodule Ribot's psychology of attention" (1890), Peirce allowed at the most the possibility of analogy.

When considering the possible difference between the parallelism of Jackson and of Ribot, it may be helpful to consider the phenomena they investigated and how these came to be regarded as relevant for psychology. For Jackson, clinical work—medical practices and medical tools available—circumscribed the field of relevancy. For Ribot, the field was circumscribed by the philosophical discourse and contemporaneous, specifically French, medical practices and their 'phenomena,' e.g., hypnotism, suggestion, and how mental illnesses were defined. In both cases, there was no 'direct approach' to 'nature,' but they adopted their differing methodological parallelist positions to discuss 'nature' through professional practices, research methods, and antecedent theorizations.

5.4.5 The Collectivity

Ribot followed Spencer on 'collectivity,' as his 'individuality' was anchored in physiology. Yet, just like Spencer's double movement between complex individuals and hierarchies of complexified collectivities, Ribot deployed SL/FNL to constitute collectivities beyond the individual (e.g., 1887, 2–3 [1885a, b]). There were three modes used for collectivizing SL/FNL memory through heredity:

(a) Through the accumulation in individuals of inherited neural patterns associated with specific habits, which, due to their adaptive utility, became evolutionarily common to the species, except in pathological cases. These patterns would be looked upon as equivalent to being innate. Already in *Eng.Psy.*, and then in *Psy. Her.*, where he states, for example, "Heredity … is to the species what memory is to the individual" (1875c, 53, 310–11 [1873a, b]), Ribot considered them to be similar to Spencer's argument on substituting Kantian categories.
(b) Patterns, considered specific to particular races or national features passed on trans-generationally, more often attributed to the early beginnings of human

groups, and notions of '*race*' and '*nation*' are used as explanatory substantives (thus as possible causes) in contemporaneous discourse. Some historians have argued that German *Volkspsychologie* had a deep impact on how collectivities were conceptualized within French psychology, starting with Steinhal and Lazarus and ending with Wundt and his double legacy. One could and should extend it to French social-sciences thinking then, e.g., the work of Celestin Bouglé, Durkheim, and Alfred Fouillé.[6]

(c) Inherited patterns of the early stages of human evolution: "we thus pass from individual to hereditary memory" (Ribot, 1882, 65). The passage from the physiological to the psychological served as the underpinning of a gradual two-way analogical transfer between memory and heredity, and thus between the evolutionary biological and the evolutionary psychological. Sommer (2016) points to Osborne's discussion of possible phylogenetic memory and relates it both to Ribot and to a quasi-genealogy going back to Jackson.

Ribot used the implied neural hierarchical order to construct an implicit psychological-hierarchical one of organic memory–volition–civilization. In it, the upper rungs would be conscious, voluntary, inherently unstable, directly given to psychological analysis, beneath which, serving as their foundation, were the lower rungs, which would also be layers of diminishing consciousness-and-autonomous-volition, exemplified in civilization-less embodied modes of behavior (else called 'primitive'), originally adaptive to environmental circumstances. These were evolutionarily "registered in the organism ... beyond the individual in the species and in all species. It comes not from above but *from below*" (1884, 38—emphasis added). These 'low modes' could surface on particular occasions, both individual and collective (1884, 333). From early on, the controversies on modes of hypnotism and on vulnerability to hypnosis were interwoven with debates on patterns of collective behavior; thus, in Ribot's later writings, these could indicate a civilizational process of the emergence of individuals from some primeval collectivity (e.g., 1887, 1). The conclusion to *Personality* was the first site where the term 'collective,' rather than 'species' or 'races,' was used to indicate non-individual inherited memory, also contrasted with both individual memorization and individually embodied inheritance (e.g., 1887, 24, 30). In *Sentiments*, Ribot tried to account for 'collective emotions' and suggested that early forms of arts related to emotion as well as to myth as 'collective' (e.g., 1896, 332, 340–41, 363).

'*Race*' was discussed within his 'collectivities' and SL/FNL framework. Already in *German Psychology*, Ribot was groping toward a generalized notion of psychology relating to the social in the aspect he later named "*psychologie des races*" (also in his *Mind* article, 1877, 2–7). Earlier, in *Eng.Psy*, he discussed how relations between the physiological and the psychological were manifested in differing races, emphasizing their fixity. Two interrelated notions and their assumed mechanisms were central in how 'race' was situated within his writing: the inheritance of acquired modifications (or characteristics) and biological heredity through (mostly

[6] Danziger (1990), Klautke (2013), Kusch (1999), and Bunzl (1996).

unconscious) memory. Questions such as the future of the 'French Nation' versus the 'German Nation,' or of 'degeneration' in French society were often discussed using them, as applied, until after the mid-1890s, to contemporary non-Western populations while deploying some anthropologists' views on races and their hierarchies. Given his various social-professional ties, Ribot too held clear-cut views on them, discussed similarly to Spencer (and many contemporaries): he related skull/brain size to cognitive abilities and concurrently denied 'lower races' any potential plasticity and future change that could be implied by the SL/FNL perspective. Biological memory was constructed in this context as transmitting cultural heredity, both racial and ancestral, serving as a framework for atavism. One could cautiously generalize that while in his writings general heredity receded in importance and environment (and sociality) became more significant around 1890, there occurred a certain modification on the issue of 'races,' by recognizing the 'lower races' capacities for morality and its attendant sentiments.

Though Ribot mentioned both Spencer and Haeckel, it seems more probable that the terms 'collective'/'collectivity' were picked from Espinas' writings. The *problématique* of individuals–collectivities came to the fore in Ribot's later work (*Sentiments*), at a time when this *problématique* became deeply controversial among sociologists (those publishing in his journal too), in particular between Durkheim and Tarde, with further sociologists and anthropologists taking part in the ongoing public debate, with discussions on 'social group' and its boundaries, on notions of 'collective solidarity.' Furthermore, in *Sentiments*, emotions were posited as having history, both evolutionary and cultural, claiming that the social facets of the affective life of individuals and their biographies should be included. Yet, Ribot also insisted that both could not be cut off from their biological roots, for example, in *Sentiments* and more explicitly in *La Logique des sentiments* (1905): "It treats a question which appears to belong to individual psychology, but is wholly collective, since human groups form and maintain themselves through community of belief" (1905, 17). Though Mucchielli (1994, 1998, 2006) and Staum (2007) argue that the rise of Durkheimian sociology impacted Ribot's views on these and other issues, perhaps the history of Ribot's relations with Tarde is relevant since both Durkheim and Tarde emphasized the 'social,' and in Tarde's case, while totally eliminating the biological; highly relevant were also Ribot's relations with Le Bon.

The civilizational hierarchical order was deployed to explain 'primitive art,' 'collective, social, anonymous' 'myths,' 'collective anonymous work of primitive humanity,' 'religious sentiments,' and particularly their 'pathologies.' The latter were distinguished from instincts, even though instincts were considered to have some measure of plasticity in humans, according to the civilizational order, already in *Psy.Her.* (1875c, 405–06). The decision to establish the Institut général psychologique occurred parallel with the fourth international congress of psychology in Paris, during the *Exposition universelle*. Mostly neo-Lamarckists as Giard and his pupil Bohn, Perrier, Bouvier, and Pieron, Binet later, were instrumental in bringing this about, as related also to the effort to establish zoological psychology (Ribot's report on the congress 1900; Thomas, 2003). 'Ecstasy,' in contradistinction to religious-sentiments pathologies, was described as an individual rather than a social state (1896, 326). Hyper-mnesias (or 'exultations of memory') were classified together with 'hypnotic' and 'hysterical' states (1881, chap. 4)—seen by Goetz and his colleagues (1995) as related to anti-religious and/or anti-clerical positions and thus political.

'Social sentiments' were not, of course, a late-nineteenth-century invention. Nevertheless, Ribot's version is of interest in interlacing them with contemporaneous evolutionary biology and current debates on the proper methodology for a smooth conceptual passage from individually based empirical results to collective consequences. Ribot's work in the late 1890s focused on affective-emotive facets of the psychological make-up and their pathologies, emphasizing the control of such facets over intellectual functioning. He looked into mechanisms of faith, defined at times as 'belief in well-fixed notions anchored in the emotional-and-organic,' i.e., in the substrate of the mind. Of note were the general context of the educational reform of the Third Republic and Ribot's affiliations with the people at the heart of that enterprise, as well as the prevalent emphasis on the close connection between civilization and individuality. Assuming a seemingly simple passage from individual to collective belief, Ribot commented on religious sentiments' 'pathologies': "none among morbid emotions has – and still more, has had – a more marked tendency to rapid propagation in epidemic form: a further proof that, in its nature, it is not so much individual as social" (1897, 327 [1896]). And in the introduction: "If the intellectual life has its roots in biology, it is only in social facts that it can find its full development. A science never gains by excessive restriction of its scope; it is better to err in the opposite direction."

These were echoed in a later book on the *Logic of Sentiments* (1905), where Ribot actually used expressions resonating with the Durkheimian group then, especially the terms 'collective' and 'social groups.' A laudatory review of that book by Durkheim appeared in *L'Année sociologique* soon after publication (1906, 156–58). It was a first, since until then no reviews of Ribot's books were published there. Durkheim suggested viewing the book as having a sociological perspective, using anthropological materials and examples that revealed collective emotions and beliefs. Indeed, Ribot attributed there a larger role to sociological considerations when dealing with higher mental functions, and his data were taken mostly not from psychopathology but from contemporary anthropology. This entanglement with the collective was also reflected in Ribot's discussion of research methods and methodology, centered primarily on the use of questionnaires and of statistics. Ribot used some form of questionnaire for his research on '*The evolution of general ideas*,' published first as an article in his journal in 1891 and then in book form in 1897.

In 1904 and in his "Psychologie" article in 1909, Ribot sharply criticized the reliance on statistics in psychological investigations. Perhaps his criticism was also directed at the legitimating rhetoric of such work by Binet (under 'democratization'). Ribot's attack could be viewed not only as suggesting that one could not directly investigate collectivities but also as a defense of the case-studies method, self-evidently used in medicine, considered the 'natural' mode of dealing empirically with pathologies, i.e., those of the individual (1891–1892, 1904). Furthermore, if one subscribed to an 'experimentalist methodology,' rather than the more general 'empirical' one, that meant a commitment to instruments/practices of measurement, which were conceived as highly individualized. Both tendencies could partially explain why within a framework of 'scientific psychology,' as the one conceived and propagated by Ribot, one resorted, and had to, to particular SL/FNL evolutionary mechanisms in order to include therein collectivities beyond the individual. This is but a component argument in an ongoing historiographical-interpretive debate I shall not enter, put in a nutshell: How to describe, analyze, and explain what seems to be the comparatively belated constitution of social psychology as an autonomous field in France.[7]

To conclude, Jackson's reversal of the Spencerian direct relation between complexity, organization, stability, and plasticity to an inverse one was projected by Ribot throughout his 'diseases' books and *Sentiments*. It allowed him to posit several *laddered continua*, which did not seem to fully overlap, except in a very general manner. This inverse version brings out the instability of those end-terms that characterized 'individualized culture' at large in Ribot's psychology, specifically:

- From 'elementary states and patterns' to sophisticated abstract ideas.
- From automatic behavior through habit to autonomous volition.
- From 'elementary' instincts to highly complex sentiments.
- From the totally determined organic unconscious states to the highly plastic, innovative, undecided conscious ones.
- From 'ego-elementary' sentiments such as fear, need for nutrition, and even anger to 'ego-less' aesthetic-intellectual ones.
- From what was inherited as collective memory to that which was individually acquired and not stable enough to be passed on inter-generationally.

Collective-memory components that were passed on through SL/NL mechanisms of heredity dealt mostly with '(human) race' and 'peoplehood.' In contrast with the racializing tendency in Ribot's work, which combined memory, heredity, collectivity, and 'race,' Freud deployed these mechanisms to posit certain universalizing all-humanity-encompassing components of the unconscious.

[7] Ash (2003), Carroy (2004), Carroy and Schmidgen (2002), Danziger (1985, 1990), and Mucchielli (1994, 2006).

5.5 Concluding Comment

When working on Ribot, I have tried to bring forth the interlacing of his two principal roles as an innovating cultural agent, laboring to fashion the intellectual public sphere, and as an original pioneering thinker who utilized the available materials in that sphere in order to attempt to fashion a new 'scientific' discipline and create an institutional setting for it.

There were a number of questions that kept coming up, to which I have tried to supply tentative multi-layered answers:

- Placing Ribot's endeavor of 'scientific psychology' within and yet outside *the positivist grid* of the second half of the nineteenth century.
- Explicating Ribot's differing uses of SL/NL as both original and yet shaped by the already existing modes of thought and patterns of practice of existing disciplines. I tried to illustrate this through the elaboration of his theorizing of 'production,' 'reproduction,' 'conservation,' and 'creation' in psychology, in particular memory heredity relations, and through his evolving conception of the predominant notion of 'individual' within the contemporaneous culture.
- Following how 'disease' was constituted in his work through its evolutionary and developmental scheme, in particular with the notions of continua and of hierarchy.

Materials from the following articles have been used by permission of:

Elsevier

- 2020 – Transfer of Lamarckisms and Emerging 'Scientific' psychologies: 19th – early 20th Centuries Britain and France, *Studies in History and Philosophy of Biological and Biomedical Sciences, 83*.

References

Ash, M. (2003). Psychology. In T. Porter & D. Ross (Eds.), *The Cambridge history of science, vol. 7: The modern social sciences*. Cambridge University Press.

Azam, E. (1887). *Hypnotisme, double conscience, et altérations de la personnalité*. J.B. Baillière et Fils.

Bacopoulos-Viau, A. (2019). From the writing cure to the talking cure: Revisiting the French 'discovery of the unconscious. *History of the Human Sciences, 32*(1), 41–65.

Baldwin, J. M. (1886). A new factor in evolution. *American Naturalist, 30*(441–51), 536–553.

Barbie, E. (1876). *L'origine des espèces au moyen de la sélection naturelle, ou La lutte pour l'existence dans la nature*. C. Reinwald.

Beaunis, H. (1909). Comment fonctionne mon cerveau, essai de psychologie introspective. *Revue Philosophique, 67*(January), 29–40.

Benichou, C. (1989). Ribot et l'hérédité psychologique. In J.-C. Fischer & C. Benichou (Eds.), *L'Ordre des caracteres* (pp. 73–94). J. Vrin.

Benrubi, J. (1933). *Les sources et les courants de la philosophie contemporaine en France*. Alcan, t. II.

Bernard, C. (1927 [1865]). *An introduction to the study of experimental medicine* (Henry C. Green, Trans.). The MacMillan Company.

Binet, A. (1908). Une enquête sur I'evolution de l'enseignement de la philosophie. *I'Année psychologique, 14*, 152–231.

Boring, E. G. (1929). *A history of experimental psychology*. The Century Co.

Bowler, P. J. (1991). *The Non Darwinain revolution*. Johns Hopkins University Press.

Brooks, J. I. (1998). *The eclectic legacy: Academic philosophy and the human sciences in nineteenth century France*. University of Delaware Press.

Bunzl, M. (1996). Review of fieldwork and footnotes: Studies in the history of European anthropology. Han F. Vermeulen and Arturo Alvarez Roldán, eds. *American Anthropologist, 98*(3), 684–686.

Canguilhem, G. (1978). *On the normal and the pathological* (C. Fawcett, Trans.). D. Reidel.

Carroy, J. (1991). *Hypnose, suggestion et psychologie: l'invention du sujet*. Presses universitaires de France.

Carroy, J. (2004). Premie'res enquêtes psychologiques françaises: L'introspection, l'individu et le nombre. *Mil neuf cent. Revue d'histoire intellectuelle, 1*(22), 59–75.

Carroy, J. (2006). Philosophie et psychologie au 19ieme siècle en France:Taine, Ribot et la Revue Philosophique. In M.-L. Pelus-Kaplan, J. Carroy, & C. Debru (Eds.), *Unité et globalité de l'homme: Des humanités aux sciences humaines* (pp. 79–91). Edition Syllepse.

Carroy, J., & Plas, R. (1996). The origins of French experimental psychology: Experiment experimentalism. *History of the Human Sciences, 9*(1), 73–84.

Carroy, J., & Plas, R. (2000). How Pierre Janet used pathological psychology to save the philosophical self. *Journal of the History of the Behavioral Sciences, 36*(3), 231–240.

Carroy, J., & Plas, R. (2006). The beginnings of psychology in France: Who was "scientific" psychologist in the nineteenth century? *Physis, 43*(1–2), 157–186.

Carroy, J., & Schmidgen, H. (2002). *Psychologies expérimentales: Leipzig – Paris 1890–1910* (pp. 1–27). Max Planck Preprint 2006.

Carson, J. (2007). *The measure of merit: Talents, intelligence, and inequality in the French and American republics, 1750–1940*. Princeton University Press.

Conry, Y. (1974). *L'Introduction au Darwinisme en France*. Vrin.

Csiszar, A. (2018). *The scientific authorship and politics of knowledge in the nineteenth century*. Chicago University Press.

Danziger, K. (1980). The history of introspection. *Journal of the History of the Behavioral Sciences, 16*, 241–2623.

Danziger, K. (1985). The origins of the psychological experiment as a social institution. *American Psychologis, 40*(2), 133–140.

Danziger, K. (1990). *Constructing the subject: Historical origins of psychological research*. Cambridge University Press.

Delay, J. (1951). Ribot et le Jacksonism. *Dialectica, 5*(3/4), 413–444.

Delage, Y., & Mary, G. (1912 [1909]). *The theories of evolution* (A. Tridon, Tran.). B. W. Huebsch.

Digeon, C. (1959). *La crise allemande de la pensée Française*. Presses Universitaires de France.

Dowbiggin, I. (1985). Degeneration and hereditarianism in French mental medicine 1840-90: Psychiatric theory as ideological adaptation. In W. F. Bynum, R. Porter, & M. Shepard (Eds.), *The anatomy of madness: Essays in the history of psychiatry* (Vol. 1, pp. 188–232). Routledge.

Dowbiggin, I. (1991). *Inheriting madness: Professionalization and psychiatric knowledge in nineteenth-century France*. University of California Press.

Ellis, J. D. (1990). *The physician-legislators of France: Medicine and politics in the early Third Republic 1870–1914*. Cambridge University Press.

Espinas, A. V. (1878). *Des sociétés animales: étude de psychologie comparée* (2iéme édition). Germer Bailliére et cie.

Fabiani, J.-L. (1988). *Des philosophes de la République*. Editions de Minuit.

Feuerhahn, W., & Trochu, T. (2016). Autoportraits de Thédule Ribot en correspondant. *Revue Philosophique, 206*(4), 521–540.

Finn, M. R. (2017). *Figures of the pre-Freudian unconscious from Flaubert to Proust*. Cambridge University Press.
Goetz, C. G., Bonduelle, M., & Gelfand, T. (1995). *Charcot: Constructingn neurology*. Oxford University Press.
Goldstein, J. (1987). *Console and classify: The French psychiatric profession in the nineteenth century*. Cambridge University Press.
Goldstein, J. (2005). *The post-revolutionary self: Politics and psyche in France, 1750–1850*. Harvard University Press.
Goldstein, J. (2013). Neutralizing Freud: The *Lycée* philosophy class and the problem of the reception of psychoanalysis in France. *Critical Inquiry, 40*(1), 40–82.
Gould, S. J. (1977). *Ontogeny and phylogeny*. Belknap Press of Harvard University Press.
Guillin, V. (2004). Théodule Ribot's ambiguous positivism. *Journal of History of Behavioural Sciences, 40*(2), 165–181.
Guillin, V. (2016). Mill, Ribot et la science du caractère. *Revue Philosophique de la France et de L'étranger, 141*(4), 489–500.
Hacking, I. (1998). *Rewriting the soul: Multiple personality and the science of memory*. Princeton University Press.
Hacking, I., Barham, P., Osborne, T., & Lynch, M. (1995). Rewriting the soul. *History of the Human Sciences, 8*(4), 107–130.
Harrington, A. (1985). Hysteria, hypnosis and the lure of the invisible: The rise of neo-mesmerism in fin-de-Siècle French psychiatry. In W. F. Bynum, R. Porter, & M. Shepard (Eds.), *The anatomy of madness: Essays in the history of psychiatry, vol. 3: The asylum and its psychiatry* (pp. 226–246). Routledge.
Harrington, A. (1987). *Medicine, mind, and the double brain: A study in 19th century thought*. Princeton University Press.
Harris, J. C. (2005). A clinical lesson at the Salpêtrière. *Archives of General Psychiatry, 62*(5), 470–472.
Henri, V. (1893). Les laboratoires de psychologie expérimentale en Allemagne. *Revue philosophique, 36*, 608–622.
Hum, P. (2014). *L'oubli 1880–1900* (PhD dissertation). Université Charles De Gaulle, Lille III.
Jackson, J. H. (1873a). Observations on the localisation of movements in the cerebral hemispheres, as revealed by cases of convulsion, Chorea and 'Aphasia.'. *West Riding Lunatic Asylum Medical Reports, 3*, 177–195.
Jackson, J. H. (1873b). On the anatomical, physiological, and pathological investigation of epilepsies. *West Riding Lunatic Asylum Medical Reports, 3*, 315–339.
Jackson, J. H. (1873c). Remarks on the double condition of loss of consciousness and mental automatism following certain epileptic seizures. *Medical Times and Gazette*, 63–64.
Janet, P. (1888). Une Chaire de psychologie expérimentale et comparée au Collège de France. *Revue des Deux Mondes, 86*(58), 518–549.
Jastrow, J. (1902). Memory (defects of). In J. M. Baldwin (Ed.), *Dictionary of philosophy and psychology* (Vol. 2, pp. 64–66). MacMillan.
Karady, V. (1979). Strategies de réussite et modes de faire valoir de la sociologie chez les durkheimiens. *Revue Française de Sociologie, 20*, 49–82.
Klautke, E. (2013). The French reception of völkerpsychologie and the origins of the social sciences. *Modern Intellectual History, 10*(2), 293–316.
Kusch, M. (1999). *Psychological knowledge: A social history and philosophy*. Routledge.
Le Bon, G. (1894). *Les lois psychologiques de l'évolution des peuples*. Felix Alcan.
Le Bon, G. (1895). *Psychologie des foules*. Felix Alcan.
Leon, X. (1917). Necrologie de Theodule. Ribot. *Revue de Métaphysique et de Morale, 24*(1), 129–132.
Leonard, J. (1992). *Médecins, malades et société dans la France de xixieme siecle*. Sciences en situation.

Lepenies, W. (1988). *Between literature and science: The rise of sociology* (R. J. Hollingdale, Trans.). Cambridge University Press.

Loison, L. (2010). *Qu'est-ce-que le néolamarckisme? Les biologistes Français et la question de l'évolution des espe'ces*. Vuibert.

Mauss, M. (1939). Ribot et les sociologues. In *Centenaire de Théodule Ribot, Jubilé de la psychologie scientifique Française1839–1889-1939* (pp. 137–138). Agen.

Mengal, P. (1994). Henri Pieron (1881-1964) et les Neo-lamarckiens français. In C. Debni, J. Gayon, & J. F. Picard (Eds.), *Les sciences biologiques et medicales en France, 1920–1950* (pp. 113–124). Editions CNRS.

Mucchielli, L. (1994). Sociologie et psychologie en France, l'appel à un territoire Commun: Vers une psychologie collective (1890–1940). *Revue de synthese IVieme s., 3–4*, 445–483.

Mucchielli, L. (1998). *La découverte du social: Naissance de la sociologie en France*. Éditions la découverte.

Mucchielli, L. (2006). Aux origines de la psychologie universitaire en France (1870–1900): enjeux intellectuels, contexte politique, réseaux et stratégies d'alliance autour de la Revue philosophique de Théodule Ribot. *Annals of Science, 55*(3), 263–289.

Nicolas, S. (1998). Aux origines de la psychologie universitaire en France (1870–1900). *Annals of Science, 55*, 263–289.

Nicolas, S. (2005). *Théodule Ribot .1839–1917*. L'Harmattan.

Nicolas, S., & Gounden, Y. (2017). Centenaire Ribot. deuxieme partie: la reception de l'œuvre de Théodule Ribot publiée chez l'éditeur Ballie're.1874-1883. *Bulletin de Psychologie, 549*(3), 179–195.

Nicolas, S., Sabourin, M., & Piolino, P. (2016). The seminal contributions of Théodule Ribot (1839-1916): The centenary of the passing of the founder of French psychology. *L'Année Psychologique, 116*(4), 519–546.

Nye, R. A. (1975). *The origins of crowd psychology: Gustave Le Bon and the crisis of mass democracy in the 3rd Republic*. Sage.

Nye, R. A. (1984). *Crime, madness, and politics in modern France*. Princeton University Press.

Ohayon, A. (2008). Psychologie clinique et psychologie expérimentale. Une union jamais consommée. *Sciences Humaines, 7*, 52–53.

Otis, L. (1994). *Organic memory: History and the body in the late nineteenth and early twentieth centuries*. University of Nebraska Press.

Parot, F. (2012). France. In L. T. Benjamin & D. B. Baker (Eds.), *Oxford handbook of the history of psychology: Global perspective* (pp. 228–254). Oxford University Press.

Peirce, C. (1890). Review of Théodule Ribot's psychology of attention. *The Nation* (June 19).

Peirce, C. S. (2010 [1890]). *Writings of Charles S. Peirce: A chronological edition Volume 8, 1890–1892*. Indiana University Press.

Perrier, E. (1881). *Les colonies animales et la formation des organismes*. Masson.

Plas, R. (2012). Psychology and psychical research in France around the end of the 19th century. *History of the Human Sciences, 25*(2), 91–107.

Prichard, J. C. (1826 [1824]). *Researches into the physical history of mankind*. John and Arthur Arch.

Rabinbach, A. (2018). *The eclipse of the utopias of labor*. Fordham University Press.

Ribot, T. (1870). *La psychologie anglaise contemporaine – école expérimentale*. Ladrange.

Ribot, T. (1873a). *English psychology*. H.S. King& Co.

Ribot, T. (1873b). *L'Heredité psychologique. étude psychologique sur ses phénoménes, ses lois, ses causes, ses consequences*. Ladrange.

Ribot, T. (1874). La psychologie physiologique en Allemagne- la measure de sensations. *Revue Scientifique, 7*, 553–563.

Ribot, T. (1875a). La Psychologie Allemande Contemporaine. *Revue Scientifique, 4*(31), 721–732, 751–760.

Ribot, T. (1875b). Psychologie physiologique, Wundt. *Revue Scientifique, 5*(31), 505–516, 544–549.

Ribot, T. (1875c). *Heredity: A psychological study of its phenomena, laws, causes, and consequences*. D. Appleton and Company.

Ribot, T. (1876a). "Introductory essay" to the 1st issue of the revue Pilosophique/originally anonymous. *Revue Philosophique de la France et de L'étranger, 1*(1), 1–4.

Ribot, T. (1876b). Review- physiologie: Hughlings-Jackson; clinical and physiological researches on the nervous system I. On the localisations of movements in the brain 1875. *Revue Philosophique, 1*, 214–216.

Ribot, T. (1876c). De la durée des actes psychiques:d'apre`s les travaux récents. *Revue Philosophique, 1*, 267–288.

Ribot, T. (1877). Philosophy in France. *Mind, 2*, 366–386.

Ribot, T. (1881). *Les maladies de la mémoire*. Librarie Generale Ballie're et Cie.

Ribot, Théodule. (1882). *Diseases of memory: An essay in the positive psychology* (William Huntington Smith, Trans.). D. Appleton and company.

Ribot, T. (1883). *Les maladies de la volonté*. G. Ballière. English edition quoted 1894, Chicago: Open Court.

Ribot, T. (1884). *The diseases of the Will* (English translation). New York, J. Fitzgerald/ Humboldt library no. 52.

Ribot, T. ([1879] 1885a). *La Psychologie Allemande contemporaine.École expérimentale*. Paris: F. Alcan] Paris: librarie Germer baillie're et cie].

Ribot, T. (1885b). *Les maladies de la personnalité*. Felix Alcan.

Ribot, T. (1886). *German psychology of to-day: The empirical school* (J. M. Baldwin, Trans.). Charles Scribner Sons.

Ribot, T. (1887). *The diseases of personality* (J. Fitzgerald, Trans.). Humboldt.

Ribot, T. ([1887] 1888). La psychologie contemporaine: leçon d'ouverture de cours de psychologie experimentale et comparée du College de France. In *Revue de l'Hypnotisme expérimental et thérapeutique* (pp. 344–47).

Ribot, T. (1889). *Psychologie de l'attention*. Felix Alcan.

Ribot, T. (1891–1892). Enquête sur les idées générales *Revue Philosophique, 376*.

Ribot, T. (1893). Heredité. In M. Berthelot, H. Derenburg, C. Defus, et al. (Eds.), *La Grande encyclopédie* (Vol. 19, pp. 1165–1169). H. Lamirault et C.

Ribot, T. (1894). *Diseases of the will* (Merwin-Marie Snell, Trans.). Open Court.

Ribot, T. (1896). *La psychologie des sentiments*. Felix Alcan.

Ribot, T. (1897). *L'évolution des idées générales*. Alcan.

Ribot, T. (1898). *Psychology of emotions*. W. Scott

Ribot, T. (1900). La psychologie de 1889 a'1900; IIIe'me Congress International de psychologie. *Revue Scientifique, 44*(12), 177–178.

Ribot, T. (1904). Sur la valeur des questionnaires en psychologie. *Journal de Psychologie Normale et Pathologique, 1*.

Ribot, T. (1905). *La logique des sentiments*. Felix Alcan.

Ribot, T. (1910). Chapter two. In H. Munsterberg (Ed.), *Subconscious phenomena* (pp. 33–39). Rebman.

Ribot, T. ([1909] 1928). Psychologie. In Henri Bouass et al. (Eds.), *De la Methode dans les sciences*, 1iere serie. Felix Alcan.

Ribot, T. (1957). Lettres a Espinas, edited by Raymon Lenoir. *Revue Philosophique, 147*, 1–14.

Ribot, T. (1962). Lettres a Espinas, edited by Raymon Lenoir. *Revue Philosophique 152*, 337–340.

Ribot, T. (1964). Lettres a Espinas, edited by Raymon Lenoir. *Revue Philosophique, 154*, 79–84.

Ribot, T. (1970). Lettres a Espinas, edited by Raymon Lenoir. *Revue Philosophique, 160*, 339–348.

Ribot, T. (2005). Letters a Henri Pieron, edited by Jacqueline Carroy and Régine Plas. *Revue Philosophique, 195*(3), 335–356.

Roger, J, Ed. (1979a). Les Néo-Lamarckiens Français. *Revue de Synthese, 95/96*, 279–469. (J. Roger, J.C. Sillard, A. Lagarde, Cl. Blanckaert, M. Viré, G. Gohau, A. Diara, K. Wellman, J.L. Fischer).

Roger, J. (1979b). Presentation. In *Les néo-lamarckiens français* (pp. 95–96), edited by Jacques Roger. Special issue of *Revue de synthèse* 3ie'me série.

Roth, M. S. (2011). *Memory, trauma and history: Essays on living with the past.* Columbia University Press.

Royer, C. (1862). *De l'origine des espèces, ou Des lois du progrès chez les êtres organisés.* Guillaumin (: A. Moulin).

Schweber, L. (2006). *Disciplining statistics: Demography and vital statistics in France and England, 1830/1885.* Duke University Press.

Shmuely, S. D. (2018). *The bureaucracy of empathy: Vivisection and the question of animal pain in Britain*, 1876–1912. PhD. dissertation. MIT.

Sommer, M. (2016). *History within: The science, culture, and politics of bones, organisms and molecules.* University of Chicago Press.

Sotirios, P. (2018). *Reappearing characters in nineteen century French literature.* Palgrave Macmillan.

Spencer, H. ([1870–1872]. *The principles of psychology*, 2nd edn. Longman, Brown, Green and Longman.

Staum, M. (2007). Ribot, Binet, and the emergence from the anthropological shadow. *Journal of the History of the Behavioral Sciences, 43*(1), 1–18.

Staum, M. (2011). *Nature and nurture in French social sciences 1859–1914 and beyond.* McGill-Queens University Press.

Taine, H. (1869). *De l'intelligence.* Hachette.

Taine, H. (1873). Review of Ribot's *Heredité. Journal des débats politiques et littéraires* (November 23).

Thomas, M. (2003). Rethinking the history of ethology: French animal behaviour studies in the Third Republic (1870–1940) (PhD dissertation). University of Manchester.

Thomas, M. (2017). Biological and social new orders: Charles Robin (1821–1885) and Comteanism in the Third Republic. *Historical Studies in the Natural Sciences, 47*(5), 653–677.

Vicenti, D. (2019). Experience and experimentation: Medicine, psychiatry and experimental psychology in Paul Janet. *Perspectives on Science, 27*(5), 704–738.

Vicenti, D. (2020). Théodule Ribot and the spiritualist tradition: The philosophical roots of scientific psychology. *British Journal for the History of Philosophy of Science, 28*(5), 1009–1030.

von Hartmann, E. (1869). *Philosophie des Unbewussten: Speculative Resultate nach inductivnaturwissenschaftlicher Methode.* Carl Duncker.

Voutsinas, D. (1988). La jubile de la psychologie scientifique française. *Bulletin de Psychologie, 49*(385), 404–469.

Wassmann, C. (2005). The science of emotion: Studying emotions in Germany, France and the US, 1860–1920, vol. 1 (PhD dissertation). University of Chicago.

Wassmann, C. (2017). Forgotten origins, occluded meanings: Translation of emotion terms. *Emotion Review, 9*(2), 163–171.

Chapter 6
Interlude: 'Hierarchy' in the Nineteenth-Century Spencerian Lamarckism/neoLamarckism and Its Transfer

I put forward a succinct comment on the significant role of the concept of hierarchy and its attendant concepts in the mid- to late-nineteenth-century evolutionary discourse, especially within diverse SL/NL frames, and their transfer to other fields.[1]

Present-day discussions of how to explain the origin of evolutionary innovation and novelty within the context of developmental processes (and vice versa), e.g., embryological ones, resort to hierarchical frames (Baedke, 2021), while the nineteenth-century SL/NL efforts that deployed 'hierarchy' looked into contemporaneous models of development for that.

Did the nineteenth-century use of hierarchy signify also the passing-on from one generation to the next of patterned whole–part relationships within the organism and those of hierarchical relations of function and structure among organisms? I think that there is no simple straightforward positive answer, not if we think along the sophisticated lines of argument indicated by the twentieth-century philosophers of biology who consider their work in some sense as revivifying the early twentieth-century organicism (e.g., Deacon, 2003; Gilbert & Sarkar, 2000), nor the present discussion of 'levels' (e.g., Brooks et al., 2021; Kostić et al., 2020).

Yet, discussions of hierarchy in Spencer, Jackson, and Ribot and in works of other contemporaries did consider those patterned relationships as transgenerational. They did so without delineating specific mechanisms besides those of SL/NL generally discussed, which should allow ample room for rising complexity as a necessary feature of evolutionary time on all levels. Complexity, even more than diversity, was at the center of their elaborations, though both were discussed.

[1] Besides the primary materials of the diverse chapters and the literature mentioned in the previous interlude, I have directly used in particular the following secondary sources on the historical angle: Droz-Mendelzweig (2010), Nicholson and Gawne (2015), Smith (1992), and Woese (2004).

S. B. Gissis, *Lamarckism and the Emergence of 'Scientific' Social Sciences in Nineteenth-Century Britain and France*, History, Philosophy and Theory of the Life Sciences 36, https://doi.org/10.1007/978-3-031-52756-2_6

Relevant here are the relations between organization, complexity, plasticity, and resilience within the nineteenth-century conceptual framework of hierarchy and levels. In his *Principles of Psychology* and *Principles of Biology*, Spencer argued that evolutionarily earlier/lower (biological) individuals were simpler and less differentiated/less specialized internally, with less coordination and integration among parts and with less interdependence among those parts, and thus they were more homogenous and less adaptive, i.e., with less plasticity in relation to environmental changes, and their modes of activity tended to be rigid and repetitive. For the Spencerian-Lamarckist, *the higher* on the evolutionary hierarchy, the *more complex*, heterogeneous, and thus *highly organized* biological units would also be endowed with *more plasticity* and would be *more resilient* under environmental change long term, because of their ability to regain their equilibrium, to innovate, in the face of such changes. The more homogenous would turn out to be more rigid, less organized, and thus less resilient in the long term. It was here that Jackson differed. For him, *the lower, less complex* (nervous centers within his view of the hierarchy of the levels of the evolutionized nervous system) were also *the more highly organized*, more rigid, and more stable, while *the higher, more complex,* were posited as more interconnected, but the *less organized*, less interdependent, more plastic, capable of *producing novel modifications*, activities, learning, etc., yet much *less resilient* in the long term. Ribot and most late nineteenth-century psychologists and neurologists, including Freud, who worked with the hierarchy model, followed Jackson's view, though at least Ribot and numerous other contemporaries were certain that they were following Spencer.

Present-day accounts of biological systems, both by those claiming mechanist and reductionist approaches and those who oppose them, agree with differing explications, on some hierarchical (and at times multi-scale as well, e.g., Fox-Keller, 2010) organization of living systems, or at least on a possible hierarchical view of evolution and on the conspicuousness of complexity as a feature of such systems. Yet, complexity rather than being an explaining feature has become that which should be accounted for and explained, often resorting to the *history* and dynamics of biological systems, moving from them also to a variety of notions—of nested levels, of controversial notions of the tree of life, e.g., Doolittle and Bapteste (2007), of other patterns, and of the concept of levels, including levels of selection. The present concept of evolutionary transitions also signifies increase of complexity, as the very nature of the hierarchical structure of life evolves or, *pace* Buss (1987), a transformation of the hierarchical level and whole–part relations. Such views also assume that such transitions also mean novel relationships between entities that became lower level and the new modes of complexity that signify the new organizational nested-ness hierarchy. Evo-devo views are presented, e.g., by Laubichler (2010), as built around a group of concepts including hierarchy, emergence, and history.

I argue that for the late nineteenth-century SL/NL characteristics of organisms were distributed along a hierarchical pattern and were thought to signify aspects of inheritance along a timeline (phylogeny), but also implied a perspective on living entities of the continuous dynamics (thus giving precedence to function over structure), assuming context/interaction/environment as fundamental. For these thinkers, such evolutionary hierarchy was in a sense 'a given,' to be elaborated, applied, and deployed, rather than the object of explanation. Jackson's description and analysis of the nervous system, including the brain, was such a case. He used the notion of 'higher level' constraining, controlling, and regulating 'lower level.' In its nineteenth-century use, however, the notion did not entail by necessity a moral–social hierarchy (as, e.g., held in R. Smith, 1992, 2020). Of significance here would be the decision whether to view organismal dynamics as under a description of top–down, or bottom–up, or an interplay between the two among levels or also within levels, and how integrative patterns were described.

Evolutionary hierarchies, though perhaps originally closer to SL/NL views, were also deployed with a variety of mixtures of contemporaneous Darwinian adaptationist–selectionist views. Hereditarian, deterministic, and environmentalist plastic versions of such hierarchies could be found in various models of degeneration and atavisms in Western Europe in the second half of the nineteenth century, particularly around the 1880s. The assumption of evolutionary hierarchies played an important role in the contemporaneous discourses on modernity and its impact on industrialization, urbanization, social and sexual deviance, mental pathology, and criminality, as well as in the political views of the middle class regarding class, nation, and 'race.' A word of caution: the specificities of locality, period, bounded communities, and demarcated scientific fields were of profound importance in the constitution, the deployments, and the transfers of these issues.

Thus, I put forward that deploying 'hierarchy' as a significant concept also implied the use of a *cluster of attendant concepts* and indicated a specific array of views on evolution, on the structure of biological nature, and on the proper methodologies to deal with these. Central among them were relations among levels, e.g., inclusion, nonreduction, part–whole, and nested relations; function; regulation, control, and constraint; evolution of organization and self-organization; complexity and complexification; continua; and emergence. I suggest that in Spencer's corpus Lamarckian assumptions were interlaced with other components to produce a view of living nature structured through this cluster of concepts, which impacted on contemporaries such as Jackson and Ribot and resonated with many others including Freud.

Using 'hierarchy' during that period as a descriptive tool indicated the possibility that living nature (or any significant part of it) could be described with some measure of precision evolutionarily and developmentally, that one did not have to know everything in order to know something. Within this frame one dealt with entities arranged in some graded series, without implying that the 'bottom' level or domain was to be considered more fundamental than others. Their relations were often reconstructed as part–whole, or linearly inclusive, or nested, while there was the background assumption that such relations were to be understood within the context of an organizational yet holistic evolutionary discourse that pertained to organisms (multicellular, in their environments) and their particular properties, their parts, and parts of parts, all mutually constitutive.

Levels in such hierarchies also indicated degrees of organizational complexity. Such degrees of complexity were differently related to plasticity, constraint, regulation, and control. The most conspicuous difference, discussed in the chapters, was the one conceptualized by Spencer on the one hand and Jackson on the other hand. These hierarchies were deemed to form graduated continua. Furthermore, given the whole–part nested relationships, mechanisms working at lower level could be considered as parts of a higher-level mechanism. Hierarchies and relations among them and their component parts did not imply assuming evolutionary transitions.

An important assumption then concerning the components of these hierarchies was that organisms were seen as self-organizing and self-maintaining, though differentially, depending on their level of (material) complexity and its accompanying features. Recent discussions coined at times as 'the return of the organism' (versus genes, populations, etc.) do deal with self-organization, developmental plasticity, and organismal agency but are rather different!

The entwinement of organisms with environments was diversely perceived and related to the discussion of passage from nonequilibrium to equilibrium states, i.e., dynamic equilibrium, as organisms were seen as continuously undergoing change. Continuous activity was attributed, even when the components of the hierarchy discussed were parts of an organism, rather than whole organisms, e.g., levels of the neural system in Jackson's theorizing. Such levels in a living entity were considered a result of evolutionary processes that the organism as a whole and its particular components had undergone. Thus, hierarchies were regarded as 'an empirical fact' to be investigated, forming part of a functional, not solely structural, explanation. Also, the time direction, i.e., the historicity of the processes, was significant.

Direction of causation came up in explicating hierarchy among the nineteenth-century figures I discuss and their contemporaries. Downward causation was implicit in notions of regulation, control, and constraint of lower levels by higher ones and often related directly or inversely also to notions of degree of organization, complexity, and plasticity. Emergence, when forming part of a particular conceptualizing of hierarchy, was related to properties of higher level that could not be reduced to those of lower levels, which also implied a basic asymmetry, that there was no upward causation. A distinction suggested by Fox-Keller (2008), following Weaver (1948), could help: between the generation (or emergence) of complexity and its organization. Furthermore, I suggest that it was implicitly assumed that components of a hierarchy—or, differently put, levels—could not be understood separately from their relations within that hierarchy, nor separately from the whole, whose hierarchy it was.

Let me end with a string of two short quotes from S.J. Gould's chapter on "Seeds of Hierarchy" in his *The Structure of Evolutionary Theory*:

> Hierarchy has resided at the heart of evolutionary theory from the very beginning, despite a temporary eclipse during the rally-round-the-flag period of strict Darwinism at the middle of the 20th century…. Call it the bad penny that keeps cropping up, or the pearl of great price always found within, but hierarchy seems unavoidable. (2002, 175, 197)

References

Baedke, I. (2021). The origin of new levels of organization. In D. S. Brooks, J. DiFrisco, & W. C. Wimsatt (Eds.), *Levels of organization in biological sciences* (pp. 153–173). The MIT Press.

Brooks, D. S., DiFrisco, J., & Wimsatt, W. C. (Eds.). (2021). *Levels of organization in biological sciences*. The MIT Press.

Buss, L. W. (1987). *The evolution of individuality*. Princeton University Press.

Deacon, T. D. (2003). The hierarchic logic of emergence: Untangling the interdependence of evolution and self-organization. In B. H. Weber & D. J. Depew (Eds.), *Evolution and learning: The Baldwin effect reconsidered* (pp. 273–308). MIT Press.

Doolittle, F., & Bapteste, E. (2007). Pattern pluralism and the tree of life hypothesis. *Proceedings of the National Academy of Sciences of the United States of America, 104*(7), 2043–2049.

Droz-Mendelzweig, M. (2010). La plasticité cérébrale de Cajal à Kandel: Cheminement d'une notion constitutive du sujet cérébral. *Revue d'histoire des sciences, 63*(2), 661–698.

Gilbert, S., & Sahorta, S. (2000). Embracing complexity: Organicism for the 21st century. *Developmental Dynamics, 219*, 1–9.

Gould, S. J. (2002). *The structure of evolutionary theory*. Belknap Press.

Keller, E. F. (2008). Organisms, machines, and thunderstorms: A history of self-organization, part one. *Historical Studies in the Natural Sciences, 38*(1), 45–75.

Keller, E. F. (2010). It is possible to reduce biological explanations to explanations in chemistry and/or physics. In F. J. Ayala & R. Arp (Eds.), *Contemporary debates in philosophy of biology* (pp. 19–31). Wiley-Blackwell.

Kostić, D., Hilgetag, C. C., & Tittgemeyer, M., Eds. (2020). Unifying the essential concepts of biological networks: Biological insights and philosophical foundations. *Philosophical Transactions of the Royal Society B, 375*(1796).

Laubichler, M. D. (2010). Evolutionary developmental biology offers a significant challenge to the neo-Darwinian paradigm. In F. J. Ayala & R. Arp (Eds.), *Contemporary debates in philosophy of biology* (pp. 199–212). Wiley-Blakcwell.

Nicholson, D., & Gawne, R. (2015). Neither logical empiricism nor vitalism, but organicism: What the philosophy of biology was. *History and Philosophy of the Life Sciences, 37*(4), 345–381.

Smith, R. (1992). *Inhibition: History and meaning in the sciences of mind and brain*. University of California Press.

Smith, R. (2020). Inhibition and top-down metaphor of organization. *Studies in History and Philosophy of the Biological and Biomedical Sciences, 80*(October), 1–9.

Woese, C. R. (2004). A new biology for a new century. *Microbiology and Molecular Biology Reviews, 68*(2), 173–186.

References

Chapter 7
David Émile Durkheim: Founding 'Scientific Sociology'

7.1 Introduction

Sociology emerged around the last third of the nineteenth century as a novel form of investigating social life, whose participants first named it 'social science.' In the 1870s, 'sociology,' coined by Comte decades earlier, did not appear in any contemporary dictionary, but by the end of the century, it occupied a space almost equal to 'economics.' The last third of the century was the arena of intense quasi-disciplinary activities in Europe and the USA: books, professional associations and journals, congresses, national and international, and gradually university chairs including 'sociology' as a component, e.g., 'history of law and sociology.' Investigating modern society was the common aim. I deal with one delimited section of this—the emergence of 'scientific sociology' promoted by Emile Durkheim, one among several in the French cultural field, with serious competitors for the position of academized 'French sociology.' Most prominent were René Worms' enterprises, Gabriel Tarde, and the two groups of Le Play followers. Until World War I, the Durkheimians seemed to gradually gain the upper hand, amid controversies on their claims, methodology, worldview, and role within French society. There is a general agreement that even though Durkheimian sociology (and sociology at large) took longer to become academized within the centralized system of French higher education, its impact on other fields of the human sciences was significant.

Evolutionary–biological rhetoric, often with medical components, was rampant in the late nineteenth-century French social/political discourses. Questions arise: Which evolutionary scheme? Was it used to legitimate the existing order, to scientifically ground a social order of inequality (social, racial, cultural), or to legitimate its reform, even revolution? Were the modes of transfer distinguished by function or by principal feature, i.e., was there an ontic identity, conceptual coherence, between the objects of the 'field of origin' and the 'target field'? For example, such a transfer, based on ontic identity, was the 'direct traffic' of

S. B. Gissis, *Lamarckism and the Emergence of 'Scientific' Social Sciences in Nineteenth-Century Britain and France*, History, Philosophy and Theory of the Life Sciences 36, https://doi.org/10.1007/978-3-031-52756-2_7

assumptions, explanatory mechanisms, and terminology from the realm of organisms to that of human societies, whereby the inner logic of the original field was kept within the 'target' one.[1]

My aim in discussing Durkheim is circumscribed—to try tracing, analyzing, and contextualizing two threads in Durkheim's activities and writings from the early beginnings till his very late writing relevant to my general argument: the endeavor to constitute a novel field, 'scientific sociology,' and turn it into an academic discipline, and the use of contemporaneous SL/FNL modes of evolutionizing for that. In my view, these two threads were intertwined in his work, and the evolving changes in both were related to each other. Furthermore, I propose that the analysis proffered in his oeuvre of the relations between collectivity and individuals played an inestimable role in both threads. Neither could be understood thoroughly without the other nor without their contextualization within the Republican regime that deployed a variety of means to mobilize, regulate, and separate social classes and political groupings. The grand multilevel educational reform emphasizing civic patriotism, moral *laicité*, and science was a component in constructing Republican national citizenship as against the Catholic Church and the 'German danger,' with a search for a new social–Republican–reformist ideology, while naturalizing class differences. Both Ribot and Durkheim participated in major polemics of the period, viewing them as an opening for establishing 'scientific' psychology and 'scientific' sociology, respectively (Fig. 7.1).

Durkheim's epistemology was aptly described by Fabiani (2003) as an epistemology of combat. Within the contemporaneous cultural field, discussions on/of 'the social' were initially attributed to the domain of philosophy, speculative metaphysics, and journalistic practices. All sociology competitors had to find their singular means to transform 'the field of the social' and its public image and status through constituting, bounding, cooperating, and co-opting with and against the already existing disciplines, especially philosophy and biology. The division of the sciences then was perceived as if constituted by their (natural) *object*, principal *assumptions*, and *methodology*, i.e., scientific method. Durkheim adopted that concept, making the role of 'the scientific' of paramount significance in establishing his version of sociology, singling out this version as reliable trustworthy producer of objective, empirical, yet rational knowledge on the social and provider of 'scientific investigation' whose results would therefore be generalizable and even universalizable.

I argue that Durkheim chose as a principal, though not unique, source of repertoire for his construction of scientific tools, practice, and rhetorical style

[1]E.g., De Lapouge, 1887; Gumplowicz, 1885, 1893; René Worms throughout the 1890s (detailed bibliography in see Gissis, 2002); von Lilienfeld, 1896. On them, Bernardini, 1997; Clark, 1984; Gissis, 2002; Leclerc, 1972; Staum, 2003, 2011; Taguieff, 1984.

Fig. 7.1 Emile David Durkheim portrait (1858–1917). Source: photographie Benque, 1903, gallica.bnf.fr

the contemporary discipline of evolutionized biology—mostly FNL and some SL. All his writings, from his early book reviews to his very late essays, were concurrently entwined and inseparable performative acts of constitution, strategic actions of legitimation, declarations and explications of 'scientific methodology,' and sociological investigations. I attempt to pull apart and reconstruct the two threads—of constituting and of the use of SL/FNL—to follow their entwined evolving and ask what remained stable throughout, what changed, into what, how, and why. In doing so, I shall adopt a 'developmental' point of view, i.e., avoid interpreting the beginnings in light of results, proceed stage by stage trying to reconstruct the repertoires within their specific cultural–historical situations and the questions, means, modes of action that were considered available and suitable then.

I divide Durkheim's work into three principal stages: early period (1882–1892), middle period (1893 until the late 1890s), and late period (the late 1890s until 1917). Given space constraints, I refrain from summarizing the contents of his works and from entering interpretive controversies. Rather my starting point is the corpus of works and whatever correspondence has become available, as well as the works of some of his contemporaries and of some members of the Durkheimian group and the FNL works of that period. I make use of the multilayered historiography of interpretations.

7.2 The Early Period: 1882–1892

7.2.1 *Spencerian Lamarckism/French neoLamarckism as Resources for a Nonbiologist*

From 1882 to 1892, Durkheim was vacillating between SL, as read in France, the evolving FNL, and the uses of biology primarily by some German social scientists. The choice to appropriate, engage with, and transfer biological mechanisms, models, and metaphors, whether SL, FNL, or Darwinian, in any new field outside evolutionized biology was consequential for that target field. The outcome reflected not only internal scientific factors but also particular juxtapositions within emerging sociologies and their position vis-à-vis contemporaneous cultural and political institutions. Thus, ongoing debates on the meaning and impact of processes of industrialization and urbanization were part and parcel of the concerns of the emergent sociologies and were fashioned through these modes of transfer. The outcome also reflected the relation of the new sociologies to and interaction with human sciences disciplines and practices perceived as adjacent and competing in the context of the Third Republic, especially philosophy, psychology, and history.

Given the (at least, initially) thorny reception of Darwin's work in France (e.g., Conry, 1974), the characteristic features of FNL were assumed to be almost self-evident. The critical gaze of recent decades, however, has revealed that there was neither any easily denoted cohesive 'neo-Lamarckian' theoretical entity nor any identifiable collective of 'NL' workers (see Gissis, 2023). This has been contentious among the principal FNL historians. Yet, the historians, whose work I have consulted,[2] would probably agree on minimal commonalities: resistance to Darwinism, yet also a modified adoption of Darwin—in both of which 'the struggle for existence' as intraspecies competition and the varied understanding of determinism versus contingency and randomness played a significant role; the impact of SL; the impact of the 'Claude Bernard model' of physiological experimental science[3]; the effect of theories on 'the economy of the organism,' especially those of Milne-Edwards. What seems clear, then, is that in the latter part of the nineteenth century, there emerged two central claims:

(a) Modifications were perceived to result primarily from adaptation to the action of the external/physical milieu/environment on the organism, i.e., the plasticity of the individual organism and individual variations. Though there had been disagreement on the mechanisms and on how to define 'need' and 'effort,' nevertheless there was a possible role for 'behavior'/'habit.'

---

[2] Beck, 2014; Bowler, 1984, 1988; Conry, 1974; De Bont, 2010, 2015; Delage and Goldsmith, 1909; Gould, 1977; La Vergata, 1996, Limoges, 1994, 2014; Loison, 2009, 2010, 2011a, b; Persell, 1999; Roger, 1979a, b. See also Loison, 2010 detailed biographies of French neo-Lamarckists; Becquemont, 2010; Blanckaert, 1979, 2004; De Bont, 2010; D'hombres, 2010; Feuerhahn, 2011; Fischer, 1979; Gohau, 1979; Thomas, 2003; Viré, 1979; Wellman, 1979.

[3] E.g., Canguilhem, 1965, 1994; Holmes, 2014: Latour, 1988; Olmsted & Olmsted, 1952; Tirard, 2013; Virtanen, 1960.

(b) Such modifications were heritable trans-generationally, i.e., the results of this plasticity were inherited.

Some FNL people (especially Jules Bonnier) tried to put the conjunction of these hypotheses to experimental test. Loison (2010, 2011a, b) argues that the neo-Lamarckists were unable to develop a systematic experimental program that combined fieldwork and laboratory work accommodating both claims, though such a program was of central importance for them. Holding both claims was, according to Loison, the root and source of the inner unsolvable tensions of FNL until the turn of the century. One could describe most, though not all, as holding some version of 'physiological positivism,' albeit heavily colored by Bernard.

There were diverse ways in which Haeckel's 'biogenetic law' (his principal works were translated in the 1870s) was integrated into this conceptual frame, differing in the ways the developmental aspect of transformism was enhanced, or the emphasis on phylogeny, or whether it served as a secondary support for theorizing on the directional complexification of organisms (Richards, 2008, Chaps. 5 and 7). Much work was invested concerning the stages of the embryo's development seen through the biogenic law, in particular the question of 'condensation' (Gould's term) or 'acceleration' in Perrier's work (which had a more important role in the writings of American NL). Note that Spencer relied on von Baer.

The centrality of the individual was conspicuous, especially in discussing growth and development issues and thus adaptation. Individual adaptive physiological variations under environmental pressures and consequently also 'function,' a central factor of evolution, were considered more significant than heredity. Yet, it was necessary to assume the efficiency of the latter for transmitting these adaptive variations trans-generationally, and the structure–function format was implied. Thus, contrary to Darwin, there was less conceptual work invested on the populational–spatial level. One should stress that environment was conceived in physical–chemical terms, including nutrition, and this allowed FNL to constitute their transformist biology as mechanist. Indeed, for Durkheim, in his transfer from FNL, all causal explanations of social processes were formulated so as to bring out that aspect deemed central in FNL, i.e., "*mais tout se passe mécaniquement*" (Durkheim, 1887b, 120). In her recent book, Riskin (2016) has argued that there were two rather than one mechanist traditions, one viewing organisms as passive and the other as acting in time, having history. The latter tradition assumed also inner-coordinated and correlated changes in/of parts (later transferred also to the view of the cell). Diversity and complexity, which had bothered both Lamarck and Darwin, related to the decision to hold the view of active mechanist organisms—"ideas of locating the source of life's agency within material nature's tendency to vary, to complexify, to progress, to revert" (Riskin, 2016, 231–32). Riskin argues that Darwin held on to the 'to-vary' part, while FLN held to the 'to-complexify-and-to-progress' parts. Due to the predominance of the environment and its diversity, the evolutionary process was conceived as a complex branching tree.[4]

---

[4] Bowler, 1984, 1988; Loison, 2009, 2010, 2011a, b.

As most FNL historians explicitly argue or imply, there did not exist any uniform particularized agreement on any of the issues mentioned above, nor on the role and extent of the significance of Darwin's arguments in *Origin*. Rather, there was, as described by the historian Conry (1974), 'a radical multiplicity of paths.' Moreover, at the center of the spectrum of positions on evolution, there was minimally a coexistence of elements, which could be defined as either Darwinian or Lamarckian.[5] Positions that, in differing degrees, adopted an exclusive stance could only be found toward both ends of that spectrum. This positioning changed with the rise of Weismann's theory and the reactions to it, combined with perceptible changes of emphasis on selection in British (neo-)Darwinism. These changes challenged FNL and sharpened both their sense of separate identity and their experimental and theoretical dilemmas (Conry, 1993, 489n6; Thomas, 2003).

Most FNL historians mention or discuss at some length the special resonance, even affinity, between this cluster of views in the biological sciences and the core tenets of the Third Republic, in particular 'mechanism' as a central mode of scientific explanation, anti-teleology, rationality, and secularism, as well as plasticity.[6] This produced within the highly centralized higher-education system a predominance of that cluster (e.g., Limoges, 1994; also, Beck, 2014; Lustig, 2004). Conry discusses political figures whose role in the predominance of FNL in academic positions was significant, e.g., Louis Liard, Paul Bert, and Leon Donnat. There was also Giard's special position at the Sorbonne since 1888 and Perrier's at the Museum d'Histoire naturelle. Both the resonance and the political connections become relevant when discussing Durkheim's very early work.

For the purposes here, an implicit consensus suffices that, above and beyond the differences, there were certain generalized features constituting 'neo-Lamarckism,' at least from the very late 1870s until soon after the turn of the century. The question of the theoretical/epistemological facets of disunity among a variety of French Lamarckists is less relevant than their methodological, epistemic, and ideological implications and their perceived status as scientific discipline. These served as a common language for the interaction between the already existing biological sciences and emerging social sciences within wider disciplinary and ideological contexts. FNL offered a double perspective on ethics and on society at large. It emphasized the overall importance of the milieu/environment in shaping present and future generations through the inheritance of acquired characters, with 'use–repetition–habit' as its principal mechanism. This implied the possibility of shaping the future: present changes could be bequeathed as prospective biological or cultural traits to be further elaborated and complexified in the future. The seeming determination of the present by the past was one consequence of such thinking, the other being the molding of the present in light of a projected future. This 'future perspective,' associated with FNL in the last decades of the nineteenth century, was of utmost importance to those social scientists who considered the emerging

[5] E.g., Alfred Giard, Edmond Perrier (Bonnier and Felix Le Dantec were somewhat later).

[6] Beck, 2014; Conry, 1974; Persell, 1999.

discipline as interwoven with social reform, assumed a continuous narrative of humanity/society, and formulated, like Durkheim, their analysis of modern society in terms of a progressively complexified division of labor.

By the mid-1890s, scientific and political–ideological reasons converged to produce a common historical–genealogical narrative of FNL with Lamarck at its center. Toward the end of that century, there appeared more synthetic overviews of the NL evolutionary theory and practice, placing it within a more general history of the rise of evolutionary thought. The utterances of all parties in the 'transformism controversies' only became more contrasting and argumentative in the 1890s (Conry, 1974, 441–60).

Thus, an interested nonbiologist could gauge some transferable commonalities between biology and the social sciences, namely (see a detailed analysis in Gissis, 2023):

- The central, all-important, causal role of environment in bringing about adaptation to changes, in the production of variations.
- The subsidiary 'negative' role of natural selection, with an emphasis on the unacceptable fortuitous character of the mechanism.
- Defining 'life,' i.e., 'organized beings,' 'organisms' by invoking the notion of protoplasm (e.g., Geison, 1969; Welch & Clegg, 2010)—a programmatic effort to emulate in physiological work the extraordinarily successful reductionist procedures of the physical–chemical sciences.
- Just as the basic entities and categories of the inanimate, physical world were a matter of theoretical decision, so was the decision to posit the individual organism, as the most elementary entity in evolutionary terms, isolated from its relational or interactionist network, as a basic epistemological category, as well as the substantive, spatiotemporal, basic entity. An implicit assumption of that procedure was the possibility of a continuous reduction to ever smaller entities. The direction of this reduction was seen as going from the whole to the parts. This was considered as enabling the bottom-to-top deterministic causal explanation, which was considered the only one acceptable scientifically. Such explanations were then perceived as mechanist.
- Concurrently, though, there were strong methodological emphases on the notion of function and on functional unity.
- There were variously formulated efforts to construct a narrative that would provide for novelty within a frame of organismal plasticity.
- Until late, heredity was dealt with in generalized terms, either as the determination catered for through the discussion of environment or else as a generalized trans-generational transfer of 'acquired features.'

To be secular, radical (in the Republican sense), to hold some form of a materialist world view, to support some mode of progressive solidarity as the moral–social ideal for one's society—these attitudes were significant features in the presentation of the self, common to many neo-Lamarckists and other members of the new elites of the *Republique*. Yet these attitudes were interlaced with concrete scientific positions adopted within the transformist camp.

7.2.2 *Durkheim: Sociology and Biology*

Durkheim's early work reveals both vacillation and gradual efforts to conceptualize and define sociology while positing it as a nascent science.[7] The 1882–1892 period encompasses the end of his studies at the ENS, the period of teaching at various provincial lycées, his journey and stay at German universities, and the very early years of his teaching post at Bordeaux University. As far as can be reconstructed, Durkheim did not have an orderly scientific education beyond what was taught at a good high school and the special obligatory science courses that became part of the reformed curriculum for philosophy students in the late 1870s to early 1880s. The course book on life sciences was authored by Alphonse Milne-Edwards. Numerous concrete biological examples in Durkheim's work seem to have been picked from this book.

As for Durkheim's image of what science was, he read Comte, and later Saint-Simon as well as Claude Bernard. He studied under Boutroux at the ENS, and his close friend Hamelin was a student and follower of Renouvier. One can surmise, based on hints in his early articles and on his meeting with Ribot in 1884—Ribot to Espinas on Durkheim complaining to him that Espinas had called all philosophy *agregés* 'neocritics': "they protested against your article. *They affirmed being evolutionists* and not at all (neo)criticists" (September 30, 1884—emphasis added)—that Durkheim followed at least to some extent controversies in the life sciences, especially those on evolution and on psychology in France, but also in some measure in Germany and in Great Britain. Thus, his criticism of teleological explanation in the social sciences was much sharper than Comte's.[8]

Durkheim fashioned his early sociological project on some hybrid model of evolutionized biology but parted company on some principal matters from the accepted commonalities and agreements of any contemporaneous 'double-barreled' (social/biological) frameworks while still using a variety of their models and metaphors. I evince this through succinct analyses of his endeavor as it evolved until the publication of *De la division du travail social-The Division of Labor in Society* (1893, *Division*).

Most of Durkheim's reviews and other articles (20 in sum) appeared in *Revue Philosophique (RP)*, some in *Revue international de l'enseignement* (*RIE*), one in the *Annales* of his faculty at Bordeaux University, and one in *Revue d'economie politique*. One could argue that an analysis of his early writings is superfluous, since after 1893 and in particular from the late 1890s onward and all the more so since the founding of *L'Année sociologique*, Durkheim wrote numerous important articles and wrote and edited numerous reviews, including reviews for the Comité des travaux historiques et scientifiques, 1903–1915 (Durkheim, 2003). It is my view however that it was a crucial period of dithering and formation, and his diverse views on the principal issues relevant to my argument concerning SL/NL can be

[7] For a detailed analysis of early Durkheim and FNL, see Gissis, 2023.

[8] Mauss, 1925; Davy, 1919, 1960; Chimisso, 2008; Fournier, 2013; Karady, 1976; Lukes, 1973a, b; Strenski, 1997, 2006.

reconstructed through such a close scrutiny. Thus, I propose to look into them not with an all-knowing retrospective gaze but 'developmentally.'[9]

A common feature of this collection of articles is the *absence of unity and consistency* along the timeline on almost any issue touched upon, except the very mission of constituting sociology as a science. Davy (1919, 1960) argued that this mission was already clear to Durkheim in the early 1880s. *Problématiques*, dilemmas, and alternative possibilities were posed, and contradictory paths were temporarily taken. The process of evolvement at times continued to include inner tensions, so that these were only gradually resolved into narrower-scope coherence. If one were to look upon this body of work over a decade, as a late nineteenth-century symphonic piece, it would become clear that it was gradually composed to sound only a few clustered notes, performatively uttered, elaborated upon, developed, and repeated: "Sociology exists; it is alive, it progresses; it has an object and a method." (Durkheim, 1886a, b, 80). Object, method, conceptual tools, projected practices, and principal p*roblématique* were deliberated upon within a double-barreled framework—social and evolutionary–biological. This new discipline would mean explicating social life not through positing the behavior of individuals as the basic unit. Rather it would look into *collectivities*, through 'scientific sociological method.' Consequently, possible positions were sifted through questions as the status of the individual within such theory and the nature of the relations among individuals and collectivities, which categories to deploy, which evolution/which tools, and, lastly, how to bound and legitimate the new field, socially and institutionally. Already at this stage there was an interdependence between contents and methodological assumptions, also relating to 'being scientific.'

7.2.2.1 The Spencerian Lamarckism/French neoLamarckism Repertoire

Durkheim's conception of 'evolutionism,' as divulged in two very early sources, was only discovered at the end of last century: the part on 'evolutionism' in his *agrégation* and notes taken mostly by Lalande of Durkheim's 1882/1883 and 1883/1884 course as a last-year high-school (lycée) teacher of philosophy: lecture 22 subsection of "Psychology" and lecture 76 subsection of "Metaphysics."[10] In both, evolutionism, though criticized along the lines of the predominant philosophical school of the day, was depicted as close to either SL or FNL positions.

The principal matters discussed in Durkheim's very early review publications mostly related to the nature and content of transfer and thus to the relations between evolutionized biology and emerging sociology, to constituting a new discipline, and to relations of sociality/collectivity and individuals. Durkheim seems to have subscribed to a view propagated by Bernard and Spencer, which separated a conceptual

[9] Relating in detail: Alexander, 1982; Giddens, 1978; Lacroix, 1981; Lukes, 1973a, b; partially Béra, 2017; Mucchielli, 1998; Wallwork, 1972.

[10] "Schmaus on Sens lectures" – http://www.relst.uiuc.edu/durkheim/reviews/Sens.Schamus.html – accessed December 19, 2000.

'building' from its 'scaffolding' in order to claim the independence of the 'building,' i.e., sociology, relegating transfer to 'scaffolding,' for example, when he wrote "as Spencer expressed it, they are a useful scaffolding, but one which masks reality from us" (1978,112 [1885]). Bernard, posited as an *exemplum* of constituting the autonomy of biology, separating it from chemistry, physics, and metaphysics, used 'scaffolding' in discussing the role of hypotheses. Durkheim distinguished between two principal modes of transfer—metaphorical and realist/literal—while explicitly promoting the deployment of metaphors, analogies, and models, e.g., 'organism,' in constructing the new field of sociology, particularly from the evolutionized-biology repertoire. For him, metaphorical transfer made available a free space of novel conceptualizations that would not otherwise have been possible (echoed in Harré & Martin, 1982).[11] Biology could and should serve as a 'model example' how to constitute an autonomous, distinct discipline, separate from fields for which it had been considered but an application. Furthermore, the social world was anchored in the world of living organisms at large (referring to Espinas and Perrier). Durkheim's further work showed continuing efforts to strike a delicate balance between benefitting from such uses and delineating precise disciplinary boundaries. From that perspective, using a repertoire from the physical sciences would entail cancelling out two crucially necessary aspects of sociological inquiry: evolution (which for him implied progress), relevant for dealing with modern societies; and history.

Durkheim's proclaimed strict adherence to ideals of empirical science also impacted on his position vis-à-vis the general presuppositions of evolution, which at that time he found lacking in specificity, though admittedly a necessary theory for any social science in its full concreteness, as: "The moralist could not reconstruct it more than a physiologist could try to recreate the organism" (1887b, 277).[12] Evolution was seen as a fundamental feature since all living entities were its products (1886b, 78), emphasizing the plasticity of interactions of living organisms and environment, the direct relation between phenomena's complexity and plasticity, and thus the plasticity of the social. Durkheim discussed the meaning of mapping evolution onto human society and human sociability, the evolutionary ordering of societies, rejecting the view of a unilinear evolution of human societies, deploying instead the image of a *branching tree* and its roots (1887a, b, c, 177), following FNL contemporaries. At a much later stage this legitimated for him using contemporaneous traditional societies as representing evolutionarily early ones, therefore the use of anthropological materials, which became later on a sore issue between him and his admired teacher Fustel de Coulanges (Momigliano, 1970). Durkheim clearly assumed FNL classifications of types, species, etc., as applied to the social, which served as grounds for distinctions that formed the conceptual condition of such complex comparative investigation. He suggested applying J.S. Mill's notion of concomitant variations (which Mill himself, in Book Six of *A System of Logic*, rejected for use in the social sciences), selecting only its technical aspect, arguing

[11] E.g., as found in Durkheim's review of Gumplowicz, 1885b, 634.

[12] "*Le moraliste n'a pas plus a' la reconstruire que le physiologiste a' refaire l'organisme*."

that social life contained many more parallel variants, whose existence served for him as a justification to use statistics.

Durkheim deployed recapitulation, prevalent among FNL (significantly but implicitly added): "Because higher organisms arise out of lower ones, the former resemble and *synopsize* the latter. The modern family *contains within itself in abbreviated form* the entire historical development of the family" (1978, 211 [1888]—emphasis added). It allowed him to claim that one could reconstruct the development of the whole from the contemporaneous vestiges of its evolutionary history. A generalized evolutionary scheme was thus offered that would locate within the same causal chain the specific phenomena to be explained, their specific environment, and their role within the general frame of the emergence of modern societies, etc. The sociology of the family served as such a specific case, where the *conjugal family* was "by virtue of a general law already observed in biology" (Durkheim, 1921, 233 [1892].

I propose to understand Durkheim's discussion of equilibrium/environment within the context of the FNL-deployed notion of *the economy of the organism* or else organismal economy, taken from Milne-Edwards, who maintained that the degree of differentiation and specialization among the elements of an organism expressed the degree of its cohering enhancement. More complex organisms had more successfully adapted themselves to their external milieu and achieved a better equilibrium in their internal milieu, termed 'the economy of the organism' (particularly so in his *Considérations sur quelques principes relatifs a la classification naturelle des animaux*, 1844, where his discussion of the 'physiological division of labor' appeared). Note that in 1833 Deshayes, Lamarck's pupil, joined Milne-Edwards in his work on his further books (Corsi, 1978). The later multivolume *Leçons sur la physiologie et l'anatomie comparée de l'homme et des animaux* had immense influence on the new generation of FNL biologists, trying to turn their research into narrowly specific laboratory work.

This notion, argues Conry (1974), served to describe and analyze internal–external coordination, so significant for FNL in their divergence from contemporaneous Darwinism. It was not conceived as a mechanist model but rather integrated certain features of both contemporary embryology and cell theory to present formation-and-functioning relations among components of the organism as the principal continual evolutionary process, where mutual coordination and adaptation to environment occurred. Plasticity was located in both individual components and organismal frame. The term 'solidarity' in FNL texts referred to the possible dynamic results of such processes. Durkheim followed Perrier's 1881 notion of plasticity combined with division of labor, leading to cohesion/solidarity, enabling adaptation to environmental pressures, whose results were to be passed on transgenerationally. Within this frame, behavior, i.e., habit/use and disuse, was considered of central significance (rather than as subsidiary, as then posited in French Darwinism). Furthermore, processes of growing differentiation, e.g., through division of labor, whether within the contemporaneous SL framework or not, were often integrated into the dynamics of the model. Thus, the notion of 'equilibration' and its derivatives became another component, reflecting ongoing active processes occurring both inside organisms and within their 'organism–environment interaction.' The 'internal milieu' of Bernard belonged to this class of models, which enabled a view of internal organismal processes as adaptation. In his later work, Durkheim drew an analogy between 'life' as a result of differential modes of association of relevant elements and 'social life,' using 'internal milieu' as a framework to discuss modes of collective functioning within society, which produced modes of solidarity, as the 'milieu' was turned into 'societal milieu.'

In some articles, particularly in the two opening lectures, Durkheim introduced the novel term 'crystallization,' which he continued to use in diverse conceptual contexts for different theoretical and descriptive purposes, enriching its uses and meaning until his very late writings. I look upon 'crystallization' during this early period as encapsulating his endeavor to explicate change/transformation, to convey a sense of social processes within an evolutionary continuum: starting with non-stable states, reaching stability/robustness, and then being challenged by novel processes undermining it. In FNL-biological terms, no social phenomenon could be looked upon as solely a mere structure or solely a mere function, transferred to the social realm.

Positing two social types as emerging, one evolutionarily earlier and a later one developing from it was related to that. Durkheim used the adjective 'organic' to characterize modern societies (1888 articles, emphatically in *Division*; no more when the specific conceptualizing supported by this transfer changed). It might have been chosen following Espinas' description of the relations of the components of the organism as both nondependent and coordinated, also in contradistinction to Tönnies' 1887 use for 'traditional similarity-based communities,' resonating with direct transfer modes found among some German and French sociologists. Though Durkheim's characterization of 'mechanical' for earlier 'homogeneous' societies was not explicated in his 1888–1889 articles, it nevertheless resonated with his implicit reasoning, which tied together homogeneity, non-difference, and physical particles. The depiction of modern societies emphasized differing webs of relations among autonomous *non-similar* individuals, forming a novel kind of collectivity. Durkheim possibly followed a Spencerian innovative inversion, where homogeneous ('simple') organisms were looked upon as less adaptive to environmental change and less resilient, in comparison with heterogenous, complex ones. These distinctions and the notion of these two types of society/solidarity were subsequently deployed in constructing the backbone of *Division*, combined with the deployment of the 'division of labor' mechanism as a producer of differentiation. Thus, his 1887–1889 articles already suggested some significant features of the general contours of a vision developed in great detail in Durkheim's first book (1893), including the use of SL/FNL repertoire and the role (picked from Spencer) of 'volume and density' of society.[13]

[13] On the Durkheim/Spencer controversy, see Jones and Perrin, 1974–1975.

Durkheim's position vis-á-vis *Darwin* was revealed in his reviews, both directly and indirectly. In discussing Wundt's work in the articles on German academia, Durkheim presented Wundt's understanding of Darwin and why Durkheim would not follow him. Both Darwin and Wundt rejected any use of explanation through ends/aims causality. Yet, the difference would be in the relations between the phenomena to be explained and antecedent ones in time and space. For Darwin, as read and understood by Durkheim through the FNL perspective, there was no connection whatsoever between physiological causes, variations in organisms, and the fact that some of these variations survived, being post-factum useful to the organism and to the species. They were random. For Wundt, customs, which served as the foundation of morality, were produced and changed without any causal connection to their post-factum ends and usefulness. Yet, in the overall scheme of morality, these were aimed at a generalized ideal of humanity and not based on the assumption of ego-centered usefulness. Thus, for Wundt and for Durkheim, there was no room for internal species conflict for the survival of organisms. One could not hold on to both the notion of random variations and causality (e.g., "Mr. Wundt absolutely refuses to apply to morals the hypothesis of natural selection. The struggle for life is a product of egoism"—Durkheim, 1887b, 126–27).[14]

Concurrently, Durkheim warned against using metaphors and analogies 'realistically,' by reification, rather than metaphorically. Already in his first review (1885c) Durkheim criticized Schaeffle for his 'direct translation' rather than (what I would call) transfer and transposition of biological categories, including 'laws,' into sociological ones. Such critique, applied to a number of figures, Comte, Spencer, Espinas, and Lilienfeld, was repeated throughout this early period. Durkheim was acutely aware of the double-edged deployment of biology as a supplier of a rich repertoire of conceptual tools, procedures, models, metaphors, analogies, and a source of legitimation for new fields aspiring to be sciences. He recommended this multifaceted transfer but concurrently cautioned repeatedly against 'direct translation,' called for distinction between fields, acutely aware that it was necessary to set sociology apart, autonomously, therefore to supply distinguishing features. This double awareness was clearly articulated especially in his discussion of analogy and its role and limits. On the one hand, as he said in his opening lesson on sociology: "for the sociologist, biology became veritable treasure trove of perspectives and hypotheses, which he doubtless had no right brutally to pillage, but which at least he might wisely exploit." On the other hand: "Social facts can be truly explained only by other social facts" (1978, 55–6 [1888]), prefiguring what he later wrote in *Rules of Sociological Method* to this effect.

Claude Bernard was deployed throughout this period, just as in FNL, as a constitutive model and *exemplum* in two important senses: for one, the emphasis on empirical work carried out with the field's distinctive toolbox, the results of which were not reducible to those of other sciences, and, second, the construction of

[14] "*M. Wundt se refuse absolutement a\` appliquer a\` la morale l'hypothèse de la sélection naturelle. La lutte pour la vie est un produit de l'égoisme.*"

'general biology' as a meta-science to sciences dealing with 'life.' Durkheim put forth the example of Bernard in order to emphasize that sociology should follow biology in its striving for scientificity, yet preserve the distinction of sociology and its autonomy, just as Bernard struggled for the separation of the biological from the physical. "In order that sociology has the right to exist, there must be in the social realm something which escapes biological investigation" (1978, 55 [1888]). In *Division* Durkheim went into great detail explicating the uniqueness of sociology vis-à-vis biology.

7.2.2.2 Constituting a New 'Scientific' Discipline

In the first two years of his university teaching, titled 'social science and pedagogy,' unprecedented within the French university system, Durkheim taught a course on sociology and then on sociology of the family, whose opening lectures were published. Both focused on the nature, genealogy, methodology, and utility of sociology and forcefully presented him as a person and a university teacher, dedicated to a mission (Lukes, 1973a, Chap. 5). The sociology opening lecture laid down general principles, while the sociology of the family one presented their implementation and elaboration in a specific subfield. There Durkheim also offered the first of numerous, ever-changing, genealogical lists of founders. There and elsewhere in his articles, he discussed two topics central to his view on 'being a science':

(a) Causes and function: Durkheim, who looked upon strict causal explanation as the scientific one, made ample use of 'function' and functional explanation, e.g., as the conditions for a cause of x. In contemporaneous contexts, such explanations were not perceived as teleological or relating to aims/ends, especially so in the wake of Comtean positivism. Durkheim thus contrasted 'functional' to 'teleological.' To understand his insistence on explicating legitimate modes of explanation, in particular when dealing with transfer, with relations of the social and the biological, one may add the growing tendency of spiritualist philosophers, e.g., Paul Janet, a spiritualist/neo-Kantian like Boutroux, to expound on teleological explanations when dealing with organisms (also Lukes, 1973a; Turner, 1986). In the age-old controversy on structures and functions, Durkheim adopted the FNL view on the priority of function, and thus of sociology as 'physiology' rather than 'morphology,' stressing the role, epistemic and ontological priority, of functional non-teleological explanations over structural ones.
(b) Scientific experimentation (or its substitutes) was central, e.g., for projected investigative practices in sociology. In the sociology of the family lecture (esp. 1978, 209–10 [1888]), Durkheim offered an analysis of 'experiment' in sociology and of experimentation in general that would enable him consider sociology an empirical science. He resorted once again (also in line with the practice of FNL members) to Bernard, in order to argue that the essential component was the ability to isolate the cause from a host of accompanying circumstances.

Such isolation could be reached either through creating artificial occasions or through the use of occurring natural changes that allowed comparisons (i.e., 'nature's experiment').

On both topics, the positions adopted were such that SL/FNL assumptions, models, and concepts deeply resonated in them.

7.2.2.3 The Object of Scientific Sociology

Durkheim's aim to posit sociology as 'scientific' depended on naturalizing its object. For him an object was to be considered natural—e.g., society—when it could not be changed according to human intention, will or whim. Societies being 'natural' were subjected to laws and open to causal explanations, as any other realm of nature; thus, he rejected the then prevalent distinction between natural and artificial societies.

The theme of individuals–collectivity relations was a foundational one, evolving throughout Durkheim's writing till the very end, closely related to that of 'scientificity.' For him, the new discipline would explicate/explain social life not through positing side by side the behavior of individuals but by investigating collectivities, considered as the basic units both methodologically and epistemologically. Given his political–moral views, Durkheim could not relinquish the autonomy of the individual (e.g., "If individual life does not have some worth, as little as that would be, then the rest is worth nothing"),[15] yet given his long-term aim and sociological views, the collectivity had methodological and epistemic priority. Three notions first articulated with evident dithering in the early reviews were individuals as shaped by their society, dependent on it; the collectivity being in some, yet undefined, way, over and beyond its individual component members and/or beyond their aggregate sum; *collective consciousness*. His conflicted position tied this argument to a call to consider the distinctiveness of the sociological inquiry vis-à-vis psychological/individualized explorations, following Boutroux's notion of the distinctness of each disciplinary realm.

Already in his first reviews 1885–1886 Durkheim posited collectivities, hence sociability, social bonds and cohesion, '*social environment*,' as the proper object of scientific sociology. Thus, he pronounced on 'morals/*moralia*' sociologically, rather than philosophically on ethics.

'Sociology' was to be scientific, not through ontological claims on the nature of its object/phenomena, but rather by virtue of specific questions asked about well-defined objects, within a field of collective patterns of behavior, needs, functions, and adaptation of collectivities, and through the specific methods deployed to investigate them in search of causal laws. Much conceptual as well as rhetorical labor was invested in constructing such methodology. Durkheim called for a clear

[15] "*Si la vie individuelle ne vaut pas quelque chose, si peu que ce soit, la reste ne vaut rien*" (1887, 140).

differentiation between questions on the nature of data (social facts) and questions on how to investigate them (sociological facts), focusing on the role of social institutions and of constraint, habit and custom, regulative and normative rules, the nature of sanctions, and social conditions of change. Deploying statistics soon became significant as a tool to make deep structures sociologically visible, especially those of collective patterns previously posited as individual/psychological or totally biological. Collectivities, as the primary object of sociology, fulfilled in Durkheim's view at this stage the enabling conceptual conditions needed for scientific investigation, e.g., stability, long-term continuity in space and time, perseverance through changes, and non-dependence on human will, and thus enabled transfer from the FNL repertoire. This made possible the use of the FNL version of evolutionized biology as a toolbox for the construction of the new 'scientific' discipline, its *problématique*, and its bounding. The role of concepts, models, etc., in the new field, transferred from biology, was supposed to be analogous (rather than identical) to their role in their source field. Thus, society was not considered an organism, but rather the role of social institutions in it was perceived as analogous to the role of the organs of the organism within it. Durkheim, following Perrier's notion of plasticity, attributed to the individual elements of the organism, combined with division of labor, leading to cohesion/solidarity, enabling adaptation to environmental pressures, whose results were to be passed on trans-generationally.

In conclusion, one could say that in these early review articles, 1882–1892 FNL was considered an important toolbox, necessary, perhaps indispensable, at that stage; nevertheless, there was also a conspicuous rhetoric of bounding and autonomy. Durkheim criticized 'realist' transfer, warning that any straightforward and 'direct translation' of the biological conceptual apparatus and practices to sociology could hamper its progress.

Was such transfer a mere rhetorical means of legitimation? A discipline-construction tool? Was it an essential explicatory mechanism? Mere scaffolding? There was no definite position in his early articles, only attempts, vacillations, and gradual evolving that in some sense never ended.

Before proceeding further, let me add a brief comment of controversy on the theme of biology–sociology in Durkheim: Not many interpreters of Durkheim have looked into the detailed makeup of the biology–sociology relations in his very early, then early, and middle periods, though all agree that he had complicated, ambivalent attitudes toward the legacies of Comte and of Spencer, both of whom dwelt at length on the biology–sociology relations. Let me *very* schematically comment on principal positions on this matter. In putting forth my argument, I would note that though *Mucchielli* (1994, 1995, 1998) in his writings on Durkheim and French sociology very generally recognizes the significance of 'biology,' mentions possible FNL influence on Durkheim and on other sociologists, and claims that the role of heredity was limited vis-à-vis that of adaptation, he nevertheless offers no discussion as to why this biology, and to what purpose. In her work on the sociology–biology issues, *Barberis* (2003) mostly presents a detailed analysis of contemporaneous organicism with the central argument that there was in fact little difference in the uses of the 'organism' among almost all sociologists of the period but hardly any analysis of FNL's role in Durkheim's work. I strongly disagree with the interpretation put forward by *Guillo* (2000, 2006). Succinctly put, though he starts only with *Division* (1893), Guillo argues categorically that a static conception of 'organization' (and not 'organism') was the concept used throughout and is to be understood as essentially *morphological*, principally based on late-eighteenth- and early-nineteenth-century mostly fixist, biological conceptualizing. Both Limoges earlier and d'Hombres more recently have dealt with 'the physiological division of labor.' *Limoges* (1994, 2014) argues that it played a central role in Durkheim's *Division* and indicates Darwin's powerful impact on him. Yet, Darwin used this concept as a metaphor (combined with the ecological division of labor), while Durkheim transferred the whole process of division to society, and it served different purposes (e.g., Gissis, 2002, 2011). Yet, I have shown above Durkheim's attitude to Darwin in the early period. In a somewhat later article, Limoges offers an interpretation of FNL that posits its main figures (in particular Giard and Perrier) as basically Darwinian, while I have tried to counter such arguments here (see also Gissis, 2023). *D'Hombres* (2010, 2012) offers a historical review and a problematized analysis of the concept of division of labor in the relations between the social and the biological around differentiation since Adam Smith, its evolving use within nineteenth-century biology, and the waning of that use toward the twentieth century. He claims that it was significant for many French neo-Lamarckists. Yet, Durkheim differed from the use of the FNL, as he deployed it to relate some version of hierarchies and progress (as in Perrier) in discussing functional plasticity.

7.3 The Middle Period: 1893 to the Late 1890s

My examination of Durkheim's writings in his middle period is tilted to a purpose; I attempt to suggest a point of view on one class of tools, methodological and conceptual, deployed in the effort to construct, legitimate, and academize a new 'scientific discipline' in a field facing other disciplines' antagonisms and competitors for a similar end. Three books, as well as some articles, written during this period are relevant for this section: *Division of Labor in Society* 1893 (*De la division du travail social—Division*), *Rules of the Sociological Method, 1895* (*Les régles de la methode sociologique—Rules*), and *Suicide 1897* (*Le suicide—Sui.*). During this period, Durkheim taught at Bordeaux University and published these three major books and a number of book reviews and articles, including a couple of programmatic essays on sociology. At the very end of this period, Durkheim founded a sociological journal, *L'année sociologique (AS)*, containing mostly reviews of publications and a couple of original articles in each volume. The first volume (1896–1897) was published in 1898, and a work collective gradually emerged around it. During the very last years of the century, Durkheim became active in the pro-Dreyfus movement.

I put forward that it is *also* within the context of transfer and transposition from the discipline of biology, FNL in particular, that we should interpret Durkheim's work during the 1890s—as a theoretician, a leader of a novel emergent field attempting to become a discipline, an observer and active agent of the educational reform, and a founder of a work collective. Moreover, Durkheim's innovative claim to explain individual and society solely in social terms was to a significant extent made possible by the sophisticated manner in which fundamental tenets, modes of conceptualizing, and terms of FNL were transferred and transposed to the domain of society. In my view, this period may be characterized as harboring certain underlying inner dithering and inner tensions, conspicuous in the early-period writings—conceptual and methodological—which found expression in volatile discrepancies between sociological practices and proclaimed programmatic positions.

The wide span of the twentieth- to twenty-first-century interpretive positions on Durkheim's 'middle' period can be sorted into those concentrating on analysis of detailed theoretical evolving, either global or of particular aspects; those striving to depict theoretical evolvement as intertwined with processes of socializing and institutionalization; and those attempting to reconstruct a sociological–historical yarn of publication, distribution, communication, and positions of teaching and research. Or else interpretive positions could be sorted according to their overall view of the dynamics of the Durkheimian corpus: those asking whether there was continuity and consistency, or at least some measure of coherence, in spite of changes; or else a gradual change in emphasis, culminating in a difference allowing claim of a hiatus between early and late; or else those suggesting there was a dramatic break or even two breaks, bringing about deeply differing sets of theoretical assumptions, of objects of investigation, of methodology, and of *weltanschauungen* (Gane [1988, 1992], Karady [1976, 1983], and Clark [1973], respectively, could serve as examples of the first and Giddens [1978], Lukes [1973a], and Alexander [1982] of the second). Contemporaneous critiques of the three middle-period books also differed widely in their understanding and in depicting the relations among them, falling roughly along disciplinary lines, political ideologies, and socio-philosophical traditions (e.g., professional philosophers and economists, e.g., Brunschvicg, Halevy, Fouillé, Tarde, Gide, Sorel, Tönnies, and Schmoller). The later fates of these three books among general readerships and their later reception among sociological readerships have been highly diverse,[16] all the more so along national lines, and so were their translations (e.g., in Germany).

7.3.1 Durkheim and Spencerian Lamarckism/ French neoLamarckism

I argue that the role of such transfer was highly significant for Durkheim during this period as well as for the Durkheimians' initial stage in the rivalry for the status of 'French sociology.' As the worldview of 'evolutionism' turned into a meta-disciplinary tenet, it colored rhetoric and epistemic assumptions and imparted an almost homogeneous appearance to a number of scientific discourses.

[16] Among sociologists, e.g., *Suicide* as a 'classic' – Marsh, 2010; Swedberg, 2019.

Durkheim held that one could not directly 'encounter evolution'; rather, one encountered specific societies that evolved independently and did not form a unilinear series. Accordingly, the presupposition of evolution should form a component of the scientificity of the field, expressed through its methodology, rather than serving as one of its philosophical characterizations, resorting to the image and metaphor of a branched-out tree for 'evolution.' "The succession of societies cannot be represented by a geometrical line; on the contrary, it resembles a tree whose branches grow in divergent directions" (1982, 64 [1895]), and earlier in *Division*: "the shape of a bushy tree, doubles with a single trunk, but with diverging branches" (1984, 100n7 [1893]). This position resonated even in Boutroux's objection to consider 'the history of philosophy' evolutionarily, as relegating it to a subsidiary role in the past of humanity.

As often argued, Durkheim's novelty was in positing a novel theorizing mode: "The causes of social phenomena are internal to the society" (1982, 141 [1895]). Unlike Spencer, Durkheim was not a straightforward bio-sociologist. Rather he intended sociology's claim to academic-and-cultural legitimation to reside in its being science. Thus, the labor of constituting a 'scientific discipline' had an indelible impact on rhetoric, on theoretical and methodological presuppositions. Yet, ideological, political, and moral commitments put constraints on choice of object, methodology, and range of possible disciplines to be considered as potential resources for a repertoire of scientific transfer (Cladis, 2020). The scientific context of the emergence of sociology was molded by sciences perceived paradigmatic in the cultural field, as examples to be emulated when constructing a system of theoretical and field-specific methods, and rules for practice when delineating the boundaries of relevant concepts and legitimate questions. Though physics still played that role for sciences in general, evolutionized biology was prominent in the cultural field. Consequently, Durkheim struggled to define proper uses of the biological repertoire. I argue that throughout the 1890s he held that while society/the social should be explained through social categories, the conceptual tools for this analysis were to be grafted on those of another science. In his middle period, Durkheim suggested that the role of the transferred/transposed (concepts, models, metaphors, analogies, etc.) in sociology would be *analogous* rather than identical to the role of the original within FNL. Transfer/transposition thus became a process that produced organization of phenomena, categories of meaning, and even a causal regime in the new field. The use of the concept of physiological division of labor was such a case.

Self-consciously, Durkheim advocated a two-layered practice: a transfer that changed the referents and dissolved the coherence in the original field, combined with a procedure through which the singular (transposed) aspects of the new field, sociology, were made conspicuous. He invested great efforts in drawing fine distinctions between this practice and other modes of transfer deemed illegitimate, in order to present the new field of 'scientific' sociology as autonomous, its object and methods unique and distinct, and its boundaries, range of legitimate problems, and modes of explanation lucidly delineated. Thus, discussion of biological transfer often served as a theoretical stage for the problematization of new objects.

Like Spencer and many of his followers in France, Durkheim deployed the contemporaneous notion of 'organism' as a necessary component in his argument about the relations between individuals and their collectivities, and vice versa. There was nowhere any working definition of 'organism.' Yet, these relationships—not the actual individuals, nor the actual collectivities—could be viewed, in some of their aspects, *as analogous* to the relation of parts of an organism to its totality. By advancing this transposed analogy, already in his early review articles and then in *Division*, Durkheim hoped to provide a semblance of observability to the collectivity. For most of his contemporaries, individuals were conceived to be epistemologically the real entities and thus the observable ones. The reality of the collectivity was often conceived as secondary, deriving from the reality of the individual, unobservable on its own. Durkheim's usage purported to make the collectivity epistemologically 'real,' to construct its 'visualization' at a time when the visibility of investigated nature had become a significant scientific issue.

To distinguish his position from Spencer's, Durkheim insisted on that very partial application of the analogy to individuals and admonished against any literal understanding of the analogy between 'individual,' 'society,' and 'organism' (on transfer in René Worms, Gissis, 2002).

Durkheim used the FNL repertoire primarily in two modes: rhetorical-legitimating and theoretical-constitutive. In the first, models and metaphors were used rhetorically to legitimate the emerging field as a science or a novel scientific discourse. In the second, models and metaphors were used to construct theory and method, to suggest strategies for further work. Durkheim's discussion of the concept of 'social life,' both in *Division* and in *Rules*, can serve to illustrate this distinction.

Rhetorical: Durkheim argued that a history of forms of social life was a necessary component for sociology just as a history of forms of life in biology was. Consequently, he incorporated sociology among those sciences that dealt with 'forms of life.' These were posited as equivalent in signification to 'forms of organization,' thus implying a methodology of 'the whole versus its parts.' Life in both significations (social and biological) depended on the association of components and on their organization, structure, and dynamics of relations. Life could not inhere in any separate component alone. Rhetorically, therefore, whatever was legitimate in biological reasoning became equivalently legitimate in sociological reasoning on social life. The distinction between the living and the nonliving within biology could serve to distinguish the subject matter of sociology from those of other disciplines and to draw its boundaries as a new field endeavoring to become a discipline.

Theoretical-constitutive: Durkheim's discussion of what should be considered the elementary unit of society—"the real social protoplasm, the germ from which all social types would have emerged" (1984, 126 [1893])—was parallel to contemporary discussions of FNL, in contradistinction to Darwinists, in search of the most elementary unit from which all life had evolved. Given Durkheim's Spencerian assumption that homogeneity was evolutionarily prior to and conditional for heterogeneity, and his own assumption that the collectivity was prior to the individual and conditioned the individual, this 'protoplasm' had to be a *homogeneous group*, i.e., an aggregate with no internal differentiation. Durkheim was fully aware that this was a hypothetical unit, a construction needed in order to draw the complete outline and genealogy of (an evolutionary) series. In FNL theorizing this methodological device was only employed when the other links in the series had already been delineated (a familiar procedure which had been affirmed even by Bernard and often resorted to also by Darwin and other evolutionists, especially in *The Descent of Man*). Durkheim's series, however, was yet to be constructed. Nevertheless, he deemed such a hypothetical construction—analogous to the FNL one—*theoretically* necessary in order to posit an evolutionary branching tree with its accumulating social traits and patterns. It meant opening up realms of social phenomena to 'scientific' sociological investigation.

I suggest to succinctly posit the particularity of Durkheim's use of 'the biological' as follows: quite a number of contemporary sociologists were Darwinian biosociologists, though perhaps with a Lamarckian subtext (Conry, 1974; Bowler, 1988), while Durkheim adhered to FNL. His position throughout the 1890s was that while society should be explained through social categories, the conceptual tools for this analysis were to be *mindfully borrowed* from another science—biology. The particular transfer and transposition suggested by Durkheim were such that the role of the transferred concepts, models, and metaphors in sociology should be analogous to the role of the original in relation to the general discipline of FNL evolutionized biology. This comes out clearly in three issues: the use of '*lutte*' (struggle) and other Darwinian terms, the use of 'division of labor,' and the way sociology was supposed to distinguish itself in its deployment of 'the biological.' Let me very briefly comment.

7.3.1.1 '*Lutte*' and Some Other Darwinian Terms

Darwin was mentioned three times in the *Division:* once in conjunction with '*densité*' (1893, I—chap. 6) and twice in connection with the '*lutte pour la vie*' (1893, II—chap. 2) and '*La division du travail social*' (1893, II—chap. 5).

Durkheim criticized the direct transfer of 'natural selection' and 'intraspecies battle for life' to society on methodological grounds and was bothered by the implicit assumption of the methodological and epistemological priority of individuals in these concepts. He held that social conflicts could not nor should be described/analyzed in these terms but only through the action of social and cultural mechanisms that were based on the collectivity as the basic unit of sociological analysis.

I argue that Durkheim could not have applied a Darwinian evolutionary model for several reasons, including his reading of it as incompatible with his political and moral views, i.e., his assumptions concerning the bonds among humans, the nature of moral solidarity in societies, and his rejection of any conception of intraspecies combat. Nor would he accept the consequences of random variations, as he understood them, for the concept of causality (e.g., Bowler, 1988; Hacking, 1983). During the 1890s, he referred to this only once: The hypotheses of Darwin, he wrote, "overlook the essential element of moral life, that is the *moderating influence that society* exercises over its members... wherever there are societies, there is altruism because there is solidarity" (1984, 197 [1893]—emphasis added). Regarding the discussion of Darwin in *Division*, I agree with Conry (1974) that Darwinian mechanisms of selection and competition were understood within a Lamarckian/neo-Lamarckian conceptual framework of the 'economy of the organism' and the 'independence and solidarity of the anatomical elements.'

7.3.1.2 The Physiological and Social Division of Labor

Durkheim did not need Darwin in order to learn the concept of the division of labor (cf. Limoges, 1994; see d'Hombres, 2010, 2016). He was familiar with SL and was a devout reader of Espinas' early work *Animal Colonies* (1877) as well as Perrier's *The Transformism* and *Animal Colonies and the Formation of Organisms* (1881).[17] Both were deeply influenced by Spencer, who borrowed from Milne-Edwards and placed his ideas within an evolutionary framework. Espinas and Perrier followed suit when adopting two of Milne-Edwards' tenets: the nature of the physiological division of labor and the relative independence of the various elements composing the organism. The cohesion of the elements in the organism was posited as a model of biological solidarity. Durkheim certainly availed himself of these FNL-elaborated conceptual tools, but in his theory *both explanandum and explanans were social.* Transferring the differentiation in the economy of the organism unto the division of labor was called 'the law of gravity of the social world' (1984, 333n1 [1893]), while the empirical measure of its differentiation in a specific society was considered an index of complexity of that society.

Thus, Durkheim presented the process of evolution, particularly in *Division*, as developing from the inside out, taking place in the 'internal milieu,' while external forces—the external milieu (other societies)—played their role. I believe Durkheim used '*évolution*' often as a polysemic term, meaning either 'development' or 'transmission' or both, context dependently. He turned the FNL 'milieu' into 'society.'[18] Among nonhuman living organisms, adaptation of the individual took place through biologically inherited characteristics, but in humans adaptation was toward culture and through culture, which led to the emergence of that which was beyond the biological and beyond biological heredity. This suggested in some contexts in all three books (*Division*, *Rules*, *Sui.*) legitimating nonmaterial explanatory categories. Explanation or description in sociology based on biological heredity was excluded: an occurrence of either biological or sociological primary processes of adaptation to environment resulted in a social/cultural practice and a social pattern. It would be transmitted as habit and custom through processes of socialization and acculturation. Adaptation of individuals could only be understood within their culture.

---

[17] On Perrier as a neo-Lamarckist, e.g., Conry, 1974; Roger and Blanckaert, 1979; Bernadini, 1997; D'Hombres, 2010; D'Hombres and Mehdaoui, 2012; Persell, 1999; Thomas, 2003; also, Gissis, 2023.

[18] "*Les causes des phénomènes sociaux sont internes à la société*" (1895, 119).

In contrast to Ribot's back-and-forth movement between the psychological and the biological, Durkheim was positing a sociological equivalent to the biological inheritance of acquired characteristic: "Man depends upon only three kinds of environment: the organism, the external world, and society" (1984, 285–86 [1893]). The molding of individuals in *Division*, more so in *Sui.*, was social–cultural and neither biological nor psychological–individual. Through the division of labor as a mechanism of change, the moral and civilizing practices and beliefs of societies could change, thereby transforming their individuals. For Durkheim (as for Spencer and Lamarck), changes and transformations applied to all members of a particular group, for these were considered as identical as far as the relevant social trait was concerned. Another layer in that transfer was the biological notion of 'balance/equilibrium,' which played a central though distinct role both for Bernard and for Spencer. It was posited, somewhat analogously to Spencer, and indicated a measure of plasticity of the organism's components, which allowed for recurring adaptational re-equilibration states. When transposed to sociology, they occurred in the internal social milieu, and human intelligence was supposed to play the role of biological plasticity (e.g., Durkheim, 1930a, b, 256 [1893]). Durkheim did not supply any social meaning to social 'external milieu,' and thus processes of change would be looked upon as internal to society.

A striking example of Durkheim's position related to the 'division of labor,' partaking in SL/FNL, was the reasoning that informed his estimation of modern society. As mentioned, high degree of differentiation and specification in the division of labor characterized modern society, resulting in complex social stratification and a sophisticated economy. Socially it produced complex networks of interdependence among individuals; smaller units in such society changed their structures when joining larger units. This situation afforded each individual a greater measure of freedom and autonomy, for the multiple networks were not overlapping, and individuals were not situated in any single totally determining network. Modern society produced thereby a new type of solidarity—which Durkheim named 'organic.' Contrary to many sociologists of his time, such as Toennies, but in line with SL, Durkheim regarded such complex societies to be more viable, stronger, and more integrated than traditional 'simple and homogeneous' societies; thus, increasing social complexity was indeed a transfer from the SL/NFL repertoire. I would like to propose that in Durkheim's work these evolutionary assumptions played a constitutive rather than a merely rhetorical role, serving as the foundation of his theory on the formation of modern societies, though a low-key presence of evolutionary assumptions compared to those in Spencer, Giddings, or Ward.

Yet, there was a certain tension during this period between Durkheim's interest in marking evolutionary stages, signaled by the use of terms such as 'low,' 'simple,' 'primitive,' 'homogeneous,' etc., and their opposites, to describe societies and/or social structures, side by side with an interest in analytically classifying social types, whereby social evolution was represented—explicitly in *Rules* and implicitly in *Sui.*—as a process of social speciation.

7.3.1.3 Distinguishing the Social from the Biological

Durkheim also used the biological repertoire to delineate the boundaries of sociology in contradistinction to biology. Transfer from one science to another could be effected when assuming, within contemporaneous scientific and/or cultural discourses, that there existed any modus of similarity or affinity between the two fields. In the process of marking boundaries, such constructed similarity could, in another context, be revealed as its opposite. I put forward that Durkheim deployed a 'double mechanism' of metaphorizing—affinity and distance—to delineate the boundaries of the new field. Thus, side by side with using models, metaphors, and analogies from that repertoire, he constructed a hiatus, a gap of dissimilarity, between the two fields, which could not have been constructed unless similarity/affinity had been presupposed. Spencer too used the same double mechanism for that purpose. In his books of the 1890s, Durkheim elaborated on a group of differences that singled out sociology. The common feature was that the biological object was characterized as being absolutely determined, while the object of sociology was presented as partially undetermined and relatively plastic (with a different 'logic' in the sense of types of order and relations, such as between functions and structures). In *Rules* and *Sui.*, the distinction, which was argued for in *Division*, was already assumed and worked into the very arguments. This distinction served to differentiate between that which could and could not be transferred over the boundary thereby constructed between biology and sociology.

7.3.2 Durkheim's Theoretical Position on the Concept of 'Social Continuum'

I suggest that Durkheim's thought during this period can be better understood methodologically as an evolving construction of classificatory social continua, differentiated by timescales and by whether they reflected individual or collective attributes, evincing the deep impact of SL/FNL. One important continuum related to 'habit,' biological or social. 'Habit' was intimately connected with Durkheim's theory of what was a social fact and how societies developed and functioned. It stretched between reflective conscious acts, informed by intentions, motives, reasoning, and acts resulting from nonreflective behavior. In my view, another continuum, and perhaps the most important one, was 'the social continuum.'[19] It stretched between acts and states of individuals and those of collective practices and institutions, as issues

[19] Durkheim, 1885b, c 1888a, b, 1893, 1895a, b, 1897, as well as 1969, 1970, 1974–75, vols.1–3. Main secondary sources: Alexander, 1982, 2005; Alexander and Smith, 2005; Besnard, 2003; Fournier, 2013; Jones-Stedman, 2001; Lukes, 1973a, b; Mucchielli, 1998; Sawyer, 2002; Schmaus, 2004; Strenski, 2006; Turner, 1993. Three close interpretations: Alexander, 1982, 2:217–19, Filloux, 1977, 107–17, Sawyer, 2002.

of the interrelation between individuals and their collectivities became radicalized and sharpened in his work.

In the last third of the nineteenth century, the relationship between the ontological and the phenomenological as to what should be the primary phenomena a science would deal with was seen as a determining factor in the choice of foundational entities and methodology. For sociology, it meant to determine the foundational entities in terms of which the nature of explanatory and descriptive relationships and categories of social life would be formulated. There were some influential, widely diffused alternative approaches, of which Durkheim was well aware, of perceiving the individual as the primary unit, delineating its relations to modes of conceptualizing organisms and collectivity/society. There was also a repertoire of collectivizing modes, primarily among German thinkers, e.g., Schaeffle, Wundt (interpreted either for 'collective' or 'national' as with Fouillé), and the somewhat earlier Moritz Lazarus and Heymann Steinthal, but also in the work of Espinas (Gissis, 2023).

For Durkheim, constituting sociology meant to provide answers: What is society? What are the features of 'social ties' (*liens sociaux*)? 'Scientific' sociological inquiry was posited as focusing on relationships, rather than on the related 'individuals,' with social relationships conceptualized within a historical-yet-evolutionary framework. "Society is not a simple collection of individuals, it is an entity which precedes those who comprise it at present and which will survive them, which acts more on them than they on it, which has its own life, own consciousness, own interests and destiny" (Durkheim, 1978, 93[1885]). His 'index methodology' was a means in the endeavor of making the field open to empirical examination. It assumed that examining external states through observation, measuring, and testing empirically, and thus objectively, would give knowledge and insight on the 'unseen/unobservable internal.' External states were thus 'indices' to internal states, processes, and structures, whose features were not observable, because social life's surface exhibited individuals, while deep structures exhibited collectivities. For example, in *Rules* "Those that are situated more deeply [*plus profondement*] are doubtless more essential, their explanatory value is greater" (1982, 75 [1895]—translation modified). Deploying statistics in *Sui.* (and in its rough forerunner "Suicide et natalité," 1888) relied on this 'index methodology' (Durkheim, 1887c).

The legal system of a society was posited as the index of its solidarity, indicating Durkheim's acute awareness of the difficulties of 'observing a collectivity' and of the need nevertheless to provide for such 'observability.' 'Solidarity' signified the degree of social cohesion and sense of collective identity and reflected a certain combination of attachment, dependency, commitment, and responsibility in the relationships among individuals. Discussion of suicide causalities was intended to achieve precisely this difficult observability by turning the causes of suicide into socially–culturally dependent ones, observable through statistical methods.[20]

[20] See Hacking, 1994, 1990, chaps 19–20; Mucchielli, 1998; Pickering, 2000; Porter, 1995.

Durkheim's notion of 'science' required, epistemologically and methodologically, 'objective,' stable-and-enduring objects of inquiry, a role which, in his view, collectivities could fulfill better than individuals. Yet, without the continuity assumption, sociology could not investigate social life and relate it to how it was phenomenologically revealed—fluid, individualized, and subjective. "We do not admit that there could be [*qu'il y ait*] such a precise point at which the individual comes to an end and the social realm commences… we pass without interval from one order of facts to the other; but this is no reason for not distinguishing them. Otherwise nothing in the world would be distinct, since there are no distinct genera and *evolution is continuous*" (2003, 277n11 [1897]—emphasis added, translation modified).

Two additional facets are relevant: the relations of the social order and its normative foundations/moral order as entangled with individual moral autonomy and the translation of 'scientific (collective) sociology cum individual moral autonomy' into civic education through sociology aiming to stabilize 'modern society' within Third Republic ideals, an aim shared by both Dreyfusards and anti-Dreyfusards. Thus, Tiryakian argues in his 2009 essay on "Durkheim and social change" that French history since the revolution was marked by state instability, which produced in most of late nineteenth-century French sociologists a view that their field should/would offer healing stabilization, social cohesion, class integration and concord, or—in late Durkheim—'civic religion.'

To my mind, one can argue that Durkheim attempted to mark a middle path resolving this dilemma at the cost of much internal tension, through the theoretical construction of continuity between individuals and collectivities. Ontologically, society existed apart from individuals, its components, but not independently of them. Properties of individuals and relations among them were considered to be real but not independent of either a specific individual or of individuals in general or of the collectivity. Facts of sociology were to be found in the relations among individuals but not solely in them. For Durkheim, there could not be, nor be perceived, an individual without a/the collectivity, since that individual had by its very existence already been socialized and collectivized. The collectivity existed through biological existence (following published biological work on marine colonies, e.g., FNL Giard) but also by virtue of the feelings and the consciousnesses of the individuals, by virtue of their joint and sharing participation in collective practices, of the cultural heritage of which they were constituent components. Rawls (1996, 1998, 2001) analyzing Durkheim's late writings stresses the significance of enacted empirical practices: "The primary unit of analysis for sociology has not been the individual but various posited levels of social organization and empirically observable social practices" (1996, 446).

Epistemologically and methodologically sociological explications should go from the collectivity and its products to the individual, the direction of historical/evolutionary process too, thus also serving as an enabling conceptual precondition to discuss 'the normal and the pathological' in social life. Features of social phenomena did not exist in their individual components, and thus social phenomena were to be considered as emergent. A mechanism providing for mutual influence/feedback loops was posited in *Rules* and later used in *Sui.*, reconciling efficient and teleological causation (Durkheim, 1895a, b 95-6n1).

The interpretation of the 'social-continuum assumption,' which I suggest, perhaps leans toward 'the principle of charity,' but I believe it does justice to the depth of Durkheim's insights I argue for. The variety of social life revealed a movement, a dynamic, *from* both states of acting and behaving, and states of thinking and feeling/experiencing that were transient, unstable, and completely unfixed, applicable for short periods of time to individuals or to limited segments of the collectivity, or to evolutionarily early collectivities, *to* currents of opinion, then to patterns that were more solidified, that last for longer periods of time, become fixed and even permanent.[21] The resonance that Durkheim found between the FNL repertoire and 'the social' related to evolutionary dynamics, following 'social facts,' rather than to 'structures' (e.g., Gane, 1994). The patterns always encompassed the entire collectivity and were reflected in the social behavior of individuals and of collectivities, e.g., systems of morals, beliefs, customs. Durkheim called this process 'crystallization,' a term indicating the effort to establish a continuity between the actions, beliefs, and experiential states of individuals and the persistent, more enduring, collective practices, patterns, and institutions, and their material products (which also constituted an integral part of that social continuum).

'Crystallizations' were manifested in systems of action, such as transportation and urbanization, but also in systems of law. The distinction between crystallized social practices that could enable categorizing/classifying, social institutions that could put constraints, impose obligations, even coerce, *and* the processes of consolidation of social phenomena, depended on the position of the phenomena along the social continuum. This would apply to less or more stable 'styles,' e.g., of thinking, of building, and of jurisdiction. Note that ontologically all crystallizations would have an equal status and that the contemporaneous conventionally-accepted demarcation between the mental and the material did not apply. A variety of aspects of the lives of individuals, such as degree of social cohesion or demographic facts, were then looked upon as 'the material foundation of collective life' and experienced as external constraints. This was explicitly articulated in Durkheim's middle period, but it had already been implied in his early writings and was implicit in his late ones. Most contemporaneous philosophical and ethical theories assumed that 'doing' involves a specific individual who was the 'doer' or 'the owner' of the doing. Durkheim's unique position vis-à-vis his contemporaries was his insistence that the 'owner' was a collectivity and that the 'doer' was a social group, supplying

[21] See, e.g., in writings before *Rules*: "*Ainsi se forment les moeurs, germe premier d'ou sont nés successivement le droit et la morale... En d'autres termes c'est comme une cristallisation de la conduite humaine. Or, les phénome'nes économiques tout comme les autres sont susceptibles de se cristalliser*" (1887a, b, 40); 1978, 66 [1888]; see also 1984, 25, 72–3 [1893]; 1982, 32–3, 51–2, 57–8, 59n4, 69–70 [1895] and also: "Thus social life consists of free ranging currents [*courants*] which are in a constant process of change. Yet we do know that social reality possesses the property of crystallizing without changing its nature... collective habits are expressed in definite forms... they constitute a fixed object... these practices are no more than social life consolidated, it is legitimate ...to study that life through these practices" 1982, 82 [1895]—translation modified; for further evidence starting in *Opening lesson* (1978 [1888]) until and including *Suicide* (1897), see also Gissis, 2002, 2018.

thus a substantive substrate then deemed necessary in philosophical discourse for relating attributes and actions. This assumption was also to serve as a methodological lever in discarding explanations given in terms of motives, intentions, meanings, and aims of individual doers and in transposing the concept of causality from the natural sciences to sociology. Clearly, the point of view that became later that of 'methodological individualism' was both methodologically and theoretically impossible for Durkheim.

Accordingly, I propose that Durkheim's '*faits sociaux*'/'social facts' should be read as *modes-de-faire*, meaning *processes-of-doing*, ways of doing, acting, thinking, representing, and feeling, i.e., social acts rather than 'things' done (Jones, Susan Stedman 2001). My reading is based on Durkheim's own usage and its implications. 'Social doings, social acts' were constituted through 'constraint,' particularly of 'externality,' i.e., collective forces. They were 'objective,' since they were epistemologically independent of a particular individual consciousness, and 'general,' as epistemologically independent of the existence of a specific individual, though not of individuals at large. 'Social acts' would thus be presented in a 'thin description,' devoid of the mode and manner in which various agents/doers experienced them.

The combination of both concepts—'social continuum' and 'social acts'—offers ways of understanding internal coherence and meaningful argumentation in the discussion of 'collectivity.' Durkheim's intention to launch a revolution in constituting 'scientific' sociology has ordinarily been interpreted as a revolution of methods. More accurately, it was *a revolution of perspective* in that the collectivity, not individuals, constituted the focus of inquiry, though throughout Durkheim had to face the difficulties of accounting for the role and status of individuals in the normative/moral order (see present-day: Douglas, 1986; Gilbert, 1989; Tollefsen, 2015). Paradoxically FNL assumptions played a role in conceptually enabling this, though they regarded 'individuals' as basic units of investigation.

7.4 The Late Period: The Late 1890s–1917

7.4.1 Durkheim's Endeavor: Bounding and Extending a Novel Scientific Field

In the following, I discuss aspects of the journal Durkheim founded, *L'Année Sociologique* (*AS*), relevant to my two threads: constituting a new 'scientific' field and the use of SL/NFL. In my view, one should consider the journal as a tool for a general mapping of the fast-growing field of sociology and a means to establish sociology as the most appropriate mode of investigating modern society and for educating the future generations of citizens of the Republic. I then show the effort invested in distinguishing sociology from philosophy and psychology while concurrently appropriating their traditional objects of research (Fig. 7.2).

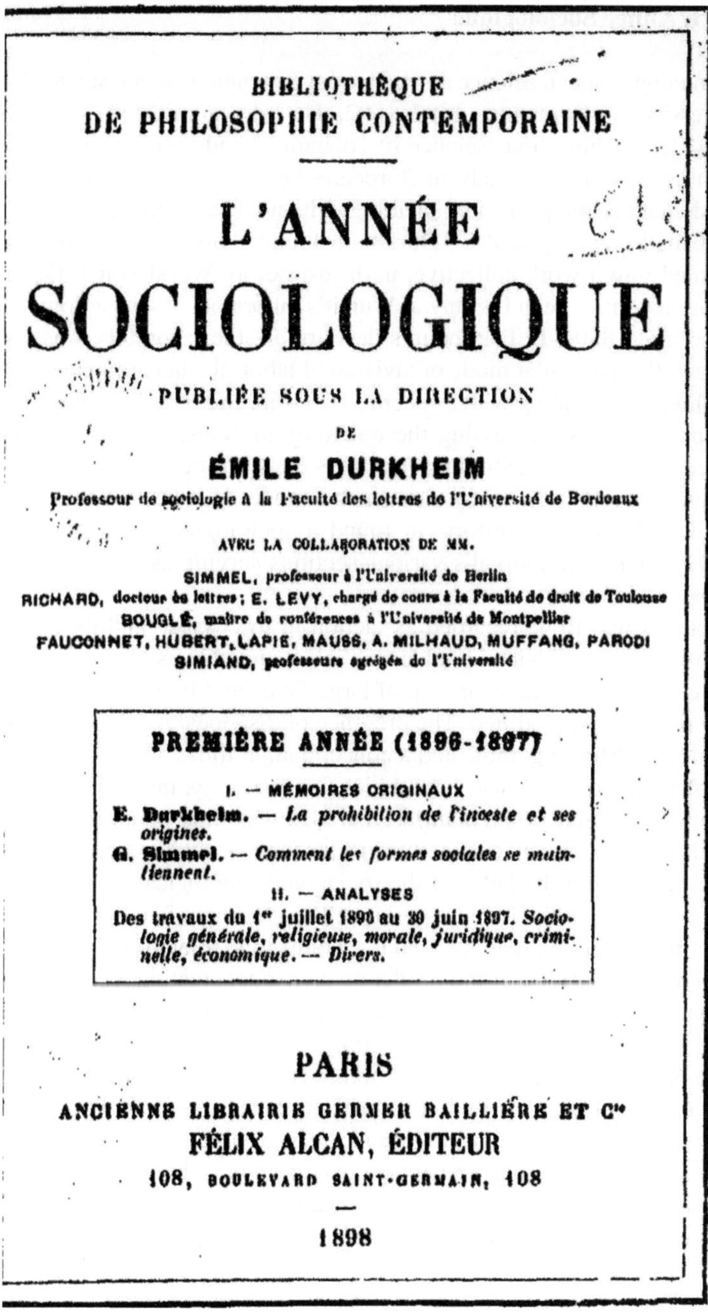

BIBLIOTHÈQUE
DE PHILOSOPHIE CONTEMPORAINE

L'ANNÉE
SOCIOLOGIQUE

PUBLIÉE SOUS LA DIRECTION
DE
ÉMILE DURKHEIM
Professeur de sociologie à la Faculté des lettres de l'Université de Bordeaux

AVEC LA COLLABORATION DE MM.
SIMMEL, professeur à l'Université de Berlin
RICHARD, docteur ès lettres; E. LEVY, chargé de cours à la Faculté de droit de Toulouse
BOUGLÉ, maître de conférences à l'Université de Montpellier
FAUCONNET, HUBERT, LAPIE, MAUSS, A. MILHAUD, MUFFANG, PARODI
SIMIAND, professeurs agrégés de l'Université

PREMIÈRE ANNÉE (1896-1897)

I. — MÉMOIRES ORIGINAUX
E. Durkheim. — *La prohibition de l'inceste et ses origines.*
G. Simmel. — *Comment les formes sociales se maintiennent.*

II. — ANALYSES
Des travaux du 1er juillet 1896 au 30 juin 1897. *Sociologie générale, religieuse, morale, juridique, criminelle, économique. — Divers.*

PARIS
ANCIENNE LIBRAIRIE GERMER BAILLIÈRE ET Cie
FÉLIX ALCAN, ÉDITEUR
108, BOULEVARD SAINT-GERMAIN, 108

1898

Fig. 7.2 Title page of the first issue of *L'Année sociologique* 1898. Source: gallica.bnf.fr

7.4.1.1 L'Année Sociologique

Durkheim taught 'social science and pedagogy,' renamed 'social science' in 1895 at the University of Bordeaux, 1887–1902. He was nominated to the Sorbonne (1902–1917), teaching first 'science of education' and then 'science of education and sociology' (1913). Already in Bordeaux he gradually gathered around him a group dedicated to the project of sociology, the nucleus of that group, around fifty, who published regularly in *AS*.[22] Durkheim founded and maintained *AS* through a group turned into a work collective, until just before World War I. One source of inspiration may have been his stay at Wundt's laboratory of physiological psychology in the mid-1880s. In Durkheim's description, the laboratory was a collective enterprise with a particular mode of division of labor: the lab group was divided into smaller units, each dealing with a specific, well-circumscribed problem, but all converged on two levels: discussing the resulting analyses, establishing generalized laws based on specific elaborations, and thus constituting the field of experimental physiological psychology. In my view, such a model of collective enterprise formed the kernel of Durkheim's efforts to found a sociological work collective of *AS* (1896–1912), with the journal's various sections serving as the laboratory and its subgroups.

The journal was to be regarded as 'action in the world,' and its changing divisions into sections of academic fields, subfields, and objects of research constituted a two-level formation—the mapping of large fields and internal sections were continuously revised and modified. The presence or disappearance of a section, e.g., on 'race' edited by Muffang, indicated a conceptual, at times also political, reorganizing, in a struggle to draw clear boundaries for sociology, indicating thus its proper objects, methods, and *problématique*. Parallel with this bounding-and-extending activity, the Durkheimians engaged in *AS* through reviewing and sectioning in an overall re-mapping of the fields of the social–anthropological–historical for a thorough sociologizing of the human sciences. The Durkheimians were not so naive as to believe that their re-mapping activity would immediately prevail, but they did hope for some reshuffling of old academic divisions, creating a new space for 'scientific sociology.' The seemingly academic polemic had far-reaching implications: sociology was posited as a solid foundation for a new morality, a substitute to religious and otherwise nonscientific 'mythologies,'[23] a vehicle for the continued effort to become part of the school 'moral-education program.' Making sociology an obligatory subject for the *agrégation* and a permissible subject for writing license thesis could have established university courses and teaching posts.

AS was published by Felix Alcan, who published Durkheim's writings, was the publisher of *RP*, later on of *RMM, L'année psychologique, Journal de psychologie*

---

[22] Primary sources: tables of contents of the *AS*; numerous review articles; Durkheim's prefaces to the first and second volumes and sections' introductions; Durkheim and Simiand (1898); Durkheim's correspondence with Mauss, in Besnard and Fournier editors (1998); his correspondence with Bouglé, in Karady (1976, II).

[23] E.g., *AS* 4 (1900):128–38; *RP* 39 (1895d):121–47.

normale et pathologique, Revue historique. As Fabiani (2003) argues, the field of publishing at that period was dominated by specializing houses, whose new-generation heads had university education and close ties with intellectual/academic worlds and also initiated collections or series (see also Hirsch, 2018).

The work on the first volume started in 1896, appearing in 1898, biannually during the first ten years, thereafter every three years, until 1913. Each volume was 500–800 pages, eight to fifteen authors, with one to three original essays (slightly less than a third of the overall volume), which appeared separately from 1906 to 1909. *AS* consisted of book reviews mostly written jointly by two or three authors, all edited by Durkheim: long and short analyses and brief notes, while two thirds of the reviewed materials were non-French, mainly from Germany, Britain, and the USA. Durkheim bore the principal burden of the enterprise—organizationally, technically, and academically, keeping continued and consistent emotional and intellectual ties and support through correspondence and if possible also meetings, with all major authors, forging relations among them, turning them into a highly productive and intense work collective, with a particular sense of joint endeavor, that produced 5000 reviews, with conspicuous cooperations especially among Durkheim, Mauss, Hubert, Simiand, Bouglé, Parodi, and Lapie. Writing in *AS* helped its authors acquire academic positions. Durkheim did not publish any book-form work between *Sui.* and *Forms*. He wrote for *AS*, six original essays, fifteen section introductions, 370 short reviews, and 300 full-length ones. None of the major figures of the collective, who survived World War I, could nor would soon continue where he left.

When it first appeared, *AS* seemed to fill a lacuna (even though sociology articles appeared in *RP* and less in *RMM*), to offer a venue claiming for scientific/methodological commitment. *AS* aimed to create a space dedicated to debate in sociology, an up-to-date potential readership, while preserving a plurality of voices and styles of reviewing. Its success was immediate, overshadowing alternatives, though its aspiration and original essays contracted sharp critique, e.g., by Tarde, Worms, and Henri Berr, and made Durkheim, also involved in the educational reform, vulnerable to the Right's attacks and demonization.

Most of Durkheim's programmatic essays on sociology, also his programmatic introductions to the second editions of his earlier books, *Division* and *Rules*, were written during the early period of *AS*, and he offered in most of them changing lists tracing the genealogy of sociology.[24] His lifelong intellectual and social–moral commitment to 'scientific methodology' and the cluster of criteria it embodied stood out in these programmatic essays and even in his reviews and recommendations (1903–1917) for the Comité des travaux historiques et scientifiques (Baciocchi & Mergy, 2003). Durkheim deployed the international dimension of *AS* as an intellectual resource but posited both sociology and *AS* as components of a (glorious) French tradition. Important indicators of the general 'cultural economy' of that period (1885–1904) were the extraordinary number of diverse new professional journals published and the numerous inter-journal debates (Belanger et al., 1970;

[24] Durkheim, 1900a, b, c 1902; Durkheim and Fauconnet, 1903; Mauss and Fauconnet, 1901.

Elwitt, 1986; Nye, 1975, 1984; Tilly, 1986; Tilly & Shorter, 1974; also, Rioux, 1991). Members of the *AS* work collective contributed to other journals, invariably to undermine the theoretical and methodological foundations of evolutionary Darwinist/organicist sociology—very careful of where they appeared (publishing with Worms indicated leaving the Durkheimian group, e.g., Richard), denoting their academia priorities through their marked preference not to participate in alternative frames of diverse novel social/political sciences colleges. The sociologists' rivalry intensified in the late 1890s (e.g., Worms' publications in the 1890s, Tarde's critique of *Suicide*, 1897), and the process by which Durkheim emerged seemingly victor and founder of 'French sociology' was due, to a large extent, not only to internal scientific developments but to a particular juxtaposition of developments in biology, sociology, anthropology, and alignments of French politics at large.

Mosbah-Natanson (2011) offers an important insight through deploying bibliometrics, showing that around the turn of the century the terms 'sociology,' 'social science,' and their declinations were used by diverse authors for a great variety of purposes, only a few of which were scientific–epistemological–professional. Looking into the use of terms in various teaching frames, learned societies, and publications at large, Mosbah-Natanson suggests a distinction between professional/semiprofessional uses and extra scholarly, also political, ones.

The need to respond to these nonprofessional uses was widely discussed, e.g., Henri Hauser's 1903 long essay, "l'Enseignement des sciences sociales – état actuel de cet enseignement dans les divers pays du monde," summing up previous years and relating to the first congress of sociology in Paris in 1900. A number of essays between 1891 and 1903 on the need for social science/sociology were published by Maurice Hauriou and Fernand Faure (law), Boutmy (economics), Marcel Bernes, Albert Sorel, and by the founder of multiple college frames, Dick May (Jeanne Weil). Frequent nonprofessional use and the above articles around the turn of the century thus pointed to the absence of clear boundaries, core autonomy, and generally defined and accepted professional identity. A gradual convergence toward a bounding of a new field emerged until World War I.

7.4.1.2 Sociology–Philosophy

I discuss two bounding efforts by Durkheim of sociology from adjacent fields—philosophy and psychology, starting with the 'ebb and flow relations' Durkheim maintained with contemporaneous philosophy as reflected in three areas: philosophy in the makeup of the Durkheimian group; Durkheim's position on the role of philosophy within the educational system and the possible role of sociology in it; and the actual relations with philosophical contemporaries—people and journals.

To distance sociology from philosophy, concurrently divesting philosophy of its authoritative position in culture and in higher education, the Durkheimians annexed to sociology some of philosophy's traditional topics—epistemology, ethics, morals, then deity, religion, and education. Even the development of institutions of knowledge was sociologized, as 'the means by which society perpetually recreates the conditions for its very existence.' Yet, numerous members of *AS* were philosophy graduates, earning their living, at least at an early stage of their careers, teaching philosophy in secondary education, some remaining in close intellectual relations with philosophy departments, journals, congresses, etc., initiated by, identified with, or highly populated with philosophers, e.g., Célestin Bouglé.

The very contours of the fields of philosophy/history of philosophy and their teaching, predominant in humanities faculties and secondary education (for males), underwent significant transformations during the last third of the nineteenth and early twentieth centuries, with the waning of undiluted spiritualism and the rise of milder varieties and of other options, e.g., neo-Kantianism. There were a couple of generalized journals publishing philosophy and three principal professional journals: Ribot's *RP* (1876), *L'Année philosophique* (1890) of Rénouvier and Pillon, and *Revue de Métaphysique et de Morale (RMM*—1893), founded and edited by Léon Brunschvicg, Xavier Léon, and Élie Halévy. The Durkheimians endeavored with little success to change that state of affairs, trying to introduce sociology/morality courses into secondary and higher education. Only after World War I sociology was introduced into the syllabus of primary-school teachers' colleges and continued to have influence since then, though hardly institutionalized in universities (Ozouf & Ozouf, 1992).

The labor of 'replacement and substitution': Sociology was posited as a substitute for philosophy, conceived earlier as a science of (lay) morality, founded in *Division* on the collectivity, while preserving/reconciling it with a moral autonomy of the individual, and in later writings with the special position of 'the human person.' From the late 1890s on, it was also posited as a collectivity-based alternative to philosophy's very construction of the categories of knowledge (particularly in *Primitive Classification* (*Classification*), 1903, and then in *The Elementary Forms of Religious Life* (*Forms*), 1912), with an overall emphasis on 'the scientific' (e.g., Chimisso, 2008; Merllie, 1993; Soulié, 2009). I propose that the labor of substitution was carried out through a highly context-sensitive movement between the language of secularization, mainly in earlier writings, and the language of sacralization in later essays, reviews, and other activities. These were intertwined with implicit membership in and close relations with a significant number of philosophers, the Paris philosophical society and its bulletin, the quasi-spiritualist journal *RMM*. The array of historiographical-interpretative positions on this is broad (e.g., the gamut between Schmaus's work and Fabiani 2020). I put forward that the Durkheimians could exercise intellectual distinctions and delineate boundaries but could not afford a clear-cut either/or positioning in the actual contemporaneous cultural and academic fields. They had to perform a complicated simultaneous double movement of distinction and separation, while signaling their belonging, resonating with Durkheim's self-awareness of his radical innovative stance concurrent with his effort to present that very endeavor as having a respected (yet constantly changing) genealogy, and thus also drawing distinctions between his sociology and socialism (esp. 1893, 1895, 1896, and in his later programmatic essays).

7.4.1.3 Sociology–Psychology

Contrary to sociology's position vis-à-vis philosophy, the issue with psychology was different.[25] Both fields were attempting to become academized, and both claimed to be scientific, attracting scholars in their early stages, especially from philosophy. To my mind, one may consider sociology's strategy toward psychology on 'how to explain behavior' as, at first, a double movement of separation and distinction on foundational assumptions and later distinction and topic-appropriation, seeking legitimation for a monopoly on the 'socio-collective' in a competitive field.[26] Durkheim was well acquainted with varieties of contemporaneous psychology, including the controversies on the unconscious hypnosis and suggestion methods, on introspection and Ribot's 'scientific psychology.' The main thrust of his work on collectivity could not allow for totally individualized 'nonsocial, non-collectivized' explanations, unless assumed within a collective framework. He did not undermine the legitimacy of psychology as a field of knowledge, but rather specific contemporary versions, perceived as subverting the legitimacy of sociology's methodological epistemological assumptions on '*faits sociaux*.' The criticized versions served to further bound sociology, pointing to possible cooperation with other versions of experimental/scientific—but not physiological—psychology (e.g., Bouglé, 1898; *AS* 1:116–23).

Klautke (2013) argues that all French social sciences in the last third of the nineteenth century were drawn to, coped with, and elaborated on the insights of *Voelkerpsychlogie*, foremost so with Wundt's view of it as a necessary extension of his individual psychology and as an alternative to Le Bon's views. Both Mauss (1908) and Durkheim (1913), however, emphasized Wundt's psychology then as individual-based. In *Division*, *Rules*, and *Sui.* Durkheim dealt with emotive phenomena, both individual and collective, positing experimental psychology as a future path to be expanded for groups, within scientific sociology. His "Representations" article (1898) argued for the irreducibility of collective representations to individual ones, claiming their independence, since though originally caused organically, they were shaped by social/collective circumstances and had existence (and efficacy).[27]

The notion that Durkheim laid the foundation for a 'purification' of sociology from evolutionary biological thinking is misplaced and should be situated within

[25] Additional secondary sources: Consolim, 2019; Cuin, 2011; Kusch, 1999; Lukes, 2018; Paoletti, 2012; Riley, 2010.

[26] E.g., Tarde, prominent sociologist of 1880–1890s, in his 1895 "Logique sociale" and "La sociologie élémentaire" versus Durkheim's "L'e'tat actuel des études sociologiques en France" (1895c).

[27] E.g., also review on Labriola (1897). Jones S. S (2001) in contradistinction to Schmaus (2000).

this context of the sociology–psychology bounding-endeavor and the wider cultural/multidisciplinary context. For example, Meloni (2016) argues that already in *Division* Durkheim evinced a 'purification' with an assumed lockstep—theoretical and biographical—between himself and Weismann. The principal bounding action in the three 1890s books—*Division*, *Rules*, *Sui.*—was that of sociology from psychology, as was explicitly pronounced in those self-same texts analyzed in Meloni's article and particularly hammered in throughout *Le Suicide*. Psychology was posited then as the principal rival of rising sociology both in academia and in the cultural field, all the more so due to its intertwining with reductive biological degeneration and criminology theories. Since Durkheim's early writings until his late ones, the individual/collectivity distinction, as applied in this bounding, was between the extant variety of psychologies and his collective stance in sociology, as signified also in the evolving role of representations. The notion of social plasticity versus biological determinism served a double purpose in those three middle-period books: first, to legitimate the transfer deployed by emphasizing human/social uniqueness (following Spencer's strategy) and, second, to tie the novel sociology both ideologically and methodologically to the vision of 'liberating/reforming modernity' and its complexified division of labor (in itself another transposition related to FNL uses). It signaled that perhaps Durkheim's way of salvaging the impact of 'collective social environment' was through the attribution of both plasticity and temporariness. Moreover, the evolving role of 'the person' in his oeuvre, and his compunction and caution throughout concerning the epistemological and ontological status of the individual in its role as an autonomous moral agent, should be taken into account (and are totally missing from Meloni's interpretation).

In 1908 Durkheim stated in answer to an inquiry, relating to the notion of 'collective representations': "*Toute la sociologie est une psychologie, mais une psychologie sui generis*" (1975, 1:61). Since the late 1890s, evolving from his new emerging *problématique* and his intellectual negotiations with members of the *AS* work collective,[28] Durkheim posited sociology as sharply different from ordinary psychology, yet as being itself group psychology or *collective psychology*, presenting a complex view of the social–psychological, such as in his essay on religious sociology in *RMM* (1909, 733–58) and his review of Ribot (*AS* 1906, 156–57). Such socio-psychology would offer explanations without resorting to individual motivations, dispositions, or even (individual) mental states at large, and the same would apply to explanations in philosophy and in history.

Mucchielli (1998) argues that Durkheim's collective sociology had a significant impact on the late 1890s psychology, yet to my mind it was a two-way affair, and his sociology had to deal in a differential manner with notions posited in the cultural field by contemporaneous psychologies, e.g., the role of emotions in social life, the unconscious. This was worked out in the double role of emotions entwined in 'the sacred' and the dynamics of sacralization, as well as in the explication of the

[28] Bouglé (later Parodi and Lapie); Durkheim's correspondence with Mauss, in Besnard and Fournier (1998), his correspondence with Bouglé, in Karady (1976, II).

normative in *Forms* (see below). As discussed above, Durkheim was attentive, in differing degrees, to the systemic necessity of positing autonomous individuals to establish morality, especially in modern societies, entwined with the central role of collectivity in the shaping the individual through the social. Strenski (2006) sees a growing emphasis of these features as resulting from the interactions among the Durkheimian group and from Durkheim's involvement in the Dreyfus affair, as was explicit in his coursework reconstructed in "Leçons de sociologie-physique de mouers (1890–1900),"[29] in the essay on intellectuals and individualism (1898), and in the complex role of the individual, the human person, in *Forms*.

7.4.2 The Late Writings: From French neoLamarckism to 'Evolution'

I have argued that transfer from FNL played a significant role in Durkheim's endeavor to constitute a new field, 'scientific sociology,' while his insistence on explaining the social, through the social, fashioned his transfer. These were instrumental in the effort to both constitute and legitimate sociology, specifically in the struggle among competing groups during the 1890s. Yet as the instrumentality of the FNL tools gradually waned, culturally and politically, and within contemporaneous biology, they were abandoned by most members of the *AS* group by 1900. I propose that though Durkheim did not use the NFL conceptual apparatus in his late writings, he conceived his sociological and epistemological theorizing such as on collectivity, religion, and knowledge, within an evolutionary framework that could not be gotten rid of.

In earlier sections, I have discussed Durkheim's particular mode of transfer and transposition, contrasted by him with 'direct/realist transfer' from the realm of organisms to that of human societies. Since Worms and his enterprises were the Durkheimians' most serious rivals from the mid-1890s on, it is not surprising that most of the attacks on Worms calling for separating sociology from direct biology transfer were led by members of the *AS* group,[30] though not directly by Durkheim. There was a noticeable difference between the polemics—style and content—around the mid-1890s and the polemics after the turn of the century. Yet, a similar difference could be found in Durkheim's own stance on the use of biological models. Also some of Worms' collaborators in the use of 'organicism' modified their biosociological position on direct transfer.

Some conspicuous changes in the cultural field around the turn of the century were particularly relevant for contextualizing Durkheim's changing use of FNL.[31]

[29] On this see H.-P. Müller's interview with Hans Joas on Durkheim's reception in Germany (2017, 399–422); Olen and Turner (2015, 954–75).

[30] For a detailed (long) list, see Gissis 2002, 101n55.

[31] Primary sources: Andler, 1932; Barrès and Maurras correspondence 1970 (1888–1923); Barrès, 1937 (1897); H. Bourgin, 1938; Lassere, 1924; Leguay, 1910; Agathon, 1911, 1913; Massis, 1931; Parodi, 1920.

For one, new positions arose in the cultural field around the turn of the century till World War I, doubting the centrality of the individual as an autonomous constructor of his/her moral and social world, the status of scientific knowledge, the meaning of 'objectivity,' and the role of rationality in culture and in sociopolitical life. These attitudes found vent in both sharp anti-German and nationalist sentiments and in attacks on positivism, thus on the 'cultural sieve' of the Republican elites and on the regime's democratic procedures at large. The lines of intellectual/political cleavage already drawn in the debate on educational reforms were turned into a political rift, with vocal (and physical) attacks on 'the new Sorbonne,' on sociology, on Durkheim, but also on the historian Lavisse and literary historian Lanson. Diverse calls against 'the regime of objectivity' were sounded by several groups gathered in the arts, philosophy, psychology, philosophy of science, positing these three figures as the epitome of the modern rationalist spirit of the Republic. Concurrently, it was the popularity height of Bergson's theorizing on the centrality of 'free will' and 'intuition,' as well as the growing popularity of the Catholic movement *Sillon*, calling for 'anti-materialism.' Both Stock (Morton) (1971) and Fournier (2013) suggest to regard it as (also) an intergenerational conflict.

The new nationalists represented the reform as a victory of 'German import' over 'Latin Spirit,' attacking 'The University,' e.g., Charles Péguy (his prose work), Georges Sorel (in 1895, a supporter of *Division*), and Pierre Lasserre. Henri Massis and Alfred De Tarde (under the pseudonym *Agathon*), among the most vociferous, were invited to join the newly established organ of the French Academy: League pour la culture française, which spans ties with the extremely active L'Action française (nationalist right-wing organization). For its members and supporters, 'reason' meant a superficial phenomenon, and 'man' essentially an emotional being, allotting a central role to the unconscious. Thus, 'sociology,' as a discipline of the new Republican university system and of secularized education reform, was depicted as an ally of the Dreyfusards, the embodiment of a democratization process threatening the imagined 'True France.' L'Action française and the nationalist groups at large annexed the figures of Renan and Taine, posthumously, to the right-wing pantheon of 'traditional French culture,' as these attitudes were diffused in the cultural discourse (e.g., Guillaume Apollinaire's 1916 "A Poet Murdered," in which a chemist incites a crowd to kill poets). Yet there were also voices calling for a national post Dreyfus-affair reconciliation, with sharpened awareness of diverse groups of 'ideals held in common' and thus a gradual emergence of the concept of '*religion laïque*': belief in science, progressist reform, and modernization of religion, alongside Republican universalism. To my mind, one may view Durkheim's late work in this context.

Secondly, evolutionized biology in France until the late 1880s contained a spectrum of views lacking a sharp demarcation, while polarized varieties emerged in the last decade of the century (persisting through at least the first third of the twentieth century).[32] Ideologically claimed oppositions among three rival groups in sociology resonated with the political divisions of the Dreyfus affair: those employing Darwinist–organicist models, those using a biological–cultural category of 'race,' and those using mediated biological transfer and solidarist/FNL idiom. An important factor of change was the publication and translation of Weismann's work as well as his somewhat later publicized polemic with Spencer and even more so the gradual realization that Weismann's theory on the total separation between somatic cells and 'material-of-inheritance' (germ) cells bore implications in biology, yet also far beyond it.[33] A polarization occurred within French communities of biologists as the predicament of FNL became acute, for they upheld a strictly empirical view of what could be regarded as scientific work. Biologists with deep ideological commitment to the Republican–solidarist ideology and/or involved in political, parliamentary, or governmental activities found it impossible to accommodate the envisioned social consequences of the new theory.

The thorniest issues were accepting absolute separation between heredity and environment since it meant giving up the notion of the direct effect of the environment through inheritance of acquired modifications/characteristics, negating the cluster of notions related to environmental plasticity and its significance for reformist social thinking. French neo-Lamarckists devised intricate mechanisms to enable them to continue their experimental scientific work without following Weismann's hypotheses (e.g., Loison, 2011a, b). The scientific status of FNL biologists therefore suffered gradual but constant erosion, both internationally and at home,[34] a case of social context contributing to the state of *problématique* and methodology. Meanwhile, the somewhat afterthought role of the figure of Lamarck in the earlier stages of FNL was transformed from the mid-1880s on but especially after the mid-1890s, as he was placed in a gallery consisting of Buffon and Etienne Geoffroy Saint-Hilaire, thus deemed significant for a French genealogy of transformism.[35] Neo-Darwinism crystallized as an exclusively selectionist, anti-FNL view of evolution, while the hallmark of FNL/NL became for many the inheritance of acquired characters.

During most of the 1890s, 'biological acculturation' of one mode or another was common to all parties and to almost all vying sociologies (with the conspicuous exception of Tarde). Before the turn of the century, the Durkheimian work collective had already distinguished itself through *AS*, publications, and other quasi-disciplinary activities, as ongoing controversies compelled them, willy-nilly, to delineate the boundaries of their purported discipline more sharply, to distinguish themselves with epistemic assumptions and rhetoric. This process converged with

[32] Additional primary sources: Delage, 1889, 1895a, b; Giard, 1889, 1890, 1904; de Lanessan, 1901, 1914; Le Dantec, 1899.

[33] Weismann, 1883, 1885, 1888, in Weismann, 1891/92; Churchill, 2015; C. Weissman, 2011; Winther, 2001.

[34] E.g., an unsigned harshly critical review (1895 *RIS* 301) of Delage's *La structure de protoplasma et les theories sur l'heredite.* and of French biological education.

[35] E.g., the relevant part in Landrieu's 1909 bibliography.

the growing cleavage within the biological field that contributed significantly to a general decrease in the transfer and transposition of the biological repertoire of tools from outside the social field and thus also in the vocabulary of the legitimating rhetoric.

Early on Durkheim also used such rhetoric to acquire general-cultural legitimation for the newly founded discipline. Its projected image as 'scientific' was one of the main sources of power of the new discipline, playing a major role in the strategy of constitution. The adoption of sophisticated transfer from FNL at one stage played an important role in the successful endeavor to create a new space for 'scientific sociology.' At another stage, its bounding as an autonomous discipline was carried out at a critical moment for its legitimation in the political and cultural context of the mid-1890s. It was instrumental in making the Durkheimians seem most suitable candidates for the role of 'French sociology,' presenting it as scientific by the very act of bounding (with FNL tools) at a moment when the scientific status of the rival organicist–Darwinist sociology was tottering and when the line dividing FNL from Republican politics within French biology was somewhat blurred. Later the outcome of the juxtaposed factors briefly discussed above was that the FNL repertoire gradually disappeared from the forefront of Durkheim's sociological works and programmatic declarations. Thus, at the height of the battle between neo-Lamarckists and neo-Darwinists within French biology, Durkheim and the *AS* work collective were moving somewhere else. This did not mean, however, that social thought and sociological activities at large in France had therefore become 'biology-free.'

Placing Durkheim within the above context can throw light on the paradoxical outcome of the bounding-endeavor in his late writings. I suggest to consider thus his effort to understand modern societies, searching for a generalized foundation of all their social and cognitive institutions, a search for both logic and historical patterns of their evolution.[36] During this period, from the very late 1890s on, there was a gradual process of transformation in the focus of interest and the features of investigated materials. While earlier on Durkheim proffered analyses of the diminishing role of religion, and thus of secularization as an inevitable normal stage in the evolution of (modern) societies, in the later period, religion, through sacralization of soci ety, gradually acquired a central role in modern society's endeavor to create a new form of social solidarity, concurrently with the centrality of the autonomous individual. In a review essay on Labriola in 1897, Durkheim claimed to have found that sought-after 'generalized foundation' in 'religion': "Religion is the most primitive of all social phenomena. It is from it that have emerged through successive transformations, all the other manifestations of collective activity – law, morality, art, science, political forms, etc. In principle everything is religious" (1978, 123–30 [1897]). In an essay on the prohibition of incest in the first *AS* volume (1898 [1896–1897], 1:70), Durkheim argued for a universalized religious institution of all primitive religions, then gradually moving to study (also together with Mauss) eth-

[36] Further secondary sources: Achimastos, 2019; Cuin, 2011; Joas, 2017; Merllié, 2017; Nungesser, 2020; Rawls, 1996, 1998, 2001; Riley, 2010; Strenski, 2006.

nological–anthropological materials, for which he had no use before, though he did find methodological justification for it already in the early and middle periods.

I shall not offer a detailed summary of the contents of his late writings, nor analysis of the historiography. Rather I examine *the use of 'evolution' and the assumption of an evolutionary framework* in this period. I ask which evolution? to what extent was it inseparable from Durkheim's evolving position on religion, its symbolic components and their action in social life? the role and status of religious emotions (on whose reality in the believers he insisted, e.g., 1914, 306–07); the nature of the sacred object at the center of (any) religion; and most significantly, the relations between individuals and collectivities— thus in collective institutions of religion and particularly in collective rituals.

Such endeavor encompassed social origins, the working of epistemology, and the emergence of social solidarity. Grossly generalizing one could claim that religious phenomena were characterized in the late work through three principal features: the dual and conflicting nature of the believer's emotions toward the object at the center of her/his belief (awe, fearful respect, and duty, on the one hand, and love, desire, and devotion, on the other hand); the kind of symbols used (collective representations) as a source of social, moral, and legal authority; and participation of individuals in varied collective institutions—most prominently in collective rituals (see Jones, 1981, 2005). Social solidarity was created and preserved through the practices of participation in the collective system of representations, the practices of collective gatherings that created and reproduced them, and the practices of collective rituals that preserved them as living collective memories (guaranteeing epistemological commonality). The collectivity was posited as the source of sacredness and the sacred/sacralized objects of religion were but the forms in which the sacralized collectivity was represented to its members. From the late 1890s on, religion was defined as a system of *beliefs and practices*, just as 'social facts' were defined earlier as manners of acting, thinking, and feeling external to the individual, consisting of *representations and actions* (Rules, 1982, 52 [1895]). Thus, religion was not only the source of social life but also a permanent phenomenon in it.

Strenski (2006, 16–17) writes in his discussion of religion–society relations: "First the familiar reductionist one that 'religion=society' meaning that religion has a social nature. But this is also to open the door to another claim—perhaps reductionist in the opposite direction—namely 'society=religion' – or that society has a religious nature." He suggests to interpret this as the inherence of the value-laden aspects, the common values, in sociability. Strenski (1997) points to the influence of French religious–liberal modernists, especially the Jewish scholars among them, e.g., Silvain Levy, who assumed social evolution. I put forward that in a very strong sense both 'secularization,' so significant in the middle period and in his involvement with the educational reform, and 'sacralization,' so central in the late period, touched upon the same unresolved dilemma of the relations between individuals and collectivity within Durkheim's sociology, and in that *they were not contrasting but complementary*.

Durkheim's analysis also turned the table on dominant contemporaneous accounts of crowd behavior and individual behavior within a crowd, i.e., Le Bon and Tarde,[37] by claiming that collective gatherings had always been the creative source of cognitive and moral ideas and the basis of social (and national) solidarity, where the individual outflowed his/her boundaries. The feared features of crowds—unruly enthusiasm, seemingly uniform irrational behavior, etc.—were redescribed/redefined as 'collective effervescence,' a lever to create collective rational content and revitalize collective cohesion, morality (normative constraint) based. The model of the French Revolution, as then presented by Mathiez, its *Fêtes* and 'crowd enthusiasms' as means of constituting novel morality, had been of significance in this innovative move.[38] This morality was embedded in a framework of religion, to a lesser degree during the years 1898–1904, but much more pronouncedly later on, most explicitly in *Forms*.[39]

I suggest to consider Durkheim's position during the late period and conspicuously so in *Forms*, as one arguing that advanced social forms of religious life developed/evolved out of evolutionarily elementary ones. This position, adhered to by Mauss as well in his participation in public debates then, emphasized developmental continuity, denying the possibility of any 'ever-presence' (e.g., conceptions of history of philosophy, of morality, etc.), as Durkheim refrained in his work from explaining the complex through the 'simple.' In his 1903 book, *La morale et la science des mouers*, Levy-Bruhl also adopted an evolutionary conception, but his three later works that argued for distinct and separate modes of 'mentality' were organized neither evolutionarily nor hierarchized as simple and complex but rather as pre-logical and logical. Durkheim's evolutionary stance was explicitly expressed in his critical review in *AS* (1913, 677–81) of Levy-Bruhl's book *Les fonctions mentales dans les sociétés inférieures* (published in 1910 in the *AS* series at Alcan, edited by Durkheim). There echoing his position in *Classification*, written with Mauss 1903, Durkheim emphasized differing rather than contrasting or even contradictory modes in 'modern culture' and 'elementary culture' as argued by Levy-Bruhl. For Durkheim these were *moments in the same continuous evolution*, which had certain components in common. He emphasized the thorough sociality of belief and of knowledge, 'of mind,' bringing in practices, thus challenging the claims of both contemporaneous philosophy and psychology, while appropriating their principal traditional objects. Note that deploying the notion of continuum was a feature of FNL evolutionism that he had already used earlier. What does come out is that

[37] E.g., de Tarde, 1890; Le Bon, 1895; Barrows, 1981; Moscovici, 1981; Metraux, 1982; Nye, 1975.

[38] See Tiryakian, 1988; indirectly Ben-Amos, 2000; also, Jones R.A. 1981, 2005, Jones S. S. 2001, 2010. Extract from P. Fontana's notes of Durkheim's 1906/07 course on the origins of religious life: "Mais *les services moraux et sociaux* que rend le culte sont indispensable et permanents tant qu'il aura des hommes, c'est-a-dire des sociétés. C'est ainsi que la Révolution française sentit *le besoin de substituer un culte nouveau* au culte ancien qu'elle voulait détruire" (Karady 1976, 2:65–122 [1907]—emphasis added).

[39] As seen in fragmentary sources, e.g., fragment of a letter to Gaston Richard, 1899; notes of a course on religion: Its origins, 1907; Lettres au Directeur de la revue neo-scolastique, 1907; Contribution to the discussion 'Science and Religion,' 1908; notes by Georges Davy of Durkheim's course 1908/1909 (1968); Durkheim L'avenir de la religion, dans: Le sentiment religieux a` l'heure actuelle, 1914; Correspondence with Mauss of those years (1998).

Durkheim insisted on an evolutionary frame. This is significant when one recalls that Mauss, Hubert, and Hertz, to whose work Durkheim resorted throughout the late period, especially to Mauss, with whom he had been closely cooperating for years, were gradually forsaking that frame in their anthropological work, e.g., already in Mauss' opening lecture in 1901.

Let me outline the general scheme of my argument, already suggested above: wavering and dithering in the very early period, yet already in the middle period Durkheim proffered a sociological discussion of 'the social' modeled on the FNL repertoire of models, metaphors, analogies, and conceptual tools. In the late 1890s and at the turn of the century, this repertoire disappeared from the foreground of his sociological work. Yet his sociology of religion was articulated within an evolutionary framework, and crucial components of it would not cohere without the assumptions of the social–environmental impact, of the accumulated effect of the collectivity on its members as passed on socially. This was the case whether considered as social or as cultural heredity, nor without the more sophisticated yet profoundly indispensable assumption of the 'social continuum.' Thus, during the late period, Durkheim indeed parted with foreground FNL, yet 'the Weismann impact' on his work was paradoxical, as he did not move to neo-Darwinian transfer, nor did he accept a severance of trans-generational social/moral/epistemological accumulation, as testified also by a posthumous last fragment prepared and published by Mauss—a search for 'the elementary forms of the moral' (1975, 2:313–31 [1917]).

One could even suggest a certain semblance between *Division* and, in as far as transformation of a social whole was discussed, the significance of close/dense social interaction emphasized, which in *Division* occurred in space only and in *Forms* took place in both space and time. Both books posited seemingly blurred or mitigated hierarchical differences. Rather, Durkheim posited a transposed conception of evolution, whose essential features, its dynamics, and its basic components were turned profoundly social/cultural. The impact of environment, the trans-generational passing-on of collective experiencing, a methodology of continua, all became social–cultural components of an overall backstage evolutionary framework. Mucchielli (1994) and even Merllié (2017) would concur with Strenski (2002, 102), stating in a negative minimal manner that one cannot evade "the residual evolutionism of Durkheim's views of social development dating from *The division of labor* to *The elementary forms of religious life*."

My argument resonates to some extent with Raymer's on American social scientists after Weismann, especially Baldwin and Ward: "In creating a new theory of social progress [I would add, in which the social environment had a central role] based on a concept called 'social heredity'" (2018, 62).

Consequently, Durkheim proceeded to examine the late-period questions through a befitting evolutionary method/order, from the simple to the more complex, starting from the 'primitive,' 'elementary,' and 'close to origins' (1912, 135). One would perhaps also note the implied notion of Spencer and Gillen (1899), i.e., uneven evolution of diverse facets, even though evolution by itself was not explicit in their descriptions of the 'Australian materials' used by Durkheim.

Such an order could presumably follow the constitution of socio-mental (secondarily emotional) classifying structures and the evolving separations that occurred (as in *Classification*) and allowed for more plasticity and complexity. This evolutionary framework became increasingly indispensable in *Forms* as the argument turned on the emergence via religion/sociality of the very categories that enabled classification of any kind (and from an evolutionary perspective of science as well).

7.5 Concluding Comment

I have asked how within the particular context of the French Third Republic a set of disciplinary theoretical assumptions and a rhetorical repertoire, namely, that of the nineteenth-century SL/FNL variety of evolutionized biology, interacted with the emerging novel *problématique* of understanding modern society and which facets of this variety had a more significant impact. I have tried to trace and follow two threads within Durkheim's oeuvre and their intertwining: constituting the novel field of 'scientific sociology' and deploying the FLN repertoire. Two topics stood out in their interdependence: collectivizing and evolutionizing, and I proposed the concept of 'social continuum' to explicate it. I suggested that the ensemble of these analyses offers another perspective from which Durkheim's endeavor can be described, analyzed, and contextualized, a perspective that points to another kind of continuity and cohesion in that endeavor. It is worth mentioning that significant facets of this perspective in my particular elaboration of them are either novel or controversial in differing degrees.

A particular connection between collectivity and evolution has been pointed to also in the chapter on Spencer's work, as well as in the one on Ribot's, and to my mind it can also illuminate a significant facet in Freud's endeavor to constitute a novel scientific field, psychoanalysis. The load of questions asked throughout the previous chapters will accompany this further journey.

Materials from the following articles have been used by permission of:

The MIT Press

2002 – Late Nineteenth century Lamarckism and French Sociology. *Perspectives on Science, 10*(1), 69–122.

Springer Journals: License number 5633100970818

2023 – The Neo Lamarckian Tools Deployed by the Young Durkheim: 1882–1892. *Journal of the History of Biology* (April) 1–37. https://doi.org/10.1007/s10739-023-09708-w

References

Achimastos, M. (2019). D'une Année à l'autre: la sociologie religieuse durkheimienne en marche. *L'Année Sociologique, 69*(1), 77–102.

Agathon (Henri Massis and Alfred de Tarde). (1911). *L'Esprit de la nouvelle Sorbonne: la crise de la culture classique, crise du Français*. Mercure de France.

Agathon. ([1913] 1995). *Les Jeunes Jens d'aujourd'hui*. Plon Nourrit.

Alexander, J. C. (1982). *The antinomies of classical thought: Marx and Durkheim*, vol. 2: *Theoretical logic in sociology*. University of California Press.

Alexander, J. C. (2005). The inner development of Durkheim's sociological theory: From early writings to maturity. In J. C. Alexander & P. Smith (Eds.), *The Cambridge companion to Durkheim* (pp. 136–159). Cambridge University Press.

Andler, C. (1932). *Vie de Lucien Herr*.

Baciocchi, S., & Mergy, J. (2003). *Emile Durkheim-L'Évaluation en comité* (S. Baciocchi & J. Mergy, Eds.). Berghahn Books.

Becquemont, D. (2010). Edmond Perrier exobiologiste. *Bulletin d'Histoire et d'Épistémologie des Sciences de la Vie, 17*(1), 91–104.

Barberis, D. S. (2003). In search of an object: Organicist sociology and the reality of society in fin-de-Siècle France. *History of the Human Sciences, 16*(3), 51–72.

Barrès, M. ([1897] 1937). *Les déracinés*. Plon.

Barrès, M., & Maurras, C. ([1888–1923], 1970). *La République ou le roi: correspondance inédite* (H. Maurras & N. Maurras, Ed.). Plon.

Barrows, S. (1981). *Distorting mirrors: Visions of the crowd in late nineteenth-century France*. Yale University Press.

Beck, N. (2014). *La Gauche évolutionniste: Spencer et ses lecteurs en France et en Italie*. Presses universitaires Franche Comté.

Ben-Amos, A. (2000). *Funerals, politics, and memory in modern France, 1789–1996*. Oxford University Press.

Belanger, J., Godechot, J., Guiral, P., & Terrou, F. (1970). *Histoire générale de la presse française*. Presses Univérsitaires de France.

Béra, M. (2017). La représentation disciplinaire du 'social' dans les références et les lectures du jeune Durkheim (1879-1894). *L'Année sociologique, 67*(2), 481–481.

Bernadini, J. M. (1997). *Le Darwinisme social en France (1859–1918)*. Editions du Centre national de la recherche scientifique.

Besnard, P. (2003). *Études Durkheimiennes*. Droz.

Blanckaert, C. (1979). Edmond Perrier et l'étiologie du 'polyzoizme'organique. *Revue de synthese, 100*, 353–376.

Blanckaert, C. (2004). *La Nature de la Société: organicisme et sciences sociales Au XIXe Siècle* (pp. 1–160). L'Harmattan.

Bouglé, C. (1898 [1896]). "Sociologie Générale" Introduction to the section. *L'Année Sociologique, 1*, 116–123.

Bourgin, H. (1938). *De Jaurès à Léon Blum, l'École Normale et la politique*. A. Fayard.

Bowler, P. J. (1984). *Evolution: The history of an idea*. University of California Press.

Bowler, P. J. (1988). *The non-Darwinian revolution: Reinterpreting a historical myth*. Johns Hopkins University Press.

Canguilhem, G. (1965). *La conaissance de la vie*. J. Vrin.

Canguilhem, G. (1994). *A vital rationalist: Selected writings* (F. Delaporte, Ed.). Zone books.

Chimisso, C. (2008). *Writing the history of the mind: Philosophy and science in France, 1900 to 1960s*. Ashgate.

Churchill, F. B. (2015). *August Weismann: Development, heredity, and evolution*. Harvard University Press.

Cladis, M. S. (2020). Durkheim's signature project: The science of morality as rational moral art. In H. Joas & A. Pettenkofer (Eds.), *Oxford handbook of Emile Durkheim*. Oxford University Press.

Clark, T. N. (1973). *Prophets and patrons: The French university and the emergence of the social sciences*. Harvard University Press.

Clark, L. (1984). *Social Darwinism in France*. University of Alabama Press.

Conry, Y. (1974). *L'Introduction au Darwinisme en France*. Vrin.

Conry, Y. (1993). Comment a-t-on pu être néo-Lamarckien en France (1843-1930)? *Nuncius, 8*, 487–520.

Consolim, M. (2019). La réception de la psychologie dans l'Année sociologique (1898-1927). *L'Année sociologique, 69*(1), 103–142.

Corsi, P. (1978). The importance of French Transformist ideas for the second volume of Lyell's principles of geology. *The British Journal for the History of Science, 11*(3), 221–244.

Cuin, C. H. (2011). *Durkheim, modernité d'un classique*. Herman.

D'Hombres, E. (2010). The 'division of physiological labor': The birth, life and death of a concept. *Journal of History of Biology, 45*(1), 3–31.

D'Hombres, E. (2012). De la différenciation biologique a` la différenciation sociale (xix-xx si'ecles): Quelques jalons historiques. *Nouvelles Perspectives en Sciences Sociales, 8*(1), 191–220.

D'Hombres, E. (2016). The Darwinian muddle on the division of labor: An attempt at clarification. *History and Philosophy of the Life Sciences, 38*(1), 1–22.

D'Hombres, E., & Mehdaoui, S. (2012). On what condition is the equation organism–society valid?' Cell theory and organicist sociology in the works of Alfred Espinas (1870s–80s). *History of the Human Sciences, 25*(1), 32–51.

Davy, G. (1919). Émile Durkheim. *Revue de Métaphysique et de Morale, 26*(2), 181–198.

Davy, G. (1960). Emile Durkheim. *Revue Française de Sociologie, 1*(1), 3–24.

De Bont, R. (2010). Organisms in their milieu: Alfred Giard, his pupils, and early ethology, 1870–1930. *Isis, 101*(1), 1–29.

De Bont, R. (2015). *Stations in the field: A history of place-based animal research, 1870–1930*. Chicago University Press.

de Lanessan, J. (1901). La Morale Scienti.que. *Revue Scientifique, 16*.

de Lanessan, J. (1914). *La crise de la République*. Felix Alcan.

De Lapouge, V. (1887). Les sélections sociales. *Revue d'Anthropologie, 2*(1), 519.

de Tarde, G. (1890). *Les lois de L'imitation: étude sociologique*. Felix Alcan.

de Tarde, G. (1895). Criminalité et santé sociale. *Revue Philosophique, 39*, 148–162.

Delage, Y. (1889). Réponse à M. Giard. *Revue Scientiaque*, 43–44.

Delage, Y. (1895a). Preface à L'Année Biologique. *L'Année biologique, 1*, vii–xviii.

Delage, Y. (1895b). *La structure du protoplasma et les théories sur l'hérédité et les grands problèmes de la biologie générale*. C. Reinwald.

Delage, Y., & Goldsmith, M. (1909 [1912]). *The theories of evolution* (A. Tridon, Trans.). B.W. Huebsch.

Douglas, M. (1986). *How institutions think*. Syracuse University Press.

Durkheim, E. (1882). L'agrégation de philosophie. *Études durkheimiennes*, fall 1993.

Durkheim, E. (1885a). Fouillée, A.: La propriété sociale et la democratie. *Revue Philosophique, 19*, 446–453.

Durkheim, E. (1885b). Gumplowicz, Ludwig: Grundrisse der Soziologie. *Revue Philosophique, 20*, 627–634.

Durkheim, E. (1885c). Schaeffle, A. Bau und Leben des sozialen Körpers: erster band. *Revue Philosophique, 19*, 840–101.

Durkheim, E. (1886a). De Greef, Guillaume: Introduction a' la sociologie. *Revue Philosophique, 22*, 658–663.

Durkheim, E. (1886b). Les Études de science sociale. *Revue Philosophique, 22*, 61–80.

Durkheim, E. (1887a). La philosophie dans les universités allemandes. *Revue internationale de l'enseignement, 23*(313–38), 423–440.

Durkheim, E. (1887b). La science positive de la morale en Allemagne. *Revue Philosophique, 24*, 33–58, 113–42, 275–84.

Durkheim, E. (1887c). Suicide et natalité: étude de statistique morale. *Revue Philosophique, 26*, 446–463.
Durkheim, E. (1888a). Cours de science sociale: leçon d'ouverture. *Revue internationale de l'enseignement, 15*, 23–48.
Durkheim, E. (1888b). Introduction de la sociologie de la famille. *Annales de la faculté des Lettres de Bordeaux, 10*, 257–281.
Durkheim, E. ([1883–1917] 1975). *Textes* (Victor Karady, Ed.), 3 vols. Les éditions du minuit.
Durkheim, E. ([1885–1912] 1978). *Émile Durkheim on institutional analysis*. (Mark Traugott, Trans.). University of Chicago Press.
Durkheim, E. ([1895] 1982). *The rules of sociological method (and other writings on sociology)* (W. D. Halls, Trans.). Free Press.
Durkheim, E. ([1893] 1984). *The division of labor in society* (W. D. Halls, Trans.). Free Press.
Durkheim, E. (1895a). L'enseignement philosophique et l'agrégation de philosophie. *Revue philosophique* xxxix, 121–147. V. Karady, Textes III. Editions de Minuit, 1975.
Durkheim, E. (1895b). Crime et santé sociale. *Revue Philosophique* xxix, 518–523.
Durkheim, E. ([1895c] 1975). L'état actuel des études sociologiques en France. V. Karady, *Textes* I. Editions de Minuit. Translation of *La Riforma Sociale, 2*(1895), 691–707.
Durkheim, E. (1895d). L'enseignement philosophique et l'agrégation de philosophie. *RP, 39*, 121–147.
Durkheim, E. ([1892] 1921). "La famille conjugale." Revue philosophique 90:1–14.
Durkheim, E. ([1893] 1930a). *De la division du travail social (étude sur l'organisation des sociétés superieures)*. Presses Universitaires de France.
Durkheim, E. ([1897] 1930b). *Le suicide: étude de sociologie*. Presses universitaires de France.
Durkheim, E. (1897). Labriola, Antonio: essais sur la conception matérialiste de l'histoire. *RP, 44*, 645–651.
Durkheim, E. ([1900a] 1953). La sociologie et son domaine scientiaque. *Rivista Italiana de Sociologia*. Translated into French. Cuvillier.
Durkheim, E. (1900b). La sociologie en France au xix siècle. *Revue Bleue, xii*, 609–613, 647–652.
Durkheim, E. (1900c). Rôle des universités dans l'education sociale du pays. In *Congrès international de l'education sociale* (pp. 128–138).
Durkheim, E. (1902). Espinas, A. 'Être ou ne pas être ou du postulat de la sociologie. *L'Année Sociologique, 5*, 127–129.
Durkheim, E. ([1903–1917] 2003). *L'évaluation en comité* (Stephane Baciocchi & Jennifer Mergy, Ed.). Berghahn Books.
Durkheim, E. ([1912] 1995). *The elementary forms of religious life* (K. E. Fields, Trans.). Free Press.
Durkheim, E. (1913). Wundt. W. Elemente der Voelkerpychologie (Leipzig,1912). *L'Année Sociologique, 12*, 50–61.
Durkheim, E. ([1914] 2011). L'avenir de la religion. *Les Cahiers de Psychologie Politique*, (18), E01–E01.
Durkheim, E. (1969). *Journal sociologique*. Edited by J. Duvignaud. Presses Universitaires de France.
Durkheim, E. ([1888–1914] 1970). *La science sociale et l'action*, introduction and presentation by J.C. Filloux. Presses Universitaires de France.
Durkheim, E. (1974–1975). Textes, edited by Victor Karady, 3 volumes. Les éditions de minuit.
Durkheim, E. (1975 [1895]). L'état actuel des études sociologiques en France. V. Karady *T, extes I*. Editions de Minuit. Translation of La Riforma Sociale 2, 691–707.
Durkheim, E. (1998). *Lettres a Marcel Mauss* (Philippe Besnard et Marcel Fournier, Ed.). Presses Universitaires de France.
Durkheim, E., & Fauconnet, P. (1903). Sociologie et sciences sociales. *Revue Philosophique, 45*, 465–497.
Durkheim, E., & Mauss, M. ([1903] 1967). *Primitive classification* (R. Needham, Trans.). University of Chicago Press.

Elwitt, S. (1986). *The Third Republic defended – Bourgeois reform in France 1880–1914*. Louisiana State University Press.
Fabiani, J.-L. (2003). Clore enfin l'e`re des généralitees. In S. Baciocchi & J. Mergy (Eds.), *L'Évaluation en comité* (pp. 151–189). Berghahn Books.
Fabiani, J.-L. (2020). Durkheim and the philosophy of his time. In H. Joas & A. Pettenko (Eds.), *The oxford handbook of emile Durkheim*. (pp.1–14). https://doi.org/10.1093/oxfordhb/9780190679354.001.0001
Feuerhahn, W. (2011). Les 'sociétés animales': un défi à l'ordre savant. *Romantisme, 154*, 35–51.
Filloux, J.-C. (1977). *Durkheim et le socialisme*. Droz.
Fischer, J. L. (1979). Yves Delages (1854-1920). *Revue de synthèse, 95–96*, 443–461.
Fournier, M. (2013). *Émile Durkheim: A biography* (D. Macey, Trans.). Polity Press.
Gane, M. (1988). *On Durkheim's rules of sociological method*. Routledge.
Gane, M. (1992). Durkheim: The sacred language. In M. Gane (Ed.), *The radical sociology of Durkheim and Mauss* (pp. 61–84). Routledge.
Gane, M. (1994). A fresh look at Durkheim's sociological method. In W. S. F. Pickering & H. Martins (Eds.), *Debating Durkheim* (pp. 66–85). Routledge.
Giard, A. M. (1889). Les facteurs d'évolution. Leçon d'ouverture – 2ieme année. *Revue scientifique, 44*, 641–648.
Giard, A. M. (1890). Le Principe de Lamarck et l'heredité des modifications somatiques. Leçon d'ouverture. *Revue scientifique*, 45–46.
Giard, A. (1904). *Controverses transformistes*. C. Naud.
Giddens, A. (1978). *Émile Durkheim*. Harvester.
Gilbert, M. (1989). *On social facts*. Routledge.
Geison, G. L. (1969). The protoplasmic theory of life and the vitalist-mechanist debate. *Isis, 60*(3), 273–292.
Gissis, S. B. (2002). Late nineteenth century Lamarckism and French sociology. *Perspectives on Science, 10*(1), 69–122.
Gissis, S. B. (2011). Lamarckism and the constitution of sociology. In S. B. Gissis & E. Jablonka (Eds.), *Transformations of Lamarckism, from subtle fluids to molecular biology*. M.I.T. Press.
Gissis, S. B. (2018). Spencer's evolutionary entanglement: From liminal individuals to implicit Collectivities. In S. Scott Lidgard & K. Lynn (Eds.), *Biological individuality, integrating scientific, philosophical and historical perspectives*. Chicago University Press.
Gissis, S. B. (2023). The Neo Lamarckian tools deployed by the young Durkheim: 1882–1892. *Journal of the History of Biology* (April) 1–37. https://doi.org/10.1007/s10739-023-09708-w
Gohau, G. (1979). Alfred Giard (1846–1908). *Revue de synthèse, 95–96*, 393–406.
Gould, S. J. (1977). *Ontogeny and phylogeny*. Belknap Press of Harvard University Press.
Guillo, D. (2000). L'empreinte de l'histoire naturelle sur la sociologie comtienne: l'héritage de l'anatomie comparée des premie'res décennies du xix sie'cle dans le cours de philosophie positive. *L'Année Sociologique, 50*(1), 195–216.
Guillo, D. (2006). La place de la biologie dans les premiers textes de Durkheim: un paradigme oublié? *Revue Française de Sociologie, 47*(3), 507–535.
Gumplowicz, L. (1885). *Grundriss der soziologie*. Manz.
Gumplowicz, L. (1893). *La lutte des races, recherches sociologiques* (C. Bayet, Trans.). Guillaumin.
Hacking, I. (1983). Nineteenth century cracks in the concept of determinism. *Journal of History of Ideas, 44*, 455–475.
Hacking, I. (1990). *The taming of chance*. Cambridge University Press.
Hacking, I. (1994). How numerical sociology began by counting suicides: From medical pathology to social pathology. In I. Bernard Cohen (Ed.), *The natural sciences and the social sciences* (pp. 101–133). Kluwer Academic Publishers.
Harré, R., & Martin, J. (1982). Metaphor in science. In D. Miall (Ed.), *Metaphors: Problems, and perspectives* (pp. 89–105). Humanities Press.
Hirsch, T. (2018). Émile Durkheim, l'Année sociologique et le 'Parti sociologique unifié': Formes et enjeux du travail collectif (1897–1913). *Mil Neuf Cent, 36*(1), 35–58.

Holmes, F. L. (2014). *Claude Bernard and Animal chemistry: The emergence of a scientist*. Harvard University Press.
Joas, H. (2017). Interview by Gregor Fitzi and Nicola Marcucci with Hans Joas on the reception of Émile Durkheim in Germany. Berlin: Humboldt University of Berlin, 6 October 2014. *Journal of Classical Sociology, 17*(4), 382–398.
Jones, R. A. (1972). Durkheim's critique of spencer's ecclesiastical institutions. *Sociological Inquiry, 44*(3), 205–221.
Jones, R. A. (1974). Durkheim's response to spencer: An essay towards historicism in the historiography of sociology. *Sociological Quarterly, 15*(3), 341–358.
Jones, R. A. (1975). Durkheim in context: A reply to Perrin. *Sociological Quarterly, 16*(4), 551–559.
Jones, R. A. (1981). Robertson Smith, Durkheim and sacrifice: An historical context for 'the elementary forms of the religious life. *Journal of the History of the Behavioral Sciences, 17*, 184–205.
Jones, S. S. (2001). *Durkheim reconsidered*. Polity Press.
Jones, R. A. (2005). Practices and pre-suppositions: Some questions about Durkheim and Les Formes élémentaires de la vie religieuse. In J. Alexander & P. Smith (Eds.), *The Cambridge companion to Durkheim* (pp. 80–100). Cambridge University Press.
Jones, S. S. (2010). Durkheim, the question of violence and the Paris commune of 1871. In S. Romi Mukherjee (Ed.), *Durkheim and violence*, Chap. 4. Wiley-Blackwell.
Karady, V. (1976). Durkheim, les sciences sociales et l'université: bilan d'un semi-échec. *Revue Française de Sociologie, 17*, 267–311.
Karady, V. (1983). The Durkheimians in academe. A reconsideration. In P. Besnard (Ed.), *The sociological domain* (pp. 71–89).
Klautke, E. (2013). The French reception of völkerpsychologie and the origins of the social sciences. *Modern Intellectual History, 10*(2), 293–316.
Kusch, M. (1999). *Psychological knowledge: A social history and philosophy*. Routledge.
La Vergata, A (1996). Giard. In P. Tort (Ed.), *Dictionnaire du Darwinisme et de la évolution* (pp. 1911–1931). Presses Universitaires de France.
Lacroix, B. (1981). *Durkheim et le politique*. Presses de la Fondation nationale des sciences politiques.
Landrieu, M. ([1809] 1909). *Lamarck, le fondateur du transformisme; sa vie, son œuvre*, a l'occasion du centenaire de la 'Philosophie zoologique.' la Société zoologique de France.
Lassere, P. (1924). *Mes routes*. Plon-Nourrit et cie.
Latour, B. (1988). *The pasteurization of France (*A. Sheridan & J. Law, Trans.). Cambridge.
Le Bon, G. (1895). *La Psychologie des Foules*. G. Masson.
Leclerc, G. (1972). *Anthropologie et colonialisme*. Fayard.
Le Dantec, F. (1899). *Lamarckiens et Darwiniens, discussions de quelques theories sur la formation d'espe'ces*. F. Alcan.
Leguay, P. (1910). *La Sorbonne*. B. Grasset.
Limoges, C. (1994). Milne-Edwards, Darwin, Durkheim and the division of labour: A case study in reciprocal conceptual exchanges between the social and the natural sciences. In I. Bernard Cohen (Ed.), *The natural sciences and the social sciences* (pp. 317–343). Kluwer Academic.
Limoges, C. (2014). A second glance at evolutionary biology in France. In E. Mayer & W. Provine (Eds.), *The evolutionary synthesis: Perspectives on the unification of biology* (pp. 322–328). Harvard University Press.
Loison, L. (2009). Les conceptions embryologiques et phylogénétiques d'Alfred Giard (1846-1908) et Edmond Perrier (1844-1921), deux appropriations de la loi biogénétique fondamental. *Bulletin d'histoire et d'épistemologie des sciences de la vie, 16*(2), 165–183.
Loison, L. (2010). *Qu'est-ce-que le néolamarckisme? Les biologistes Français et la question de l'évolution des espe'ces*. Vuibert.
Loison, L. (2011a). French Roots of French Neo-Lamarckisms, 1879–1985. *Journal of the History of Biology, 44*(4), 713–744.

Loison, L. (2011b). The Notions of Plasticity and Heredity among French Neo-Lamarckians (1880–1940): From Complementarity to Incompatibility. In S. Gissis & E. Jablonka (Eds.), *Transformations of Lamarckism* (pp. 67–76). MIT Press.

Lukes, S. (1973a). *Émile Durkheim: His life and work: A historical and critical study*. Allen Lane.

Lukes, S. (1973b). *Individualism*. Blackwell.

Lukes, S. (2018). Durkheim and the new sociology of morality. In H. Joas & A. Pettenkofer (Eds.), *The Oxford handbook of Émile Durkheim*. Oxford University Press. https://doi.org/10.1093/oxfordhb/9780190679354.013.30

Lustig, A. J. (2004). Ants and the nature of nature in Auguste Forel, Erich Wasmann, and William Morton wheeler. In L. Daston & F. Vidal (Eds.), *The moral authority of nature* (pp. 282–307). University of Chicago Press.

Marsh, R. M. (2010). Merton's sociology 215-216 course. *American Sociologist, 41*, 99–114.

Massis, H. (1931). *Evocations: Souvenirs 1905–1911*. Librairie Plon.

Mauss, M. (1908). L'Art et le mythe d'apres M. Wundt. *Revue Philosophique, 66*, 47–79.

Mauss, M. (1925). In memoriam, l'oeuvre inédite de Durkheim. In V. Karady (Ed.), *Marcel Mauss œuvres* (Vol. 3, pp. 473–488). Les éditions de minuit.

Mauss, M., & Fauconnet, P. (1901). Sociologie. In *La Grande Encyclopedie* (Vol. 30). Société anonyme de la Grande Encyclopédie.

Meloni, M. (2016). The transcendence of the social: Durkheim, Weismann, and the purification of sociology. *Hypothesis and Theory, 1*, 11.

Merllié, D. (1993). Les rapports entre Revue philosophique et la Revue de métaphysique et de morale-MM, Leon Xavier, Theodule Ribot, Lucien Levy Bruhl. *Revue de Métaphysique et de Morale, 1–2*, 59–108.

Merllié, D. (2017). Correspondance d'Émile Durkheim avec Lucien Lévy-Bruhl. *Cahiers Vilfredo Pareto, 55*(2), 105–168.

Metraux, A. (1982). French crowd psychology: Between theory and ideology (chapter 11). In W. R. Woodward & M. G. Ash (Eds.), *The problematic science: psychology in nineteenth-century thought*. Praeger.

Momigliano, A. (1970). The ancient city of Fustel de Coulanges. In *Essays in ancient and modern historiography* (pp. 325–344). Chicago University Press.

Morange, M. (2010). What history tells us: The French neo-lamarckians. *Journal of Bioscience, 35*(4), 515–517.

Mosbah-Natanson, S. (2011). La sociologie comme 'mode'? Usages éditoriaux de label 'sociologie' en France a` la fin du XIX siècle. *Revue Française de Sociologie, 52*(1), 103–132.

Moscovici, S. (1981). *L'Âge des foules: Un traité historique de psychologie des masses*. Fayard. Motabe 2020.

Mucchielli, L. (1994). Sociologie et psychologie en France, l'appel à un territoire Commun: Vers une psychologie collective (1890–1940). *Revue de synthese IVieme s., 3–4*, 445–483.

Mucchielli, L. (1995). Pourquoi réglementer la sociologie? Les interlocuteurs de Durkheim. In M. Borlandi & L. Mucchielli (Eds.), *La Sociologie et sa méthode* (pp. 15–48). L'Harmattan.

Mucchielli, L. (1998). *La découverte du social: Naissance de la sociologie en France*. Éditions la découverte.

Nungesser, F. (2020). The sociality of mind: Key arguments, inner tensions and divergent appropriations of Durkheim's sociology. In H. Joas & A. Pettenkofer (Eds.), *The Oxford handbook of Émile Durkheim*. Oxford University Press.

Nye, R. A. (1975). *The origins of crowd psychology: Gustave Le Bon and the crisis of mass democracy in the 3rd Republic*. Sage.

Nye, R. A. (1984). *Crime, madness, and politics in modern France*. Princeton University Press.

Olen, P., & Turner, S. (2015). Durkheim, sellars, and the origins of collective intentionality. *British Journal for the History of Philosophy, 23*(5), 954–975.

Olmsted, J. M. D., & Olmsted, E. H. (1952). *Claude Bernard and the experimental method in medicine*. H. Schuman.

Ozouf, J., & Ozouf, M. (1992). *La republique des instituteurs*. Julliard.

Paoletti, G. (2012). *Durkheim et la philosophie: représentation, realité et lien social*. Classiques Garnier.
Parodi, D. (1920). *La philosophie contemporaine en France: essai de classification des doctrines*. Felix Alcan.
Perrier, E. (1881). *Les colonies animales et la formation des organismes*. Masson.
Perrin, R. G. (1975). Durkheim's Misrepresentation of Spencer: a Reply to Jones' "Durkheim's Response to Spencer". *Sociological Quarterly, 16*(4), 544–550.
Persell, S. M. (1999). *Neo-Lamarckism and the evolution controversy in France, 1870–1920*. The Edwin Mellen Press.
Pickering, W. S. F. (2000). Representations as Understood by Durkheim- an Introductory Sketch. In *Durkheim and Representations*, edited by W. S. F. Pickering, 11–23. Routledge.
Porter, T. M. (1995). Statistical and social facts from Quetelet to Durkheim. *Sociological Perspectives, 38*(1), 15–26.
Rawls, A. W. (1996). Durkheim's epistemology: The neglected argument. *American Journal of Sociology, 102*(2), 430–482.
Rawls, A. W. (1998). Durkheim's challenge to philosophy: Human reason explained as a product of enacted social practice. *The American journal of sociology, 104*(3), 887–901.
Rawls, A. W. (2001). Durkheim's treatment of practice: Concrete practice vs. representation as the foundation of reason. *Journal of Classical Sociology, 1*, 33–68.
Raymer, E. J. (2018). Human progress by human effort: Neo-Darwinism, social heredity, and the professionalization of the American social sciences, 1889–1925. *History and Philosophy of the Life Sciences, 40*(4), 1–23.
Ribot, T. (1900). *Essai sur l'imagination créatrice*. Felix Alcan.
Richards, R. J. (2008). *The tragic sense of life: Ernst Haeckel and the struggle over evolutionary thought*. University of Chicago Press.
Riley, A. T. (2010). *Godless intellectuals? The intellectual pursuit of the sacred reinvented*. Berghahn Books.
Rioux, J.-P. (1991). *Chronique d'une fin de siècle: France, 1889–1900*. Seuil.
Riskin, J. (2016). *The restless clock: A history of the centuries-long argument over what makes living things tick*. Chicago University Press.
Roger, J, Ed. (1979a). Les Néo-Lamarckiens Français. *Revue de Synthese, 95/96*, 279–469. (J. Roger, J.C. Sillard, A. Lagarde, Cl. Blanckaert, M. Viré, G. Gohau, A. Diara, K. Wellman, J.L. Fischer).
Roger, J. (1979b). Presentation. In *Les néo-lamarckiens français* (pp. 95–96), edited by Jacques Roger. Special issue of *Revue de synthèse* 3ie'me série.
Sawyer, R. K. (2002). Durkheim's dilemma: Towards a sociology of emergence. *Sociological Theory, 20*(2), 227–247.
Schmaus, W. (2000). *On Sens lectures: Downloaded*. http://www.relst.uiuc.edu/durkheim/reviews/Sens.Schamus.html.
Schmaus, W. (2004). *Rethinking Durkheim and his tradition*. Cambridge University Press.
Simiand, F. (1898). Sociologique- L'année 1897. *Revue de Métaphysique et de Morale, 6*(5), 608–653.
Smith, R. (2005). Does reflexivity separate the human sciences from the natural sciences? *History of the Human Sciences, 18*(4), 1–25.
Soulié, S. (2009). *Les philosophes en République-l'aventure intellectuelle de la revue de métaphysique et de morale et de la société française de philosophie.1891–1914*. Presses universitaires de Rennes.
Spencer, B., & Gillen, F. J. (1899). *The native tribes of Central Australia*. Macmillan.
Staum, M. J. (2003). *Labeling people: French scholars on society, race and empire, 1815-1848*. McGill-Queen's University Press.
Staum, M. (2011). *Nature and nurture in French social sciences 1859–1914 and beyond*. McGill-Queens University Press.

Stedman Jones, G. (2002). The new social history. In C. Jones & D. Wahrman (Eds.), *The age of cultural revolutions: Britain and France 1750–1820* (pp. 94–136). University of California Press.
Stock-Morton, P. (1971). Students versus the University in Pre-World War France. *French Historical Studies, 7*, 83–110.
Strenski, I. (1997). *Durkheim and the Jews of France*. University of Chicago Press.
Strenski, I. (2002). Durkheim, disciplinarity, and the 'sciences' religieuses (chap. 5). In A. Anderson & J. Valente (Eds.), *Disciplinarity at the fin de siècle* (pp. 153–173). Princeton University Press.
Strenski, I. (2006). *The new Durkheim*. Rutgers University Press.
Swedberg, R. (2019). How do you make sociology out of data? Robert K. Merton's Course in Theorizing (Soc 213–214). *The American Sociologist, 50*, 85–120.
Taguieff, P. A. (1984). Georges Vacher de la Pouge 1854–1936. In D. Huisman (Ed.), *Dictionnaire des Philosophes*. Presses Univérsitaires de France.
Thomas, M. (2003). Rethinking the history of ethology: French animal behaviour studies in the Third Republic (1870–1940) (PhD dissertation). University of Manchester.
Tilly, C. (1986). *The contentious French*. Belknap Press.
Tilly, C., & Shorter, E. (1974). *Strikes in France 1830–1968*. Cambridge University Press.
Tirard, S. (2013). Claude Bernard et les trois formes de vie. In F. Duchesneau, J. J. Kupiec, & M. Morange (Eds.), *Claude Bernard: la méthode de la physiologie* (pp. 49–62). Éditions Rue d'Ulm.
Tiryakian, E. A. (1979). L'ecole Durkheimienne a la recherch de la societe perdue: La sociologie naissante et son milieu culturel. *Cahiers Internationaux de Sociologie, 66*, 97–114.
Tiryakian, E. A. (1988). Durkheim, Mathiez, and the French Revolution: The political context of a sociological classic. *Archives européennes de sociologie. European Journal of Sociology, 29*(2), 373–396.
Tiryakian, E. A. (2009). *For Durkheim: Essays in historical and cultural sociology*. Ashgate.
Tollefsen, D. P. (2015). *Groups as agents*. Polity Press.
Turner, S. P. (1986). *The search for a methodology of social science: Durkheim, Weber, and the nineteenth century problem of cause, probability and action*. D. Reidel.
Turner, S. P. (1993). *Emile Durkheim: Sociologist and moralist*. Routledge.
Virtanen, R. (1960). *Claude Bernard and his place in the history of ideas*. University of Nebraska Press.
Von Lilienfeld, P. (1896). *La pathologie sociale*, preface R. Worms. V. Giard & E. Brière.
Wallwork, E. (1972). *Durkheim: Morality and milieu*. Harvard University Press.
Weissman, C. (2011). The first evolutionary synthesis of august Weismann and the origins of neo-Darwinism (PhD dissertation). Tel-Aviv University.
Welch, G. R., & Clegg, J. S. (2010). From protoplasmic theory to cellular systems biology: A 150-year reflection. *American Journal of Physiology: Cell Physiology, 298*(6), C1280–C1290.
Wellman, K. (1979). Felix Le Dantec et le néo lamarckisme français. *Revue de Synthese, 95–96*, 423–442.
Weismann, A. (1891–1892). *Essays upon heredity and kindred biological problems*. In E. B. Polton, S. Schonland, & A. E. Shipley (Eds.), 2 vol. Oxford University Press.
Winther, R. G. (2001). August Weismann on germ plasm variation. *Journal of the History of Biology, 34*, 517–555.

Chapter 8
Sigmund Freud, a neoLamarckist – Very Short Coda

8.1 Introduction

"I must, however, in all modesty confess that nevertheless I cannot do without this factor in biological evolution" (Freud, 1975, 23:99–100 [1939])—a sentence taken out of the often-quoted paragraph where Freud deliberated on the theoretically unavoidable use of SL/NL tools and assumptions, which has been deployed by detractors, apologists, and present-day interpreters positing Freud as some prefiguration of epigenetic explanations of trauma (e.g., Malabou, 2018). Indeed, by the 1930s, the processes of purging evolutionary biology in general from the social sciences were predominant, but the option of soft inheritance in evolutionary biology was still there in many ways. Thus, it seems that all above interpretations have misplaced their positional load[1] (Fig. 8.1).

Most of Freud's writings and the evolving of the novel field he constituted, psychoanalysis, intended as a scientific discipline as well as a therapeutic practice, fall outside the time span of the book. Yet I include a brief coda on him, for I consider him to be a member of this 'group of SL/NL users–founders,' though only in some limited yet meaningful aspect of his work. Thus, the short coda is a continuation of the book's main thrust, in which I propose to view Freud's oeuvre from the same specific historical perspective employed in the previous chapters. This is the perspective of the entwining of his system with the transfer from the SL/NL framework as a significant tool in constituting a new field, its theoretical foundations, research methodology, and scientific objects, positing it as a scientific discipline and hoping

[1] Since Freud's diverse portraits are so very familiar, I present here a marionette created by Sophie Taeuber-Arp for special performances of the eighteenth-century play by Carlo Gozzi *The Stag King*. The play was updated as a humorous take on psychoanalysis. The magician in the original play became a character called Freud analyticus. Writers and artists associated with the Dada movement praised these marionettes.

S. B. Gissis, *Lamarckism and the Emergence of 'Scientific' Social Sciences in Nineteenth-Century Britain and France*, History, Philosophy and Theory of the Life Sciences 36, https://doi.org/10.1007/978-3-031-52756-2_8

Fig. 8.1 "Freud Analyticus" (1856–1939). Sophie Tauber-Arp, marionette from the play "The King Stag" 1918. Photograph Ernst Linck. Source: Stiftung Hans Arp und Sophie Tauber-Arp e.V

to have it acknowledged and legitimated in academia. Two principal components of SL/NL considered significant for such an endeavor are relevant here: the back-and-forth mechanism of inheritance between the biological and the social/psychological and perceiving the SL/NL frameworks as more congenial to greater plasticity yet open to determinism as well.

I propose to analyze Freud's use of 'archaic memory,' 'memory traces,' the inheritance of modifications of acquired trauma, and phylogeny–ontogeny narratives, through this lens. I begin by explicating Freud's position as a user of SL/NL and then describe and analyze in more detail his practice of transfer. I put forward that it was deployed as a tool in the more general effort to constitute a 'universal' individual and that similarly to Spencer and Ribot, Freud also deployed it to bridge between individual and collectivity. Repeatedly and diversely deploying the SL/NL repertoire helped the metaphors, analogies, and models used to accrue more meaning, wider systemic anchoring, and explanatory generativity.

In my argument, I disagree with two principal interpretive biological-reduction positions on Freud's deployment of the evolutionized-biology repertoire—Lucille Ritvo and Frank Sulloway. Let me clarify while greatly simplifying both positions for my purpose here. Ritvo argues that whatever Lamarckian features found in

Freud's oeuvre, they are actually the Lamarckian features found in Darwin (including the notion of continuum). These significantly include the inheritance of acquired characteristics/modifications, which, she claims, are of ancillary importance for both Darwin and Freud. Furthermore, Ritvo argues that whatever is attributed to the 'impact of the environment' is solely derived from Darwin, as transmitted to Freud (particularly via Claus) since in her depiction of Lamarckism 'environment' is of secondary significance in every way (Ritvo, 1990).

In claiming Freud for the history of biology, rather than for that of psychology, Sulloway (1992) presents a complex and nuanced picture. He invests much effort in situating Freud's ideas within what Sulloway conceives as the evolutionized biology of Freud's time. He posits Freud as a user of this biology for the purpose of constructing fully fledged 'scientific laws,' as *biological* answers to three foundational psychological questions: What was the nature of pathological repression? Why the centrality of sex? How did the choice of neurosis occur? This is carried out, so Sulloway claims, especially through the pervasive direct use of components from Darwin's major works (and some late ones published by Romanes, especially after 1882). Positing 'conflict' as foundational to Freud is considered as another major Darwinian feature, as well as Freud's mode of understanding of some of his contemporaries such as Richard von Krafft-Ebing and Henry Havelock Ellis. Sulloway argues that Freud's dynamic–genetic model of the mind and his preoccupation with the instinctual and irrational prove him as first and foremost a Darwinian.

Yet, Sulloway (1992) also discusses at some length and detail what he considers as 'Lamarckian influences' and Freud's use of recapitulation. Yet, Freud read Lamarck's work, according to his own testimony, in 1917 and gradually abandoned it during that selfsame year, as attested in his correspondence with Ferenczi during the years of World War I (detailed, e.g., in A. Young, 2006). That means that he read Lamarck's *Philosophie zoologique* after he had already written about and deployed SL/NL mechanisms and assumptions. Thus, I argue that he read Lamarck through SL/NL conceptualizing, with which he had already been familiar and had already deployed (combined with recapitulation).

Early on, in the 1890s, Freud wrote about the theoretical–methodological need to assume a universal archaic memory,[2] and later he brought forth two significant works deploying similar mechanisms for their principal narrative. Therefore, representing Freud as adopting specifically Lamarck's understanding of the use-inheritance mechanism ignores or even reads backward into Lamarck the important Spencerian modifications and additions, adopted by late nineteenth century, also by some neo-Lamarckists, consisting in the back-and-forth movement between the social/psychological and the biological/neurological. It was this 'Spencerian addition' that performed the intellectual alchemy needed for the particular deployment of the use-inheritance mechanism, first by Spencer, then by Ribot, and later by Freud. Furthermore, Freud used a combination of the SL/NL notion of inheritance of acquired modifications together with the Haeckelian 'biogenetic law' of

[2] Grubrich-Simitis, 1987, 91–92, discussing Freud's correspondence.

recapitulation, which he transferred from evolutionized biology to the description and explanation of the psychological. This particular combination of components was a hallmark feature of neo-Lamarckism adherents, especially in France at that time, whether used in evolutionized biology or in transfer to emerging social sciences.

Freud came to hold these views, as expressed in his extensive correspondence, in particular with Wilhelm Fliess, already in the 1890s, and then it appeared in his correspondence with Karl Abraham and with Sandor Ferenczi in the years after the turn of the century and well into World War I. It is significant that Freud spent a few months in Paris when Charcot was at the apex of his fame (1885), as well as Bernheim in Nancy—a meaningful period of ascending FNL. It has been suggested by some psychoanalysis historians that Freud might have met Ribot while there, e.g., in the Charcot sessions at the Salpêtrière hospital. It can be reasonably assumed that Freud was familiar with the main argument in some of Ribot's works, which indeed deployed the SL/NL perspective, particularly *The Diseases of Memory* (1881), written when Spencer was still deeply influential in France. Freud was certainly familiar with the SL perspective as deployed by Jackson, whom he quoted profusely in those years (e.g., *On Aphasia*, 1891). There were also some significant SL/NL figures within his familiar German-speaking context, e.g., Carl F.W. Claus.

In the controversy on correlations between physiological/neurological descriptions of events and mental/psychological ones as indicating causality, etiology, etc., analyses of Freud's 1893 'Project' and its abandonment and withdrawal from publication, and his ongoing comments on methodological issues, have served all parties. If he had been a straightforward parallelist, there would have been no scientific reason for him to discuss, to gesture toward, or to offer an integration of the two modes of description, 'the physiological/neurological' and the 'mental/psychological' (e.g., Tauber, 2010), but rather follow a path of concomitance. Yet, accumulated evidence throughout his writings and correspondence seems to point to another possibility, namely, that Freud did believe on the one hand in there being an ontological gap (between the neurological/physiological and the psychological/mental) and on the other hand in *eventually* establishing psychoanalysis on *evolutionized*-biology foundations at some point in the indefinite future; he did use and transfer evolutionary mechanisms and assumptions, though declaratively and manifestly applying them to 'the mental/psychological' rather than to the physiological/neurological (e.g., *Beyond the Pleasure Principle*, 1920, 1975, 18:60). Ridding himself of the theoretical need for 'a physiological correlate' allowed him to constitute this new field aspiring to be a scientific academic discipline. In that he mirrored in obverse manner Jackson, who strove to rid himself of 'the mental' in order to constitute the clinical–scientific discipline of neurology. I suggest that holding on to both positions, perhaps with differing theoretical emphases at different periods, or through deploying a future projection, made it possible to relate the new field to established sciences, thereby clothing it with a scientific garb.

Already around 1900 Freud became familiar with anthropological writings of Lubbock and Tylor (in particular *Primitive Culture*), and perhaps somewhat later McLennan, and Frazer, Wundt in his *Völkerpsychologie* writings, and the sociological–anthropological ones of Spencer.[3] Time was presented there, especially in Spencer, from the evolutionary-developmental perspective rather than as either 'neutral time' or 'historical time' (Fabian, 1983). Thus, to provide a description and analysis how X evolved was posited as the proper answer to the question 'what X was.' Insights gained, materials and methodological suppositions used, such as assuming a parallel between the psychology of 'normal savages' and that of 'neurotics,' under the same causal-explanation frame, and explaining the evolutionarily 'higher' through the 'lower' were highly instrumental in constructing the Freudian constitutive quasi-evolutionary 'sequence' narrative to be discussed here.

Freud adopted, transferred, and elaborated on Jackson's view of the nervous system. Thus, it and the abstract site of 'the mental' (in later Freud) were layered hierarchically, both structure-wise and function-wise. It meant that levels of greater complexity were thus considered as later evolutionary stages in the development of nervous, psychological, or mental systems, and pathology was defined as either a *developmental* regression, repression, or fixation along that scale (e.g., Stengel's 1954 interpretation, also on aphasias). Following Jackson (and possibly Ribot), Freud too construed lower levels as more stable and less plastic (tending to automatic action; also, Solms & Saling, 1986).

8.2 Spencerian Lamarckism/neoLamarckism, Memory, Individual, and Collectivity

Methodologically following Claude Bernard (as well as Jackson, Ribot, and contemporaneous French psychiatry), Freud assumed a continuity between normal and pathological states of the organism or any system within it, applied to the psychological realm. Furthermore, having adopted the particular SL/NL version of the mechanism of inheritance of acquired characteristics, Freud implicitly assumed the Spencerian back-and-forth movement. Thus, already from the early 1890s on, there were hints, in published writings and in correspondence with Fliess, of Freud's view that one had to presuppose, methodologically and epistemologically, an archaic heritage, an archaic memory, common to all humankind.[4]

This idea further evolved into a *sequence of social–sexual conflictual traumatic events*, evolutionarily primeval, then repetitive, which became phylogenetically constitutive through recapitulation. It formed within each individual a necessary component of the load of memory and of heredity. Thus, Freud deployed an already

[3] Quoted in *Totem and Taboo*, while anthropologists mostly connected with the 'Australian materials,' e.g., Spencer and Gillen, were barely mentioned.

[4] For example, already in 1893; letters toward the end of 1896; in a letter of January 30, 1899, the two kinds of memory, not just the later repressed memories (1985, 338–39); in a letter of December 21, 1899, the archaeological metaphor of Schliemann excavating Troy 'once again' appeared in a somewhat different context, of searching for primal infant memory (1985, 391–92); work on 'the contents of the unconscious,' 1912 (1975, 12:255–66).

known analogy, prevalent earlier among some biologists and psychologists as a one-way analogy between memory and heredity, that evolved into a two-way analogical mapping between heredity and memory, brought to its extreme in Semon's work (Semon, 1921, 1923; Schachter, 1996, 2001). It could be considered an elaborate variation on Spencer's notion, used by Jackson, of past modes of experiencing passed on trans-generationally, further endowing them with a spatial dimension in Freud's 'topographic,' layered view of the psychological makeup of the individual. In Spencer's psychology, though, this load was dynamic, inhering yet modifying the modes of experiencing of coming generations.

In the 1910s and 1920s, especially in the first years of World War I, Freud's evolving view of memory and of 'the psychological' at large followed SL/NL conceptualizing as applied both to structures and to functioning, as attested through the intense letter exchange with Ferenczi. Accordingly, Freud laid down a classification of the neuroses based on an evolutionary narrative (of such sequence), already elaborated on and developed in *Totem and Taboo* (called by Hughes 1961 'anthropological fantasy'). This more detailed view was constructed in the unpublished metapsychology essay "Overview of the transference neuroses" (1915), now titled *A phylogenetic fantasy* (edited by Grubrich-Simitis, 1987), presenting a scheme of parallel evolutionary time lines of primeval sequences of group events and of developmental appearance order of two classes of neuroses in individuals. The archaic/phylogenetic memory and/or its expository role of the order-of-neuroses was mentioned briefly and/or more broadly in writings from then on, even though the 1915 paper arguing for it in detail was never published.

The principal ones were the following: In 1917, in the *Introductory Lectures on Psychoanalysis*, the part on the general theory of the neuroses, Freud discussed children's fantasies, the 'libido neuroses,' and the 'ego neuroses' as if it was self-evident that they were of 'phylogenetic origin,' 'abbreviated recapitulations' of primeval mostly traumatic development passed on trans-generationally, common to all humans (1975, 16:354, 370–71). In 1918 and 1919, the phylogenetic dimension in children's experiencing was discussed, e.g., a paragraph added in 1919 to a 1900 text: "Behind this childhood of the individual we are promised a picture of a *phylogenetic childhood*—a picture of the development of the human race, of which the individual's development is in fact an *abbreviated recapitulation* influenced by the chance circumstances of life… Dreams in neuroses seem to have preserved more mental antiquities than we could have imagined possible; so that psychoanalysis may claim a high place among *the sciences* which are concerned with the reconstruction of the earliest and most obscure periods of *the beginning of the human race*" (1975, 5:548–49—emphasis added).[5] Then again it was discussed in the *Group Psychology and the Analysis of the Ego*, 1921 (1975, 18:122ff). It was assumed in *The Future of an Illusion*, 1927 (1975, 21:42), indirectly relating this explanatory pattern also to Freud's explication of the dynamics of repression and

---

[5] Also, *From the history of an infantile neurosis*, 1918, possibly 1914 (1975, 17:17); *A child is being beaten*, 1919 (1975, 17:203–04).

civilization in *Civilization and Its Discontent*, 1930, and finally leading up to yet another detailed sequence portrayal in *Moses and Monotheism*, 1937–1939.

There were three principal varieties of this sequence: *Totem and Taboo; Overview of the Transference Neuroses; Moses and Monotheism.* I present the first as a model repeated with variations later on.

Totem and Taboo (1912–1913)—Freud's main sources were the later work of James G. Frazer, to some extent Wundt, even while disagreeing with him. In *Totem and Taboo*, Freud elaborated on the sequence narrative itself in section iv (1975, 13:100–61); the sequence's *kernel* first appeared as a primeval one—remembered, repeated, transformed, and socially transforming (in 13:140–46); and the issues of trans-generational inheritance were summarized in the last subsection (13:155–61). Freud assumed there the universality of totemism, as interpreted by him, as the foundation of 'morality, society, religion'—one should add 'civilization.' The first subsections of *Totem and Taboo* (13:1–99) consisted of an overview of diverse contemporaneous approaches to totemism, as an introduction to the discussion of the sequence narrative Freud offered later in that work, where the figure of a sexually monopolizing and otherwise tyrant father played a foundational role. This, so he claimed, allowed him to relate to totemism central psychoanalytic concepts, in particular the Oedipus complex, applied to contemporaneous children and thus to children in general (Brunner, 1995, esp. chap. 16).

Robertson-Smith's notion of the 'totemic meal' as a sacrificial/festive ritual served as a frame of reference for developing a narrative which combined that notion with a murder of the (sexually) tyrant father (later substituted in the rituals by the totem animal) by a band of male sexually and otherwise frustrated and threatened sibling-sons. These siblings were consequently filled with ambivalence toward their deed, their murdered father, and the possible fulfillment of their sexual desires for his wives (their mothers, sisters, etc.). That fulfillment was perceived after the murder as endangering their achievement of power as a group and equal freedom as individuals. The ensuing strictures on murder of kin (later on humans at large) and prohibition of incest were posited as a result of this sequence of events, repeated in primeval times and echoed forever in the Oedipus complex universally and necessarily in each individual child and in the collectivity of humanity at large (Winter, 1999). Freud argued that subsequent elaborations and repetitions of this sequence and its ritualistic memorialization should be viewed as significant foundations of the concept of god and of religion in general, as well as of the transformation of society into patriarchal societal structure. In the last, seventh subsection, he attempted to provide an explanation and justification for the assumed continuity of these psychological structures and of their trans-generational inheritance. He therefore resorted to two, not quite compatible explications: the assumption of collective mind, somehow resonating in individual ones, which supposedly served as a repository which guaranteed that continued existence, and inheritance of 'psychical (unconscious) dispositions,' i.e., individually.

Freud pointed to the similarity between 'savage' (*naturvölker*) and 'neurotic' as a support for this assumption (while implicitly also primeval ancestors and children formed elements of the same construction).[6] I would argue that Freud manifestly and intentionally did not deploy the possibility of that narrative structure passing on *solely socially*, as a component of tradition, custom, ritual, etc. In order to be able to argue that it functioned continuously, trans-generationally, through the unconscious, he implicitly assumed the SL mechanism of back-and-forth inheritance of acquired modifications. 'The psychological' was implicitly inherited biologically—though there was a marked absence of biological–evolutionary terms of any kind—so that Freud could argue that it appeared and functioned in the unconscious, though in fact it could have appeared in the next generation either as biological or as psychological. While Spencer, Ribot, and Jackson used SL/NL to locate mental mechanisms as physiologically inherited and expressed mentally, Freud used it to map the mental in such a way so as to avoid passage between the biological and the mental/psychological and to imply alternative dynamics. Freud's further elaborations show that providing a convincing account of continuity and trans-generational inheritance of this sequence kept bothering him till the very end.[7]

Overview of the Transference Neuroses (1915) was a very rich *dual-sequenced narrative stretching over two generations*, presenting adaptive patterns to both natural catastrophe and human-produced constraints and prohibitions, incorporating the one already delineated in *Totem and Taboo*. The order of neuroses relied on previous work distinguishing between neuroses based on differentiating between their focus on 'ego' or on 'libido' and their relation to 'reality,' also relating the libidinal ones via recapitulation to vertebrates in general. Thus, the ego-narcissistic neuroses were posited as uniquely human, appearing in later stages of the individual's development. This indicated there being two stages of the development of 'humanity' evolutionarily (see Gould, 1992, 2002), and of the particular biographies of individuals, who might go through these passage stages, or else failed to change their formerly suitable adaptive mechanism, and thus stumbled in individualized paths through regression, repression, fixation, i.e., became neurotic.

Freud continued to use the framework and common components in later works both individually and collectively, especially two: trauma memorized and continuously inherited memory. Within his psychoanalytic framework, Freud resorted to SL/NL to perform for both components the labor of memory as the labor of heredity and vice versa.

Moses and Monotheism (1937–1939) has been highly controversial, primarily, though not solely, because of the hypothesis of an Egyptian Moses in the origins of Judaism, origin of religion, and explanation of anti-Semitism.[8] Here too, while deploying contemporaneous (racialized) anthropology, Freud strove even more

---

[6] The full title included: "some points of agreement between the mental lives of savages and neurotics."

[7] Freud–Jung relations are not discussed.

[8] Primarily: Assmann (1997), Bernstein (1998), Ginsburg and Pardes (2006), Marinelli and Mayer (2006), Sharvit et al. (2018), and Yerushalmi (1991) but also Hacking (1998).

conspicuously to present a universalist anthropo-psychological frame, blurring differences and variations. Both individual's and collectivity's (traumatic) experiences and trans-generational inheritance were more emphatically posited as evolving in tandem (e.g., 1975, 23:94). Though here Freud rejected any collective-mind explanation, the occurrences, repressed and inaccessible in the unconscious, were capable of returning, e.g., "by a recent real repetition of the event" (1975, 23:101).

Freud considered this sequence a keystone of that foundational part of the individual's psychic makeup, 'the unconscious,' 'the Id.' He depicted it as structured, active, and 'faculty like.' Yet, most of its descriptive and analytic attributes would have been considered 'negative' by contemporaneous thinkers, because the items comprising it were neither logically nor temporally ordered. Hence, this narrative component, the sequence, was constituted in order to play a significant role in the metapsychological, psychoanalytic, and therapeutic aspects of the evolving endeavor of constituting psychoanalysis. In its evolved dual-sequence form (versions 2 and 3), it was construed as a working out of the central axis of the Freudian structure—the Oedipus complex—turning it into a multigenerational, quasi-familial, quasi-societal collectivity-narrative, both phylogenetic narrative of the ontogeny of each and every individual singly, and a psychic evolutionary classifying schedule of the neuroses through individual's life span. Being modeled on a SL/NL back-and-forth mechanism, it could offer either a deterministic plot or one of plasticity. It provided universality and necessity to the analysis of the mental/psychic apparatus of individuals, thus components of 'a science,' while concurrently endowing these individuals with all-humanity collective attributes. Sociality was posited in some of Freud's middle and late works as external-repressive to the individual, to his/her needs. Concurrently, this collective-memory-narrative sequence was an internal attribute of the individual's psychological makeup. Yet, Freud's analysis of the phenomenal actuality of being a member of a collectivity beyond the family was mediated either through socially diffuse notions such as 'the masses' or through rigid collective frameworks such as the army and the church.

SL/NL served as a conceptual–metaphorical bridge from the contemporary individual to the collectivity, to sociality, but it resulted in explaining the individual, in spite of the randomness of personal histories, *through* archaic traumatic experiencing, and its trans-generational memory, a memory that was in its diverse manifestations—primeval horde, sibling group, etc.—primarily collective, common to 'the human race.' Freud did not provide any connecting links, neither as mechanism nor as narrative pieces, that would explicate possible modifications between those primeval events, their archaic memory formation, and later history of both individuals and collectivities (except partially in the constitution of the religion narrative). Thus, the sequence in its three versions was assumed throughout Freud's oeuvre as crucial, yet it was not generatively articulated into the evolving dynamics of the conceptual and therapeutic frames. In that sense, 'historical time' was annulled, as these narratives were posited as psychological–foundational for morality, religion, and civilization; yet these primeval formations could be considered through their biological assumptions, implicit or explicit, also as part of 'nature.'

Though at times calling it a 'myth' and at other times 'a fantasy,' Freud always characterized this (double) sequence as *scientific*. His individuals contained the collectivity within them through SL/NL evolutionary mechanisms, i.e., as emergent products of evolutionary-developmental processes inherited and inscribed in the 'evolutionary-developmental memory' of all individuals as members of the human species, inhering in the common-species unconscious. It thus formed an indispensable component in the explanatory narrative of both civilizational history and individual life history, normal and pathological.

Contrary to Spencer (and Durkheim), however, Freud did not expect his *therapeutics* to have social utopian/redemptive dimensions. It remained basically within the boundaries of the individual, with the modest hope of reduction of suffering and increase in the person's autonomy. Could one, though, look upon the *theory* of psychoanalysis as collectively emancipatory in the Enlightenment sense, in combatting misguided hopes, unraveling the hold of religion on humans, to make them realize that they can improve humanity's lot in this life?

Materials from the following articles have been used by permission of:

Elsevier

2020 – Transfer of Lamarckisms and Emerging 'Scientific' psychologies: 19th – early 20th Centuries Britain and France, *Studies in History and Philosophy of Biological and Biomedical Sciences, 83*.

References

Assmann. (1997). *Moses the Egyptian: The memory of Egypt in Western Monotheism.* Harvard University Press.

Bernstein, R. (1998). *Freud and the legacy of Moses*. Cambridge University Press.

Brunner, J. (1995). *Freud and the politics of psychoanalysis*. Blackwell.

Fabian, J. (1983). *Time and the other: How anthropology makes its object.* Columbia University Press.

Freud, S. (1975). *The standard edition of the complete psychological works of Sigmund Freud.* Hogarth Press.

Freud, S. (1985). *The complete letters of Sigmund Freud to Wilhelm Fliess, 1887–1904* (J. Mussaieff-Masson, Trans. and Ed.). Belknap Press.

Ginsburg, R., & Pardes, I. (Eds.). (2006). *New perspectives on Freud's "Moses and Monotheism"*. Conditio Judaica.

Gould, S. J. (1992). Ontogeny and phylogeny revisited and reunited. *BioEssays, 14*(4), 275–279.

Gould, S. J. (2002). *The structure of evolutionary theory*. Belknap Press.

Grubrich-Simitis, I. (1987). An essay on Freud. In I. Grubrich-Simitis (Ed.), *A phylogenetic fantasy*. Belknap.

Hacking, I. (1998). *Rewriting the soul: Multiple personality and the science of memory*. Princeton University Press.

Malabou, C. (2018). Is psychic phylogenesis only a phantasy? New biological developments in trauma inheritance. In G. Sharvit, K. Feldman, et al. (Eds.), *Freud and monotheism* (pp. 177–198). Fordham University Press.

Marinelli, L., & Mayer, A. (2006). Editors' introduction: Forgetting Freud? For a new historiography of psychoanalysis. *Science in Context, 19*(1), 1–13.
Ribot, T. (1881). *Les maladies de la mémoire*. Librarie Generale Ballie're et Cie.
Ritvo, L. B. (1990). *Darwin's influence on Freud: A tale of two sciences*. Yale University Press.
Schachter, D. L. (1996). *Searching for memory: The brain, the mind, and the past*. Basic Books.
Schachter, D. L. (2001). *Forgotten ideas, neglected pioneers: Richard Semon and the story of memory*. Psychology Press.
Semon, R. (1921). *The Mneme*. George Allen and Unwin.
Semon, R. W. (1923). *Mnemic psychology*. G. Allen & Unwin, Ltd.
Sharvit, G., Feldman, K. S., Assman, J., Berenstein, R., Gotschel, W., & Hendel, R. (Eds.). (2018). *Freud and monotheism: Moses and the violent origins of religion*. Fordham University Press.
Solms, M., & Saling, M. (1986). On psychoanalysis and neuroscience: Freud's attitude to the localizationist tradition. *International Journal of Psycho-Analysis, 67*, 397–416.
Stengel, E. (1954). The significance of Freud's study of aphasia for psychoanalysis. *Psyche (Stuttgart), 8*(1), 17–24.
Sulloway, F. J. (1992). *Freud, biologist of the mind: Beyond the psychoanalytic legend*. Harvard University Press.
Tauber, A. (2010). *Freud, the reluctant philosopher*. Princeton University Press.
Winter, S. (1999). *Freud and the institution of psychoanalytic knowledge*. Stanford University Press.
Yerushalmi, H. (1991). *Freud's Moses: Judaism terminable and interminable*. Yale University Press.
Young, A. (2006). Remembering the evolutionary Freud. *Science in Context, 19*, 175–189.

Chapter 9
Interlude: 'Collectivity' in the Nineteenth Century Between the Biological and the Social

During this period, the concepts of collectivity and of the individual, besides reflecting deep metaphysical commitments, pointed to what was perceived as a profound dichotomy, which some would argue still exists. It resulted in an unresolved tension in the perception of living nature and of sociality and their description and analysis.

It seems that there has been a certain conceptual 'rapprochement' in present-day discourses of both social and biological sciences and their philosophies. Principal nineteenth-century terms used in dealing with 'individual' were 'action' and 'intention' (indicating 'agency'), yet now these are qualified not only by adjectives emphatically individualizing but also indicating diverse degrees of pertaining to collectivity: shared, joint, coordinated, cooperative, collaborative, social, normative, institutional, and collective. There has been a convergence in the diversity and multiplicity of recurring analyses of humans as collective constructors and selectors of their collective culture, attending to looping processes of interacting, coordinating, cooperating, collaborating, of communicative intentionality, of interdepending roles, as well as discussions of the Spencerian metaphor of 'superorganism' nature of (modern) human societies. In these analyses humans are posited learning, spreading, and maintaining adaptive innovations, goals between individuals, among individuals in a group, among groups, of groups as a special kind of 'collective individual.'[1]

---

[1] A rough mapping of contemporary resources would yield such adjectives as found in the work of present-day principal theoreticians on the spectrum of collectivity as well as the phenomenological perspective. Much work has been invested especially in recent decades from the 'cultural evolutionary' perspective, with sharp differences in presuppositions and possible resources for investigation, e.g., infants, human groups (their artifacts and norms), apes, other mammals, and birds, bio-archaeological findings, language evolution, genetic and epigenetic findings, also discussions of specific stages, mechanisms, and whether they lead to an evolutionary transition of individuality (most recently the two clusters of articles in Philosophical Transactions of the Royal Society B during 2023 (Carmel et al., 2023; Ioannou & Laskowski, 2023)).

S. B. Gissis, *Lamarckism and the Emergence of 'Scientific' Social Sciences in Nineteenth-Century Britain and France*, History, Philosophy and Theory of the Life Sciences 36, https://doi.org/10.1007/978-3-031-52756-2_9

I have argued that throughout the nineteenth century, emerging fields dealing with the social and the psychological, all the more so when aspiring to be considered 'scientific,' were deeply influenced by the paradigmatic example of contemporaneous evolutionized biology, transferring a repertoire prevalent in its diverse branches. Though there were certain notable, complicated, yet significant and influential exceptions in the biological disciplines in Britain and France (e.g., Espinas, Perrier), generally, and particularly within SL/NL frames, the 'individual'—a multicellular organism—was perceived as an entity in terms of which all else had to be described, analyzed, and explained. As such it was regarded as a focus of causation, a substantive spatiotemporal basic evolutionary entity. Individuals were also seen as having plasticity and the capacity to change over time and, therefore, as endowed with the capacity of adaptation to novel contexts—those in which they found themselves and for the SL/NL also those with which they interacted, or formed.

Individuals were perceived as the epistemologically real, observable, and visualizable entities. Through such transfer from evolutionized-biology frames, isolatable individuals were regarded within the new social sciences as foundational, decoupled, and sharply distinct from collectivities, yet seen as operating within a definite, given collectivity/social contexts. Individuals were thus looked upon as the most fundamental elements of the newly emerging disciplines that deal with the social and the mental/psychological. The descriptions, analyses, and explanations of collectivity-situations became those of the aggregated reactions of each solitary individual to a state of affairs deemed common, such as in probabilistic accounts of births, deaths, marriages, and suicides (e.g., Hacking, 1990, 1994, 101–33; Turner, 1986). Only by employing statistical methods could one construct a 'visualization' of a collectivity. Some versions of individualism posited 'collectivity properties' as supervenient on the individual or at times as emergent products of processes of individuals, that is, collectivity properties and processes were consequent upon individuals. By these overall deployments in these fields, 'the individual' became naturalized/came to be seen as a natural entity (on individualism, e.g., Lukes, 1973a, b).

I use a nineteenth-century conceptual space in which 'population' referred to the broadest characterization of the individuals being delineated. British, French, and German users had therefore different understandings of aggregated statistical population, none of which was the later 'Mendelian' one used by eugenicists. As discussed in the Durkheim chapter, 'society' was regarded then as coextensive with the nation-state. The term has to be contextually qualified by the diverse contemporaneous discourses on sociality/sociability. There were diverse attempts to elucidate what ties human beings together and transforms them from disparate entities into entities living within some form of grouping of meaningful higher-level bonds and interactions, in particular groupings not enforced from above. When the nineteenth-century users referred to 'collective' (as noun or as adjective), they more often than not assumed that such belonging occurred by 'implanting it inside the individual' as biologically rooted in humans, thus material, universal, and necessary (rather than through the godly gaze of either reason or sympathy in the seventeenth and eighteenth centuries). It mitigated the individual's discreteness, thus facilitating investigation of groupings through individuals and their relationships.

I use 'collectivity' as that which bounds a much narrower group than the above use of 'population,' yet more amorphic than the 'society' as nation-state.

In what follows, my argument is that transfer from the SL/NL framework served to bridge that dichotomy between individual and collectivity, so that through the use of particular mechanisms of evolutionizing, individuals became fundamentally related to collectivities. Furthermore, this framework played a crucial role in enabling the 'spectrum of the roles of collectivity.'

Nineteenth-century general questions relating to the social were heavily colored by the gradually vanishing theological answers to contemporary issues, such as the following: What connects humans together so that they can live in groups? What are the fundamental features of these relations? What are the foundations of sociality? Such general questions were formulated more specifically in the search for models of societal organization, a much looked-for formula of cohesiveness in and of society, of solidarity in and of the nation-state, or of the possibility of some material universal foundation of being human. Gradually, there emerged a range of answers, mostly from an individualistic perspective, some of which relied on transfer between the biological and the social. The individualist perspective suggested that one could integrate the 'collectivity' and the 'sociality' components into the physiological, even neural, makeup of the individual and thus into the 'experience' of relations among individuals, whereby investigating the individual would provide sufficing answers for the above questions.

Yet, some contemporaries pointed out that such perspective was insufficient and that there would still arise difficulties in any attempt to explain social institutions, social norms, or sociality at large. One way posited then of dealing with such difficulties was a symmetrically opposing perspective, whereby individuals, or fundamental aspects of them, would be constituted through their partaking in a collectivity, being shaped and organized also in their personal identity by/through social institutions and social norms, and be dependent on them. Furthermore, collectivities bore attributes at the group level, as containing individuals, that would not fit the individuals within these collectivities separately. If society/culture advanced progressively, in a long-term cumulative manner, then no single individual could have embodied/possessed that, and progress depended on the collectivity as a whole. Collectivity-belonging, characterizing modern societies in such a view, produced complex networks of relations and interactions, creating human 'density,' whereby each individual partook in a variety of nonoverlapping sub-collectivities, differentially central to her/his social/individual existence. In this view, modern society was resilient by virtue of being both complex and heterogenous. This was Durkheim's generalized position. Present-day collectivity theorists convey quite close positions through terms such as 'we mode' versus 'I mode,' etc. (e.g., Gilbert, 1989, 2014; Ingold, 1986a, b; Tollefsen, 2002, 2015; Tomasello, 2009, 2014; Tuomela, 2007, 2013; even Searle, 1995).

The predominance of an individualist ontology in this period impelled biologists, sociologists, and psychologists, who did not adopt the 'collectivity mode,' yet were aware of the inherent difficulties in this ontology, to look for ways and means to cope methodologically and epistemologically with issues related to collectivities and collectivity frameworks, i.e., to sociality. I suggest that in the emerging fields I deal with it was not carried out through assuming a common *cultural*-quasi-chronological *history* of the human species. Rather the SL/NL perspective on adaptation, evolution, and heredity, together with the assumption of memory as heritable (often as material), served to link individuals harboring memory common to the human species or else to sub-collectivities within it. This feat was achieved by embedding in the individuals a universally or at least generally shared memory structure and content, which related both to individual development and to general *evolutionary* and developmental history of humans (seen as part of the evolution of 'living nature'). The individual was described and analyzed through its ontogenetic development.

Yet, its very development relied on there being indispensable inherited components, binding the individual evolutionarily and developmentally to some collectivity, minimally phylogenetically to human species. These components could be inherited evolutionized modes of acting, of experiencing, of perceiving, and *of memory*, i.e., modes of accumulated acquired modifications. These modes could be either universal or specific to a subsection of the human species. For Spencer perceptual experience and learning processes very gradually formed and fashioned biological–psychological competences—a mechanism which operated via a collectivity's biological heredity. Experiencing and behavior manifested their collective hereditary transmission origins, working equally on all possible generations of individuals of the relevant group. In their turn, these competences were the enabling preconditions for further individual accumulation of experience and thus for learning and communication, also allowing for hierarchical levels of individuals. This heredity was conceptualized as being liminal between the individual and the collectivity (the 'species,' the 'race'), forming an enabling condition for sociality, as an all-important component of both the species and the individual. For Spencer heredity made collectivities into necessary *relational* components in the construction and constitution of individuals. There were a number of extant versions on the nature and role of memory/heredity after Spencer's writings of the 1850s and 1860s, conspicuously Ribot, but especially Hering (1870), Butler (1880), and later in the early twentieth century the work of Semon (1922 [1908]; see, e.g., Schacter, 2001).

During the later years of the nineteenth and the early twentieth centuries, the scientific and cultural contexts, in which this particular transfer from the evolutionary biological to the psychological occurred (memory as inherited, memory likened to heredity), helped turn it into a two-way traffic, whereby memory gradually became the metaphor for evolutionary heredity, and evolutionary mechanisms were deployed as 'internalized bridges' between 'individual' and 'collectivity.' Note that in these contexts there arose also a search, as in Spencer's work, for a broader causal repertoire, e.g., relations between whole and parts, looping interactions, and emergence. Thus, when discussing the impact of varieties of Lamarckism on the new 'scientific' social sciences, Spencer offered a mode of moving from individuals to collectivities through SL evolutionizing. Components of that mode were partially adopted, explicitly or implicitly, by members of the new 'scientific' sociology and psychology communities, most conspicuously the notion of the inherited impact of a collectivity's past generations. As shown in the chapters on Spencer and Ribot, 'races' were one of the categories of collectivity explicated and elaborated on in that manner.

In Spencer the bridging mechanism was effectively present in the notion of collectivities inhering within individuals through the back-and-forth mechanism between biological heredity and cultural–psychological inheritance of acquired characteristics. He deployed it in discussing both his psychology and his sociology and became a collectivist in disguise. Ribot fashioned his bridging mechanism between individuals and various kinds of collectivities, including 'races' on that model. Freud too used the SL/NL memory model to fashion his particular evolutionary bridging between individuals and forms of collectivity.

Thus, transfer of SL/NL evolutionizing models, especially those that related memory to inheritance of acquired modifications, served to construct bridges over the nineteenth-century dichotomy between individuals and collectivity. Durkheim, who was a conspicuous exception, as he considered collectivities as the basic units of his system, mostly applied the SL/NL 'bridging' repertoires to the relations among individuals, using these repertoires to constitute the individuals within the collectivities shaping them.

Gradually, from the turn of the twentieth century on, evolutionized biology ceased to play that bridging role, in part because of the crisis over the explanation of heredity in SL/NL biology versus Weismann/Weismannism/neo-Darwinism, starting in the 1890s and also resulting in a clear-cut separation between cultural–psychological inheritance and biological heredity.

In sum, transfer from SL/NL repertoires was used as a powerful tool to stretch and overcome the focus of contemporaneous frameworks on dichotomous individuals, seeking answers outside them, when attempting to formulate novel problématiques relating together individuals and collectivity. These transfers were perceived to supply 'thickness' to these relationships and helped posit individuals as natural entities, concurrently constructing the novel fields as universalizing social sciences.

References

Butler, S. (1880). *Unconscious memory*. Ballantyne Press/Hanson and Co.

Carmel, Y., Shavit, A., Lamm, E., & Szathmáry, E. (Eds.). (2023). Human socio-cultural evolution in light of evolutionary transitions. *Philosophical Transactions of the Royal Society B, 378*(1872).

Gilbert, M. (1989). *On social facts*. Routledge

Gilbert, M. (2014). Joint commitment: How we make the social world.

Hacking, I. (1990). *The taming of chance*. Cambridge University Press.

Hacking, I. (1994). How numerical sociology began by counting suicides: From medical pathology to social pathology. In I. Bernard Cohen (Ed.), *The natural sciences and the social sciences* (pp. 101–133). Kluwer Academic Publishers.

Hering, E. (1870). *Über das gedächtniss als eine allgemeine function der organisirten materie*. K.K. hofund staatsdruckerei.

Ingold, T. (1986a). *The appropriation of nature: Essays on human ecology and social relations*. Manchester University Press.
Ingold, T. (1986b). *Evolution and social life*. Cambridge University Press.
Ioannou, I., & Laskowski, K. L. (2023). A multi-scale review of the dynamics of collective behaviour: From rapid responses to ontogeny and evolution. *Philosophical transactions of the royal society B, 378*(1874).
Lukes, S. (1973a). *Émile Durkheim: His life and work: A historical and critical study*. Allen Lane.
Lukes, S. (1973b). *Individualism*. Blackwell.
Schacter, D. L. (2001). *Forgotten ideas, neglected pioneers: Richard Semon and the story of memory*. Psychology Press.
Searle, J. R. (1995). *The construction of social reality*. MIT Press.
Semon, R. W. ([1908] 1922). *The Mneme* (Louis Simon, Trans.). G. Allen and Unwin.
Tollefsen, D. P. (2002). Challenging epistemic individualism. *ProtoSociology, 16*, 86–117.
Tollefsen, D. P. (2015). *Groups as agents*. Polity Press.
Tomasello, M. (2009). *Why we cooperate*. MIT Press.
Tomasello, M. (2014). *A natural history of human thinking*. Harvard University Press.
Tuomela, R. (2007). *The philosophy of sociality: The shared point of view*. Oxford University Press.
Tuomela, R. (2013). *Social ontology: Collective intentionality and group agents*. Oxford University Press.
Turner, S. P. (1986). *The search for a methodology of social science: Durkheim, Weber, and the nineteenth century problem of cause, probability and action*. D. Reidel.

Chapter 10
Concluding Reflection

In the beginning of the introduction, I have asked how and why a particular kind of evolutionized biology became significant for constituting novel scientific fields. My answers, in the plural, have been given by way of describing and analyzing case studies of discipline founders in their particular contexts and deployment of evolutionized biology.

Yet, the book is not simply a series of case studies on individuals but rather an account of the emergence of 'scientific' social sciences and their entwinement with varieties of Lamarckism, pointing thus to broader contemporaneous currents. In order to expose the commonalities of an implicit conceptual frame underlying these diverse fields, I have emphasized particular aspects of their theories and practice, centering them rather than others. I have focused on these aspects to articulate the relationship between SL/NL and the founding endeavors which, I suggest, attempted to produce new fields, new objects, and possibilities for new disciplines and their methodologies, with their new scientific personae. I have tried to detail the intricate interactions, contemporaneous and intergenerational, which were crucial for the emergence of the new fields. The book presents individualized processes, since for severe constraints of space I have addressed eclectically rather than thoroughly elaborated on these individuals' situatedness within communities and the situatedness of those communities within cultures and their values and societal and institutional pressures and needs.

Some of these founders thought that the repertoire of SL/NL evolutionized biology could be of use in formulating and providing an adequate account of the complexity of specifically human knowledge—individual and collective—in experiential, contextual, reflexive, historical terms. In differing ways, and by relating to each other's work, all of them believed in 'enlightenment,' for some even 'progress,' to be brought about by science. Thus, they sought to create 'a network of objectivity' for and of their fields through endowing them with necessity and

S. B. Gissis, *Lamarckism and the Emergence of 'Scientific' Social Sciences in Nineteenth-Century Britain and France*, History, Philosophy and Theory of the Life Sciences 36, https://doi.org/10.1007/978-3-031-52756-2_10

universality, employing those yarns of scientific knowledge on individuals and collectivities 'for the good of humankind.'

Spencer related to Lamarck; Jackson, Ribot, Durkheim, and Freud related to Spencerian Lamarckism and to neo-Lamarckism; and within that network Ribot and Freud related to Jackson. Lamarck aspired to constitute a novel 'science' of living entities; Spencer, Ribot, Durkheim, and Freud deployed the SL/NL repertoire to constitute nonbiological disciplines dealing with humans, aspiring to achieve a status of 'science' and of 'scientists' for their endeavors and for those laboring within them. Jackson held a somewhat similar aspiration for a section of medicine.

They used SL/NL in diverse ways, privileging both plasticity and evolutionized hierarchy as exemplified in their modes of metaphorizing and analogizing. As argued throughout, all of them deployed metaphors, analogies, models, and mechanisms, from the broadly construed SL/NL repertoire as important tools, proving by the very rhetoric of their endeavor that they considered the literal–metaphorical continuum as a useful one. Even when claiming to see these uses as 'scaffoldings,' some components did become an integral part of their 'scientific' edifice. Other components were deployed as interim answers to difficult questions, as props of complex assumptions in the absence of an existing relevant terminology, and in general as meaningful toolbox in the labor of constructing their science. Metaphors were used both as creators–constructors of affinities and as pointers of distance, often the same ones situated in different theoretical contexts and deployed for different purposes, in particular in order to claim the autonomy and separateness of the field being constructed through the use of those metaphors.

Their myriad affiliations with and belonging to various communities and to institutions of knowledge and their moving among them were significant for their ability to make both intermediary and radical conceptual transitions and transformations. In multiple, complex, direct and indirect ways, their search channeled and helped focus diffuse as well as more concentrated intellectual disaffection and search within relevant contemporaneous communities.

Concurrently, their entangled ties, their positions as simultaneously insiders and outsiders, central in one sense and at least temporarily marginal or peripheral in another crucial sense, were significant. This enabled them to locate and question contradictions, conceptual insufficiencies, inabilities to deal with new objects within existing frameworks, from the inside and from the outside, to seek answers through posing questions that stretched existing frameworks. This questioning process then compelled them to seek answers outside those frameworks, to gradually construct novel ones, positing new objects, establishing new repertoires of practice. Thus, they labored to research phenomena posited as new, revealed through the deployment of conceptual repertoires not used earlier in their fields, transferred from SL/NL, but at times extant within their societal contexts. They devised new vocabularies and strenuously performed boundary work for legitimation and in some cases strove to constitute new institutional frameworks. Thus, diverse hybrids emerged that only gradually became coherent novel frameworks. Only then, retrospectively, could those hybrids be looked upon as intermediary stages.

There were social–political facets to the SL/NL framework stemming from the focus on organism–environment interactions, a view of inheritance that would emphasize plasticity (but also determinism), a role for behavior/learning, an implicit focus on processes and continua. By virtue of these, and given the predominant view of the centrality of the 'individual' in the biological and uniquely human disciplines, SL/NL was deployed by some of these founder figures in complex ways to bridge the deemed ontic or epistemic hiatus between the collectivity and the individual. The diverse uses of SL/NL by these figures demonstrated that within their contemporary social/cultural context, this repertoire could be deployed to advocate social reform or reinforce existing racist dogmas, at times through that very act of bridging.

In the early twentieth century, the human sciences had undergone a process discussed by historians of these fields as 'separation' of the social/historical and the biological, or as a 'purging/purification' or 'denaturalization,' regarded then also as emancipation. This process has also been conceived, in a manner relevant to this book, as relying on the drastic separation of genetic/hard heredity from the modifications acquired through life course, considered as soft heredity (e.g., Meloni, 2016, 2019; Kronfeldner, 2009). Throughout the twentieth century, there occurred a number of attempts to establish newly 'combined disciplines' in the human sciences, most prominently sociobiology and early evolutionary psychology.

In the recent generation of scholars, there have been works either pointing to a new de facto intertwining (e.g., Rabinow's bio-sociality) or manifestly urging for a novel 'interdisciplinarity' of the socio-cultural and the evolutionary biological, either by virtue of an increased role of the environment, the growing legitimation of epigenetics within 'the modern synthesis,' or because of new conceptions of agency, directionality, intention, and even sociality in evolutionary biology. These attempts have been carried out through metaphorizing or analogizing nature–culture, evolutionary biology–social–human sciences, through the creation of new facilitating biosocial hybrids posited as 'boundary objects' (Star & Griesemer, 1989; Star, 2010; Strathern, 2004, 2007).

The elaborate studies of earlier endeavors, presented in this book, may perhaps be of some comparative use in probing into recent novel ones, which assume the in-principle or in-practice usefulness/relevance/productivity/indispensability of the biological for the explanation and understanding of the social–cultural. In which ways can one discuss such interdisciplinarity taking place between groups (not only in the work of particular individuals)? What ways are suggested to overcome the nature/culture structurally and institutionally entrenched lines of power relations, with conspicuous hierarchies among disciplines and knowledge areas and inside them, where 'Science,' with a capital S, exacts higher value than culture, humanities, and social sciences (all the more so with rampant balkanization of disciplinary territories)? What means are called for to create a common shared space, resulting in a coherent new theoretical alignment, between two distinct epistemic cultures?

Proponents of such novel production (e.g., Meloni, 2016; Ingold & Palsson, 2013) or even those differing positions that support diverse notions of cultural evolution (e.g., Boyd & Richerson, 2005; Henrich, 2016; Jablonka & Lamb, 2014;

Mesoudi, 2016; Sterelny, 2012 and 2021, a whole issue of the *Royal Society Philosophical Transaction B* in 2021 dedicated to "Foundations of Cultural Evolution," (vol. 376/issue 1828) and once again in 2023 "Human Socio-Cultural Evolution in Light of Evolutionary Transitions" (vol. 378/issue 1872)) do offer answers. While some view 'the social' as wholly grounded in the authority and logic of the biological, varied positions of the cultural evolution spectrum propose to apply diverse evolutionary explanatory toolboxes to human collectivities, to complex and constructive culture, and to the study of language, the development of social norms, etc. Some do so even when adhering to any version of distinct channels of biological and cultural inheritance (discussed, e.g., in Kronfeldner, 2018, 2021). Opponents of any such moves question, in somewhat weaker (e.g., R. Smith, 2005, 2007) or stronger (Collingwood, e.g., 1961, and his disciples) versions, whether a sufficing account can be given of what they consider as the specifically human, either because of the two-way interaction between the emic and the etic in human cultures, or their self-reflexivity, beyond the issues of storage/transfer and transmission, or because of the role of ideas and modes of learning in individuals and among people, which Evelyn Fox Keller (2016: 38) termed the 'material powers of mind' in the dynamics of change. In both versions, reflexive-knowledge situations in the human sciences are seen as 'historical entities,' and 'nature' is not seen as a given through evolutionary processes.

The founder figures discussed in this book assumed that their endeavors acquired with the help of their transfer practices significant empirical advantage over other contemporary founding endeavors. Present-day positions ask how 'evolution' enabled the emergence of consciousness, and with it ever-changing human life, while their opponents ask how consciousness and reflexivity have enabled the possibility of knowledge, including knowledge of evolution, so that when self-description categories change, human life changes too.

References

Boyd, R., & Richerson, P. J. (2005). *The origin and evolution of cultures*. Oxford University Press.

Collingwood, F. J. (1961). *Philosophy of nature*. Prentice Hall.

Henrich, J. (2016). *The secret of our success: How culture is driving human evolution, domesticating our species, and making us smarter*. Princeton University Press.

Ingold, T., & Palsson, G. (Eds.). (2013). *Biosocial becomings: Integrating social and biological anthropology*. Cambridge University Press.

Jablonka, E., & Lamb, M. (2014). *Evolution in four dimensions: Genetic, epigenetic, behavioral, and symbolic variation in the history of life*. The MIT Press.

Keller, E. F. (2016). Thinking about biology and culture. *The Sociological Review Monographs, 64*(1), 26–41.

Kronfeldner, M. E. (2009). If there is nothing beyond the organic...: Heredity and culture at the boundaries of anthropology in the work of Alfred L Kroeber. *Naturwissenschaften, Technik und Medizin, 17*(2), 107–133.

Kronfeldner, M. E. (2018). The right to ignore: An epistemic Defense of the nature/culture divide. In R. Joyce (Ed.), *The Routledge handbook of evolution and philosophy* (pp. 210–224). Routledge.

Kronfeldner, M. (2021). Digging the channels of inheritance: On how to distinguish between cultural and biological inheritance. *Philosophical Transactions, B.376/1828*. https://doi.org/10.1098/rstb.2020.0042

Meloni, M. (2016). The transcendence of the social: Durkheim, Weismann, and the purification of sociology. *Hypothesis and Theory, 1*, 11.

Meloni, M. (2019). *Impressionable biologies: From the archaeology of plasticity to the sociology of epigenetics*. Routledge.

Mesoudi, A. (2016). Cultural evolution: A review of theory, findings and controversies. *Evolutionary Biology, 43*(4), 481–497.

Smith, R. (2005). Does reflexivity separate the human sciences from the natural sciences? *History of the Human Sciences, 18*(4), 1–25.

Smith, R. (2007). *Being human: Historical knowledge and the creation of human nature*. Columbia University Press.

Star, S. L. (2010). This is not a boundary object: Reflections on the origin of a concept. *Science, Technology, and Human Values, 35*(5), 601–617.

Star, S. L., & Griesemer, J. R. (1989). Institutional ecology, 'translations' and boundary objects: Amateurs and professionals in Berkeley's Museum of Vertebrate Zoology, 1907-39. *Social Studies of Science, 19*(3), 387–420.

Sterelny, K. (2012). *The evolved apprentice: How evolution made humans unique*. The MIT Press.

Sternly, K. (2021). *The Pleistocene social contract: Culture and cooperation in human evolution*. Oxford University Press.

Strathern, M. (2004). *Commons and borderlands: Working papers on interdisciplinarity, accountability and the flow of knowledge*. Sean Kingston Publications.

Strathern, M. (2007). Interdisciplinarity: Some models from the human sciences. *Interdisciplinary Science Reviews, 32*(2), 123–134.

References

[illegible]

Appendix: Concise Biographical Portraits

Jean-Baptiste Pierre Antoine de Monet, chevalier de Lamarck, usually named **Jean-Baptiste Lamarck** (1744–1829), was born in Picardy, France. He removed the nobility tag of his impoverished family from his name early in the first year of the French Revolution. As a teenager, he served in the French army until injured in 1768. Already while in the army he developed a keen interest in botany. After leaving the army, he studied medicine for four years and worked as an accountant to make ends meet. While studying (including with the well-known naturalist Jussieu at the Jardin du Roi, the distinguished natural history institution), he wrote a classificatory book of French flora, shunning Linnaean epistemology. His *Flore Françoise* attracted Buffon's attention, and Buffon helped to have it published in 1778. Through the ministrations of Buffon, Lamarck became a tutor of his son, became part of the Jardin du Roi staff, though non-salaried, and was also elected to the botanical section in the *Académie des sciences*. Later on he wrote the botanical section of the *Encyclopédie méthodique.* During the French Revolution, the Jardin du Roi was turned into *Muséum national d'Histoire naturelle*, and Lamarck was nominated as one of its permanent members. The *Académie* was turned into the *Institut national des Sciences et des Arts*, and Lamarck became a member of its sciences section, as well as a member of some of the active learned societies of the period. Yet, since he had no previous economic resources of any kind to rely on, as he came from an impoverished family, and had a large family of his own, his economic situation was very shaky all throughout his many scientific activities and until his death. During the first decade of the French Revolution (1789) and the rule of the directory (from 1795), and almost until Napoleon declared himself First Consul (1801) and then emperor (1803), Lamarck was trying to develop a more general system of dynamic natural history of the earth, including chemistry and geology. These efforts were not well received, and his status within the museum was further diminished with the rise of Cuvier to central positions of power there and in the field of sciences in general in France. Lamarck's prolonged attempt to develop a

S. B. Gissis, *Lamarckism and the Emergence of 'Scientific' Social Sciences in Nineteenth-Century Britain and France*, History, Philosophy and Theory of the Life Sciences 36, https://doi.org/10.1007/978-3-031-52756-2

science of meteorology was also undermined by Cuvier. Lamarck taught and wrote on living entities, including humans, on fossils, and on natural history classifications, until the late teens of the nineteenth century, i.e., during the rule on Napoleon as emperor and under the restored crown (in 1814) which also changed hands within the next ten years, until he became completely blind. He died in dire poverty, and his library was sold soon after.

Herbert Spencer (1820–1903) was born in Derby to a middle-class dissenter (Wesleyan Methodist and later on Quaker) well-educated wider family. He was the only surviving child in his nuclear family. He spent his adolescent years at his uncle's house, who became an Anglican minister in a place near Bath. Both his father and his uncle were connected to groups of provincial religious intellectuals. From 1837 to 1848, somewhat intermittently, Spencer worked in railway engineering, i.e., during the construction of the great lines. During that period, he was involved in writing for their professional magazines. He also read then both Lyell's *Principles of Geology* and phrenological writings. In the 1840s, he became deeply engaged in radical political circles in Birmingham. In 1842 he wrote for *The Nonconformist* on spheres of governing. In 1844 he became a subeditor and writer in a radical paper, *The Pilot*, and then went back to engineering until he moved to London in 1848 to become a subeditor in *The Economist.* He entered into intellectual circles in London and struck up friendships that lasted all his life. In 1850 his first book, *Social Statics*, was published. A series of seminal articles followed and then in 1855 the first edition of *Principles of Psychology. First Principles* followed in 1862, then *Principles of Biology* (1864–1867) and the second, much enlarged and changed edition of *Principles of Psychology* (1870/1872), and in 1874 *Study of Sociology.* Spencer turned quite quickly into a well-known, established writer, contributor to journals, a prominent intellectual radical figure, and a member of the X Club, which included some of the highly influential and innovative scientists of the period. He continued writing and publishing essays and his *Principles of Sociology*, and later on *Data of Ethics*, but also revised second and third editions of his other books, as well as a two-volume autobiography. His renown was possibly at its height from the 1860s to the late 1880s, when his work was widely translated and disseminated in Europe, USA, and Russia but also in Asia and Latin America. From the 1890s on, his elaborate system lost much of its appeal and audiences.

John Hughlings Jackson (1835–1911) was born in Yorkshire to a family whose economic state worsened dramatically during his early childhood. His three elder siblings migrated to New Zealand. Early on he adopted Hughlings, his Welsh mother's surname, as his middle name. He studied at York Medical School, where he was also apprenticed to a central local physician, and was a student of Thomas Laycock. While there he also observed mental disease at the York Retreat under Dr. Tuke. He spent a year at St. Bartholomew's Hospital under Dr. James Paget, London, which enabled him to become a member of the Royal College of Surgeons. Thereafter, he passed his examination for the MD degree at the University of St. Andrews (1860). He had further practice as a resident medical officer at York Dispensary. In London he first worked at the Metropolitan Free Hospital, then at the Islington Dispensary, and at Moorfields Eye Hospital. Starting in 1862 he worked at the National Hospital for the Relief and Cure of the Paralysed and Epileptic at Queen Square, working his way through the positions of assistant physician and then physician, and also at the London Hospital. He investigated diseases of the nervous system during 1866–1876 at the West Riding Lunatic Pauper Asylum, Wakefield, then under JamesCrichton-Browne. He married his cousin in 1865; she died in 1876. They had no children. For a while during the 1860s he was employed together with Dr. Jonathan Hutchinson as medical reporter on hospital practice for J. Churchill's *Medical Times and Gazette,* which allowed both of them privileged access to hospitals. Hutchinson, who came from a well-established London family, was of great significance in enabling Jackson, who came from a provincial medical school, to enter the London hospital system. In 1868 Jackson became a fellow of the Royal College of Physicians and in 1878 a fellow of the Royal Society. He worked in the hospital until his retirement in 1894. He continued publishing short papers of medical cases in *Lancet* until 1909. He was honored by various universities, e.g., Glasgow, Edinburgh, Bologna, Leeds, Yorkshire, and the Neurological Society instituted an annual Hughlings Jackson lecture in 1897, but no state honors.

The principal sites where the work I relate to was conceptualized and carried out were as follows: West Riding Lunatic Pauper Asylum, Wakefield, under James Crichton-Browne, primarily during 1866–1876, and the National Hospital at Queen Square, London (1862–1906), originally under Charles-Édouard Brown-Séquard. In accordance with Jackson's will, all his drafts, notes, letters, etc., were burned after his death.

Théodule Armand Ribot (1839–1916) was born in a small community in Brittany and studied in a school of a more central town there. He then worked there for two years as an officer in the local branch of the Ministry of Finance. In 1862, after trying twice, he was accepted to the Ecole Normale. He studied philosophy mostly with Caro, Lemoine (of the French eclectic tradition), and Lachelier (of the idealist tradition). Relations with some of his cohort students were of immense significance later. He then taught philosophy at provincial lycées—Vesoul and then Laval—both under strong Catholic influence. In 1870 he published his first book *English Psychology*, where Spencer was significant. The book was an indirect but sharp criticism of figures of the French eclectic tradition. In 1872 he handed in his PhD dissertation in philosophy, *Psychological Heredity*, and the Latin language one, then obligatory, on David Hartley. In spite of its strongly positivist anti-spiritualist message, the former was approved in 1873. Ribot could not get a job in the higher education system, took a leave of his lycée teaching, and enrolled in the faculty of medicine in Paris while spinning a network of connections with prominent positivist philosophers in England and in France. His book on psychological heredity, based on his PhD dissertation, came out in 1873. Together with his lifelong friend Espinas, he translated Spencer's *Princ.Psycho.*ii. In 1874 he published a book on Schopenhauer. In 1876 he founded an independent journal *Revue philosophiquede la France et de l'étranger* and published *German Psychology* in 1878. In 1885/1886 he taught a complementary philosophy course at the Sorbonne on 'experimental psychology.' In 1888–1902 he moved on to teach an open course on 'experimental and comparative psychology' at the College de France. In 1889 he was on the organizing committee of the first International Congress of Psychology. From 1881 to 1885, he published a series of three books on the diseases of memory, will, and personality. In later years, he published on the psychology of attention, sentiments, creative imagination, and some more. He edited the journal almost until his death.

Emile David Durkheim (1858–1917) was born in Épinal, in that part of Lorraine that remained under French rule after the 1871 defeat of France by Germany. He was born to a Jewish orthodox family, and his father (like all males three generations back) was a rabbi, originally under the French Jewish consistory system, which was abolished in that area in 1871 (adversely affecting his father's salary). There were close ties with other members of both sides of the family (Marcel Mauss was his nephew). On his mother's side, the women in her family had high school education. He studied at the École normale superieure (ENS) where he stood out in his studies of philosophy; his PhD dissertation became his first book, *De la division du travail social* (*Division*). There he met some of his lifelong friends and teachers who left a deep impact on him. Soon after graduating, Durkheim started teaching philosophy in diverse provincial lycées and read copiously English and French as well as some German authors, especially philosophy, 'moral sciences,' and law/politics. Between 1885 and 1896, he published book reviews in Ribot's journal, *Revue philosophique*. In 1886, Durkheim received a scholarship intended to broaden the views of promising humanities scholars by learning at German model research universities. Once there he spent a few months in Wundt's famous experimental psychology laboratory. Upon his return, Durkheim was nominated to teach a pedagogy and social science course at the University of Bordeaux. Just before moving there, he married a wife from a well-to-do Jewish family. In 1896 he organized a work collective around a newly founded journal, *L'année sociologique*, which was published until the war in 1914. In 1902 Durkheim was nominated to teach at the Sorbonne in Paris (which was united with the ENS into one school in 1903). He was a staunch supporter of the Third Republic, its institutions, civic aims, and educational reform. Durkheim had clear socialist leanings but did not join the party, though close to Jean Jaures, presumably since he thought it might compromise his position on sociology. He was politically engaged and active in the League for Human Rights during the Dreyfus affair in France and for a while then rather isolated in his own university and faced anti-Semitic harassment. He was a secular person but kept his Jewish family ties. Gradually, Durkheim turned into one of the prominent intellectuals of the Third Republic and was involved in numerous controversies and educational enterprises. Once World War I started, he became involved in the French propaganda war effort. The death during the war of some of his close students and members of the *AS* group (nine people altogether) and the death of his own son in late 1915 undermined his health and brought about his early death in 1917.

References

Achimastos, M. (2019). D'une Année à l'autre: la sociologie religieuse durkheimienne en marche. *L'Année Sociologique, 69*(1), 77–102.

Ackerknecht, E. H. (1967). A plea for a "Behaviorist" approach in writing the history of medicine. *Journal of the History of Medicine and Allied Sciences, 22*(3), 211–214.

Adam, A. E. (1988). *Spontaneous generation in the 1870S: Victorian scientific naturalism and its relationship to medicine* (PhD). Sheffield City Polytechnic http://shura.shu.ac.uk/2695/.

Adriaens, P., & De Block, A. (2010). The evolutionary turn in psychiatry: A historical overview. *History of Psychiatry, 21*(2), 131–143.

Agathon (Henri Massis and Alfred de Tarde). (1911). *L'Esprit de la nouvelle Sorbonne: la crise de la culture classique, crise du Français*. Mercure de France.

Agathon. ([1913] 1995). *Les Jeunes Jens d'aujourd'hui.* Plon Nourrit.

Agathon (Henri Massis and Alfred de Tarde). ([1913] 1995). *Les Jeunes jens d'aujourd'hui* (J. J. Becker, Ed.). Imprimerie nationale.

Agulhon, M. (1979). *Marianne au combat: l'imagerie et la symbolique républicaine de 1789 à 1880*. Flammarion.

Alexander, J. C. (1982). *The antinomies of classical thought: Marx and Durkheim*, vol. 2: *Theoretical logic in sociology*. University of California Press.

Alexander, J. C. (1986a). Rethinking Durkheim's intellectual development I: On 'Marxism' and the anxiety of being misunderstood. *International Sociology, 1*, 91–107.

Alexander, J. C. (1986b). Rethinking Durkheim's intellectual development II: Working out a religious sociology. *International Sociology, 1*, 189–201.

Alexander, J. C. (2005). The inner development of Durkheim's sociological theory: From early writings to maturity. In J. C. Alexander & P. Smith (Eds.), *The Cambridge companion to Durkheim* (pp. 136–159). Cambridge University Press.

Algazi, G. (2016). Exemplum and Wundertier: Three concepts of scholarly persona. *Low Countries Historical Review, 31*(4), 8–32.

Allen, G. E. (1985). Heredity under an embryological paradigm: The case of genetics and embryology. *Biological Bulletin, 168*, 107–121.

Amundson, R. (1996). Historical development of the concept of adaptation. In M. R. Rose & G. V. Lauder (Eds.), *Adaptation* (pp. 11–53). Academic Press.

Andermann, A. A. J. (1997). Hughlings-Jackson's deductive science of the nervous system: A product of his thought collective and formative years. *Neurology, 48*(2), 471–481.

S. B. Gissis, *Lamarckism and the Emergence of 'Scientific' Social Sciences in Nineteenth-Century Britain and France*, History, Philosophy and Theory of the Life Sciences 36, https://doi.org/10.1007/978-3-031-52756-2

Anderson, W. (2013). The case of the archive. *Critical Inquiry, 39*(3), 532–547.
Andler, C. (1932). *Vie de Lucien Herr.*
Ankeny, R. (2011). Using cases to establish novel diagnoses: Creating generic facts by making facts travel together. In M. Morgan & P. Howlett (Eds.), *How well do facts travel? The dissemination of reliable knowledge* (pp. 252–272). Cambridge University Press.
Annales du Muséum national d'Histoire naturelle. (1802, 1803, 1804, 1805, 1806). Paris: Museum national d'Histoire naturelle. Les frères Levrault.
Appel, T. A. (1987). *The Cuvier-Geoffroy debate: French biology in the decades before Darwin*. Oxford University Press.
Ash, M. (2003). Psychology. In T. Porter & D. Ross (Eds.), *The Cambridge history of science, vol. 7: The modern social sciences*. Cambridge University Press.
Assmann. (1997). *Moses the Egyptian: The memory of Egypt in Western Monotheism.* Harvard University Press.
Baciocchi, S., & Mergy, J. (2003). *Emile Durkheim-L'Évaluation en comité* (S. Baciocchi & J. Mergy, Eds.). Berghahn Books.
Bacopoulos-Viau, A. (2019). From the writing cure to the talking cure: Revisiting the French 'discovery of the unconscious. *History of the Human Sciences, 32*(1), 41–65.
Baczko, B. (1992). La constitution de l'an iii et la promotion culturelle du citoyen. In F. Azouvi (Ed.), *L'Institution de la raison: la révolution culturelle des idéologues* (pp. 21–40). J. Vrin.
Baczko, B. (1994). Enlightenment and the institution of society: Notes for a conceptual history. In W. Melching & W. Wyger (Eds.), *Main trends in cultural history: Ten essays* (pp. 95–120). Rodopi.
Baedke, I. (2021). The origin of new levels of organization. In D. S. Brooks, J. DiFrisco, & W. C. Wimsatt (Eds.), *Levels of organization in biological sciences* (pp. 153–173). The MIT Press.
Baker, K. M. (1973). Politics and social science in eighteenth-century France: la Société de 1789. In F. Bosher (Ed.), *French government and society 1500–1850: Essays in memory of Alfred Cobban* (pp. 208–230). Athlone Press.
Baker, K. M. (1990). *Inventing the French revolution: Essays on French political culture in the eighteenth century*. Cambridge University Press.
Baldwin, J. M. (1886). A new factor in evolution. *American Naturalist, 30*(441–51), 536–553.
Baldwin, J. M., Ed. (1901–1905). *Dictionary of philosophy and psychology*, 3 vols. McMillan.
Baldwin, M. (2015). The successors of Club X? Late Victorian naturalists 1869-1900. In B. Lightman & G. Dawson (Eds.), *Victorian scientific naturalism: Community, identity, continuity* (pp. 288–308). Chicago University Press.
Bandinelli, A. (2013). Le Origini Chimiche della Vita:legami tra la rivoluione de Lavoisier e la biologia di Lamarck. *Biblioteca di Nuncius, Studi e Testi, 71*, 133–200.
Bange, C. (1997). Théorie et practique de la taxonomie végétale chez Lamarck et ses continuateurs. In G. Laurent (Ed.), *Jean-Baptiste Lamarck 1744–1829* (pp. 207–225). Editions de CTHS.
Barberis, D. S. (2002). Moral education for the elite of democracy: The 'Classe de Philosophie' between sociology and philosophy. *Journal of the History of the Behavioral Sciences, 3*(4), 355–369.
Barberis, D. S. (2003). In search of an object: Organicist sociology and the reality of society in fin-de-Siècle France. *History of the Human Sciences, 16*(3), 51–72.
Barbie, E. (1876). *L'origine des espèces au moyen de la sélection naturelle, ou La lutte pour l'existence dans la nature*. C. Reinwald.
Barclay, J. (1992). *A caring community: A centenary history of the National Society for epilepsy and the Chalfont Centre, 1892–1992*. National Society for Epilepsy.
Barrès, M. ([1897] 1937). *Les déracinés*. Plon.
Barrès, M., & Maurras, C. ([1888–1923], 1970). *La République ou le roi: correspondance inédite* (H. Maurras & N. Maurras, Ed.). Plon.
Barrows, S. (1981). *Distorting mirrors: Visions of the crowd in late nineteenth-century France*. Yale University Press.

Barsanti, G. (1995). La naissance de la biologie: observations, theories, metaphysiques en France 1740-1810. In C. Blanckaert, J. L. Fischer, & R. Rey (Eds.), *Nature, histoire, Société: Essays en homage a` J. Roger* (pp. 197–227). Klinchsiech.

Barsanti, G. (1997a). Lamarck et la naissance de la Biologie. In Goulven Laurant (Eds.), *Jean-Batiste Lamarck 1744–1829* (pp. 349–68). Editions du CTHS.

Barsanti, G. (1997b). Le 'Systeme de la nature' de Lamarck (1794): Analyse d'un ambitieux projet avorté d'apres un manuscript oublié. In C. Blanckaert, C. Cohen, P. Corsi, & J.-L. Fischer (Eds.), *Le Muséum au premier siècle de son histoire* (pp. 219–228). Éditions du Muséum national d'hisotire naturelle.

Barsanti, G. (2000). Lamarck: Taxonomy and theoretical biology. *Asclepio, 52*(2), 119–132.

Bartha, P. F. A. (2013). *By parallel reasoning: The construction and evaluation of analogical arguments*. Oxford University Press.

Barthelmy-Madaule, M. (1982). *Lamarck, the mythical precursor: A study of the relations between science and ideology* (M. H. Shank, Trans.). MIT Press.

Barton, R. (1990). An influential set of chaps: The X-club and royal society politics 1864–85. *The British Journal for the History of Science, 23*(1), 53–81.

Barton, R. (1998). Huxley, Lubbock, and half a dozen others: Professionals and gentlemen in the formation of the X-Club, 1851-1864. *Isis, 89*(3), 410–444.

Bartrip, P. W. (1992). The British medical journal: A retrospect. In W. F. Bynum, S. Lock, & R. Porter (Eds.), *Medical journals and medical knowledge: Historical essays*. Routledge.

Bassiri, N. (2015). Epileptic insanity and personal identity – John Hughlings-Jackson and the formation of the neuropathic self. In N. Bassiri & D. Bates (Eds.), *Plasticity and pathology: On the formation of the neural subject* (pp. 65–111). Fordham University Press.

Bateson, P., & Gluckman, P. (2011). *Plasticity, robustness, development and evolution*. Cambridge University Press.

Beaunis, H. (1909). Comment fonctionne mon cerveau, essai de psychologie introspective. *Revue Philosophique, 67*(January), 29–40.

Bechtel, W., & Richardson, R. C. (2010). *Discovering complexity: Decomposition and localization as strategies in scientific research*. The MIT Press.

Beck, N. (2014). *La Gauche évolutionniste: Spencer et ses lecteurs en France et en Italie*. Presses universitaires Franche Comté.

Becquemont, D. (2010). Edmond Perrier exobiologiste. *Bulletin d'Histoire et d'Épistémologie des Sciences de la Vie, 17*(1), 91–104.

Delage, Y. (1889). Réponse à M. Giard. *Revue Scientiaque*, 43–44.

Delage, Y. (1895a). Preface à *L'Année* Biologique. *L'Année biologique, 1*, vii–xviii.

Delage, Y. (1895b). *La structure du protoplasma et les théories sur l'hérédité et les grands problèmes de la biologie générale*. C. Reinwald.

Belanger, J., Godechot, J., Guiral, P., & Terrou, F. (1970). *Histoire générale de la presse française*. Presses Univérsitaires de France.

Ben-Amos, A. (2000). *Funerals, politics, and memory in modern France, 1789–1996*. Oxford University Press.

Ben David, J. ([1960] 1991). Roles and innovations in medicine. In Gad Freudenthal (Ed.), *Scientific growth by J. Ben David* (pp. 33–48). University of California Press.

Benichou, C. (1989). Ribot et l'hérédité psychologique. In J.-C. Fischer & C. Benichou (Eds.), *L'Ordre des caracteres* (pp. 73–94). J. Vrin.

Benrubi, J. (1933). *Les sources et les courants de la philosophie contemporaine en France*. Alcan, t. II.

Bernadini, J. M. (1997). *Le Darwinisme social en France (1859–1918)*. Editions du Centre national de la recherche scientifique.

Bequemont, D., & Mucchielli, L. (1998). *Le cas Spencer: Religion, science et politique*. Presses Universitaires de France.

Béra, M. (2014). *Émile Durkheim a` Bordeaux (1887–1902)*. Éditions confluences.

Béra, M. (2017). La représentation disciplinaire du 'social' dans les références et les lectures du jeune Durkheim (1879-1894). *L'Année sociologique, 67*(2), 481–481.

Béra, M. (2019). Aux sources de la sociologie. *Archives des sciences sociales des religions, 188*(4), 111–130.
Béra, M., Marcel, J.-C., & Mosbah-Natanson, S. (2019). L'histoire de l'Année sociologique: de l'ère des pionniers à la normalisation. *L'Année sociologique, 69*(1), 11–19.
Berg, M., & Bowker, G. (1997). The multiple bodies of the medical record: Toward a sociology of an artifact. *The Sociological Quarterly, 38*(3), 513–537.
Berlant, J. L. (2007). On the case- introduction. *Critical Inquiry, 33*(4), 363–372.
Berlucchi, G., & Buchtel, H. A. (2009). Neuronal plasticity: Historical roots and evolution of meaning. *Experimental Brain Research, 192*, 307–319.
Bernard, C. (1927 [1865]). *An introduction to the study of experimental medicine* (Henry C. Green, Trans.). The MacMillan Company.
Bernardini, J. M. (1997). *Le darwinisme social en France (1859–1918)*. Editions du Centre national de la recherche scientifique.
Bernstein, R. (1998). *Freud and the legacy of Moses*. Cambridge University Press.
Berrios, G. E. (2012). Memory disorders and epilepsy during the nineteenth century. In A. Zeman, N. Kapur, & M. Jones-Gotman (Eds.), *Epilepsy and memory*. Oxford University Press.
Besnard, P. (Ed.). (1979). Les Durkheimiens. *Revue Française de Sociologie, 20*.
Besnard, P. (1983). The Année Sociologique Team. In P. Besnard (Ed.), *The sociological domain: The Durkheimians and the founding of French sociology* (pp. 11–39). Cambridge University Press.
Besnard, P. (2003). *Études Durkheimiennes*. Droz.
Biener, Z., & Schliesser, E. (Eds.). (2014). *Newton and empiricism*. Oxford University Press.
Bilder, R. M., & Frank LeFever, F. (Eds.). (1998). *Neuroscience of the mind on the centennial of Freud's project for a scientific psychology*. New York Academy of Sciences.
Binet, A. (1908). Une enquête sur l'evolution de l'enseignement de la philosophie. *l'Année psychologique, 14*, 152–231.
Birenbaum, P. (2020). The Dreyfus affair and Durkheim's experience of anti-semitism. In H. Joas & A. Pettenkofer (Eds.), *The Oxford handbook of Emile Durkheim*. Oxford University Press.
Black, M. (1962). *Models and metaphors: Studies in language and philosophy*. Cornell University Press.
Black, M. (1979). *On metaphor*. University of Chicago Press.
Blanckaert, C. (1979). Edmond Perrier et l'étiologie du 'polyzoizme'organique. *Revue de synthese, 100*, 353–376.
Blanckaert, C. (1988). Variations sur le Darwinisme epistmologique et transfert lexical. In M. Groult (Ed.), *Transfert de Vocabulaire dans les sciences*. Editions du Centre national de la recherche scientifique.
Blanckaert, C. (Ed.). (2001). *Les politiques de l'anthropologie:Discours et pratiques en France.1860–1940*. L'Harmattan.
Blanckaert, C. (2002). Le moment 'naturaliste' des sciences des homme. *Revue d'histoire des Sciences Humaines, 3*(2), 117–160.
Blanckaert, C. (2004). *La Nature de la Société: organicisme et sciences sociales Au XIXe Siècle* (pp. 1–160). L'Harmattan.
Blanckaert, C. (2009). *De la race à l'évolution: Paul Broca et l'anthropologie française 1850-1900*. L'Harmattan.
Boring, E. G. ([1929] 1957). *A history of experimental psychology*. Appleton-Century-Croft.
Boring, E. G. (1929). *A history of experimental psychology*. The Century Co.
Borlandi, M. (1993). Durkheim lecteur de Spencer. In P. Besnard, M. Borlandi, & P. Vogt (Eds.), *Division du travail et lien social: la thèse de Durkheim un siècle après* (pp. 67–112). Presses Universitaires de France.
Borlandi, M. (2006). Durkheim, Rauh et la part qui revient à l'individu dans la genèse des phénomènes sociaux. *Durkheimian Studies, 12*(1), 17–29.
Bouglé, C. (1898 [1896]). "Sociologie Générale" Introduction to the section. *L'Année Sociologique, 1*, 116–123.

Bouglé, C. (1901). Le proce's de la sociologie biologique. *Revue philosophique, 52*, 121–146.
Bouglé, C., & Andler, C. (1896). Sociologie, psychologie, histoire. *Revue de Métaphysique et de Morale, 1*, 144.
Bourgin, H. (1938). *De Jaurès à Léon Blum, l'École Normale et la politique*. A. Fayard.
Bowler, P. J. (1983). *The eclipse of Darwinism, anti-Darwinian evolution theories in the decades around 1900*. Johns Hopkins Press.
Bowler, P. J. (1984). *Evolution: The history of an idea*. University of California Press.
Bowler, P. J. (1988). *The non-Darwinian revolution: Reinterpreting a historical myth*. Johns Hopkins University Press.
Bowler, P. J. (1989). *The invention of progress: The Victorians and the past*. Johns Hopkins University Press.
Bowler, P. J. (1991). *The Non Darwinain revolution*. Johns Hopkins University Press.
Bowler, P. J. (1996). *Life's splendid Drama: Evolutionary biology and the reconstruction of life's ancestry, 1860–1940*. University of Chicago Press.
Bowler, P. J. (2009). *Science for all: The popularization of science in early 20th century Britain*. Chicago University Press.
Bowler, P. J. (2015a). From agnosticism to rationalism: Evolutionary biologists, the rationalist press association, and early twentieth-century scientific naturalism. In B. Lightman & G. Dawson (Eds.), *Victorian scientific naturalism: Community, identity, continuity* (pp. 309–326). Chicago University Press.
Bowler, P. J. (2015b). Herbert Spencer and Lamarckism. In M. Francis & M. W. Taylor (Eds.), *Herbert Spencer legacies* (pp. 203–241). Routledge.
Boyd, R., & Richerson, P. J. (2005). *The origin and evolution of cultures*. Oxford University Press.
Brazier, M. A. B. (1988). *A history of neurophysiology in the 19th century*. Raven Press.
Bret, P. (1991). L'emergence d'une histoire des sciences du moment revolutionnaire. In M. Vovelle (Ed.), *Recherches sur la Révolution* (pp. 169–176). La Decouverte.
Bret, P. (2002). *L'état, l'armée, la science: l'invention de la recherche publique en France (1760–1830)*. Presses Universitaires de Rennes.
Bristowe, J. S., & Holmes, T. (1863). On the Hospitals of the United Kingdom. Appendix 15 to the *Privy Council Medical Officer's Report*, London.
Broadbent, W. (1903). Hughlings Jackson a Pioneer in nervous physiology and pathology. *Brain, 26*(3), 305–356.
Broman, T. (1998). The Habermasian public sphere and science in the enlightenment. *History of Science, 36*, 123–149.
Brooks, J. I. (1998). *The eclectic legacy: Academic philosophy and the human sciences in nineteenth century France*. University of Delaware Press.
Brooks, D. S., DiFrisco, J., & Wimsatt, W. C. (Eds.). (2021). *Levels of organization in biological sciences*. The MIT Press.
Brunner, J. (1995). *Freud and the politics of psychoanalysis*. Blackwell.
Brunschvicg, L. (1893). La philosophie de l'inconscient. *Revue de Métaphysique et de Morale, 1*(4), 405–410.
Buklijas, T. (2018). Histories and meanings of epigenetics. In J. Maurizio Meloni, D. F. Cromby, et al. (Eds.), *The Palgrave handbook of biology and society* (pp. 167–188). Palgrave Macmillan.
Bunzl, M. (1996). Review of fieldwork and footnotes: Studies in the history of European anthropology. Han F. Vermeulen and Arturo Alvarez Roldán, eds. *American Anthropologist, 98*(3), 684–686.
Burkhardt, R. (1995). *The Spirit of system: Lamarck and evolutionary biology*. Harvard University Press.
Burlingame, L. J. (1973). Lamarck. In C. C. Gillispie (Ed.), *Dictionary of scientific biography* (Vol. 7, pp. 584–594). Scribner.
Burlingame, L. J. (1981). Lamarck's chemistry: The chemical revolution rejected. In H. Woolf (Ed.), *The analytic spirit* (pp. 37–64). Cornell University Press.
Buss, L. W. (1987). *The evolution of individuality*. Princeton University Press.

Butler, S. (1879). *Evolution, old and new; or, The theories of Buffon, Dr. Erasmus Darwin, and Lamarck, as compared with that of Charles Darwin*. Fifield.

Butler, S. (1880). *Unconscious memory*. Ballantyne Press/Hanson and Co.

Bynum, W. F. (1990). The nervous patient in 18th and 19th century Britain: The psychiatric origins of British neurology. In R. M. Murray & T. H. Turner (Eds.), *Lectures on the history of psychiatry: The Squibb series* (pp. 115–127). Gaskell.

Bynum, W. F., & Wilson, J. C. (1992). Periodical knowledge: Medical journals and their editors in nineteenth century Britain. In W. F. Bynum, S. Lock, & R. Porter (Eds.), *Medical journals and medical knowledge: Historical essays*. Routledge.

Bynum, W. F., Porter, R., & Shepard, M. (Eds.). (1985–1988). *The anatomy of madness: Essays in the history of psychiatry*, 3 vols. Routledge.

Cabanis, P. J., & Georges. (1805). *Rapports du physique et du moral de l'homme* (2nd ed.). Crapart, Caille et Ravier.

Cameron, L., & Low, G. (1999). *Researching and applying metaphor*. Cambridge University Press.

Campbell, D. T. (1975). On the conflicts between biological and social evolution and between psychology and moral tradition. *American Psychologist, 30*(12), 1103–1126.

Canguilhem, G. (1965). *La conaissance de la vie*. J. Vrin.

Canguilhem, G. (1978). *On the normal and the pathological* (C. Fawcett, Trans.). D. Reidel.

Canguilhem, G. (1994). *A vital rationalist: Selected writings* (F. Delaporte, Ed.). Zone books.

Cannon, W. B. (1900). The case method of teaching systematic medicine. *The Boston Medical and Surgical Journal, 142*(2), 31–36.

Carmel, Y., Shavit, A., Lamm, E., & Szathmáry, E. (Eds.). (2023). Human socio-cultural evolution in light of evolutionary transitions. *Philosophical Transactions of the Royal Society B, 378*(1872).

Carneiro, R. L. (1973). Structure, function and equilibrium in the evolutionism of Herbert Spencer. *Journal of Anthropological Research, 29*(2), 77–95.

Carpenter, W. B. (1861). *Researches on the foraminifera*. Taylor and Francis.

Carrithers, D. (1995). The enlightenment science of society. In C. Fox, R. Porter, & R. Wokler (Eds.), *Inventing human science: Eighteenth century domains* (pp. 232–270). University of California Press.

Carroy, J. (1991). *Hypnose, suggestion et psychologie: l'invention du sujet*. Presses universitaires de France.

Carroy, J. (2004). Premie'res enquêtes psychologiques françaises: L'introspection, l'individu et le nombre. *Mil neuf cent. Revue d'histoire intellectuelle, 1*(22), 59–75.

Carroy, J. (2005). L'etude de cas psychologique et psychanalytique (XIX' siecle-debut du xx' siecle). In J.-C. Passeron & J. Revel (Eds.), *Penser par cas* (pp. 201–228). Éditions de l'EHESS.

Carroy, J. (2006). Philosophie et psychologie au 19ieme siècle en France:Taine, Ribot et la Revue Philosophique. In M.-L. Pelus-Kaplan, J. Carroy, & C. Debru (Eds.), *Unité et globalité de l'homme: Des humanités aux sciences humaines* (pp. 79–91). Edition Syllepse.

Carroy, J., & Plas, R. (1996). The origins of French experimental psychology: Experiment experimentalism. *History of the Human Sciences, 9*(1), 73–84.

Carroy, J., & Plas, R. (2000). How Pierre Janet used pathological psychology to save the philosophical self. *Journal of the History of the Behavioral Sciences, 36*(3), 231–240.

Carroy, J., & Plas, R. (2006). The beginnings of psychology in France: Who was "scientific" psychologist in the nineteenth century? *Physis, 43*(1–2), 157–186.

Carroy, J., & Schmidgen, H. (2002). *Psychologies expérimentales: Leipzig – Paris 1890–1910* (pp. 1–27). Max Planck Preprint 2006.

Carroy, J., Ohayon, A., & Plas, R. (2002). *Histoire de la psychologie en France* (pp. xix–xx). La Découverte.

Carroy, J., Ohayon, A., & Plas, R. (2015). La 'psychologie' au college de France. *Revue Philosophique, 140*(2), 225–228.

Carroy, J., Feuerhahn, W., Plas, R., & Trochu, T. (2016). Les entreprises intellectuelles de Théodule Ribot. *Revue Philosophique, 141*(4), 451–464.

Carson, J. (2007). *The measure of merit: Talents, intelligence, and inequality in the French and American republics, 1750–1940*. Princeton University Press.
Casper, S. T. (2014). *The neurologists: A history of a medical specialty in modern Britain c.1789–2000*. Manchester University Press.
Casper, S. T. (2016). The currency of consciousness: Neurology, specialization and the global practices of medicine. *Canadian Bulletin of Medical History, 33*(2), 321–363.
Cazelles, E. M. (1870). *Introduction a la traduction de les premiers principes par Herbert Spencer I-LXXX*. Felix Alcan.
Ceccarelli, D. (2019). Between social and biological heredity: Cope and Baldwin on evolution, inheritance, and mind. *Journal of the History of Biology*, 161–194.
Censer, J. R., & Hunt, L. A. (2001). *Liberty, equality, fraternity: Exploring the French revolution*. Pennsylvania State University.
Chamberlain, J. E., & Gilman, S. L. (Eds.). (1985). *Degeneration: The dark side of Progress*. Columbia University Press.
Charland, L. C. (2010). Reinstating the passions. In P. Goldie (Ed.), *The Oxford handbook of philosophy of emotion* (pp. 237–259). Oxford University Press.
Charle, C. (1987). *Les élites de la République: 1880–1900*. Fayard.
Charle, C. (1990). *Naissance des "intellectuels" 1880–1900*. Les éditions de minuit.
Charle, C. (1994). *La république des universitaires, 1870–1940*. Seuil.
Charlton, D. G. (1984). *New images of the natural in France: A study in European cultural history, 1750–1800*. Cambridge University Press.
Charon, R. (1992). To build a case: Medical histories as traditions in conflict. *Literature and Medicine, 11*(1), 115–132.
Cherkaoui, M. (1983). Education and social mobility: Paul Lapie's pathbreaking work. In P. Besnard (Ed.), *The sociological domain* (pp. 217–230). Cambridge University Press.
Cheung, T. (2006). From the organism of a body to the body of an organism: Occurrence and meaning of the word 'organism' from the seventeenth to the nineteenth centuries. *The British Journal for the History of Science, 39*(3), 319–339.
Chimisso, C. (2008). *Writing the history of the mind: Philosophy and science in France, 1900 to 1960s*. Ashgate.
Chirimuuta, M. (2019). Synthesis of contraries: Hughlings Jackson on sensory motor representation in the brain. *Studies in History and Philosophy of Biological and Biomedical Sciences, 75*, 34–44.
Churchill, F. B. (1987). From heredity theory to Vererbung: The transmission problem 1850-1915. *Isis, 78*, 3.
Churchill, F. B. (1991). The rise of classical descriptive embryology. In S. Gilbert (Ed.), *Conceptual history of modern embryology* (pp. 1–29). Plenum.
Churchill, F. B. (2007). Living with the biogenic law: A reappraisal. In M. D. Laubichler & J. Maeinschein (Eds.), *From embryology to Evo Devo* (pp. 37–82). MIT Press.
Churchill, F. B. (2015). *August Weismann: Development, heredity, and evolution*. Harvard University Press.
Cladis, M. S. (2020). Durkheim's signature project: The science of morality as rational moral art. In H. Joas & A. Pettenkofer (Eds.), *Oxford handbook of Emile Durkheim*. Oxford University Press.
Clark, T. N. (1973). *Prophets and patrons: The French university and the emergence of the social sciences*. Harvard University Press.
Clark, L. (1984). *Social Darwinism in France*. University of Alabama Press.
Clark, M. J. (1985–1988). Morbid introspection unsoundness of mind and British psychological medicine 1830-1900. In W. F. Bynum, R. Porter, & M. Shepard (Eds.), *The anatomy of madness: Essays in the history of psychiatry* (vol. 3, pp. 71–102). Routledge.
Clarke, E., & Jacyna, L. S. (1987). *Nineteenth-century origins of neuroscientific concepts*. University of California Press.
Class, M. (2014). Introduction: Medical case histories as genre: New approaches. *Literature and Medicine, 32*(1), vii–xvi.

Clauzade, L. (2018). Comte's positive biology. In M. Bourdeau, M. Pickering, & W. Schmaus (Eds.), *Love, order, and Progress: The science, philosophy, and politics of Auguste Comte* (pp. 93–127). University of Pittsburgh Press.

Cohen, I. B. (Ed.). (1981). *The conservation of energy and the principle of least action*. Arno Press.

Collingwood, F. J. (1961). *Philosophy of nature*. Prentice Hall.

Conry, Y. (1974). *L'Introduction au Darwinisme en France*. Vrin.

Conry, Y. (1981). Une lecture Newtonienne de Lamarck, est elle possible? In J. Schiller (Ed.), *Lamarck et son temps, Lamarck et notre temps* (pp. 37–64). Vrin.

Conry, Y. (1993). Comment a-t-on pu être néo-Lamarckien en France (1843-1930)? *Nuncius, 8*, 487–520.

Conry, Y. (1994). Lamarck, penseur de frontiere. *Nuncius, 9*(2), 559–592.

Consolim, M. (2019). La réception de la psychologie dans l'Année sociologique (1898-1927). *L'Année sociologique, 69*(1), 103–142.

Corsi, P. (1978). The importance of French Transformist ideas for the second volume of Lyell's principles of geology. *The British Journal for the History of Science, 11*(3), 221–244.

Corsi, P. (1982). Models and analogies for the reform of natural history. Features of the French debate 1790-1800. In W. Bernardi & A. La Vergata (Eds.), *Lazzaro Spallanzani e la biologia del Settecento: Teorie, esperimenti, istituzioni scientifiche: Atti del convegno, marzo 1981* (pp. 381–396). L.S. Olschki.

Corsi, P. (1987). Julien Joseph Virey, le premier critique de Lamarck. In S. Atran et al. (Eds.), *Histoire du concept d'espece dans les sciences de la vie* (pp. 176–187). Fondation Singer Polignac.

Corsi, P. (1988). *The age of Lamarck: Evolutionary theories in France 1790–1830* (J. Mandelbaum, Trans.). University of California Press.

Corsi, P. (1997a). Le Muséum et l'Europe. In C. Blanckaert, C. Cohen, P. Corsi, & J. L. Fischer (Eds.), *Le Muséum au premier sie'cle de son histoire* (pp. 635–647). Editions du Muséum d'histoire naturelle.

Corsi, P. (1997b). Les élèves de Lamarck: un projet de recherche. In G. Laurent (Ed.), *Jean-Baptiste Lamarck 1744–1829* (pp. 515–526). Editions de CTHS.

Corsi, P. (1998). A Devil's Chaplain's calling? *Journal of Victorian Culture, 3*(1), 129–137.

Corsi, P. (2005). Before Darwin: Transformist concepts in European natural history. *Journal of the History of Biology, 38*(1), 67–83.

Corsi, P. (2006). Biologie. In P. Corsi, J. Gayon, G. Gohau, & S. Tirard (Eds.), *Lamarck philosophe de la nature* (pp. 37–64). Presses Universitaires de France.

Corsi, P. (2012). Idola Tribus: Lamarck, politics and religion in the early nineteenth century. In S. Fasolo (Ed.), *The theory of evolution and its impact* (pp. 23–25). Springer.

Corsi, P. (2018). Systèmes de la nature and theories of life: Bridging the eighteenth and nineteenth centuries. *Republics of Letters, 6*, 1.

Corsi, P. (n.d.). *General Editor of the Lamarck Site on the Web*http://www.lamarck.cnrs.fr/index.php?lang=fr

Creager, A. N. H. (2016). Paradigms and exemplars meet biomedicine. In R. J. Richards & L. Daston (Eds.), *Kuhn's 'structure of scientific revolutions' at fifty: Reflections on a science classic*. Chicago University Press.

Creager, A. N. H., Lunbeck, E., & Wise, N. (2007). Introduction. In A. N. H. Creager, E. Lunbeck, & N. Wise (Eds.), *Science without laws: Model systems, cases, exemplary narratives* (pp. 1–21). Duke University Press.

Critchley, M. (1960). Hughlings Jackson, the man: And the early days of the national hospital. *Proceedings of the Royal Society of Medicine, 53*(8), 613–618.

Critchley, M., & Critchley, E. A. (1998). *John Hughlings Jackson: Father of English neurology*. Oxford University Press.

Csiszar, A. (2010). Seriality and the search for order: Scientific print and its problems during the late nineteenth century. *History of Science, 48*(3/4), 399–434.

Csiszar, A. (2018). *The scientific authorship and politics of knowledge in the nineteenth century*. Chicago University Press.

Csiszar, A. (2019). *The scientific journal*. University of Chicago Press.

Cuin, C. H. (2011). *Durkheim, modernité d'un classique*. Herman.

Cuvier, G. (1805). *Leçons d'anatomie comparée ,recueillies et publiées sous ses yeux par C. Duméril, an VIII [1800]-an XIV--1805*. Baudouin.

Cuvier, G. (1810). *Rapport historique sur les progrès des sciences naturelles depuis 1789, et sur leur état actuel....* l'Imprimerie impériale.

Cuvier, G. ([1815] 1989). *Rapport à l'Empereur sur le progrès des sciences, des lettres et des arts depuis 1789. 2. Chimie et sciences de la nature*. Edited by Yves Laissus. Belin.

Cuvier, G. (1860). *Éloges historiques: précédés de l'éloge de l'auteur par M. Flourens*. E. Ducrocq.

D'Hombres, E. (2010). The 'division of physiological labor': The birth, life and death of a concept. *Journal of History of Biology, 45*(1), 3–31.

D'Hombres, E. (2012). De la différenciation biologique a` la différenciation sociale (xix-xx si'ecles): Quelques jalons historiques. *Nouvelles Perspectives en Sciences Sociales, 8*(1), 191–220.

D'Hombres, E. (2016). The Darwinian muddle on the division of labor: An attempt at clarification. *History and Philosophy of the Life Sciences, 38*(1), 1–22.

D'Hombres, E., & Mehdaoui, S. (2012). On what condition is the equation organism–society valid?' Cell theory and organicist sociology in the works of Alfred Espinas (1870s–80s). *History of the Human Sciences, 25*(1), 32–51.

Damasio, A. R. (1999). *The feeling of what happens: Body and emotion in the making of consciousness*. Harcourt Brace.

Damasio, A., & Damasio, H. (2006). Minding the body. *Daedalus, 135*(3), 15–22.

Dantec, L., & Alexandre, F. (1903). *Les limites du connaissable: La vie et les phénomènes naturels*. Alcan.

Danziger, K. (1980). The history of introspection. *Journal of the History of the Behavioral Sciences, 16*, 241–2623.

Danziger, K. (1982). Mid-19th century British psycho-physiology: A neglected chapter in the history of psychology. In W. R. Woodward & M. G. Ash (Eds.), *The problematic science: Psychology in 19th century thought* (pp. 119–146). Praeger.

Danziger, K. (1985). The origins of the psychological experiment as a social institution. *American Psychologis, 40*(2), 133–140.

Danziger, K. (1990). *Constructing the subject: Historical origins of psychological research.* Cambridge University Press.

Darwin, C. ([1859] 1964). *On the origin of species*. A facsimile of the First Edition. Harvard University Press.

Darwin, C. ([1872] 1998). *The expression of the emotions in man and animals*, 3rd edn. Oxford University Press.

Daston, L. (1978). British responses to psychophysiology 1860-1900. *Isis, 69*(2), 192–208.

Daston, L. (1990). Nationalism and scientific neutrality under Napoleon. 'Commentary', Inge Jonsson. In T. Frängsmyr (Ed.), *Solomon's house revisited* (pp. 95–132). Nobel Symposium 75: Science History Publications.

Daston, L. (2000). The coming into being of scientific objects. In L. Daston (Ed.), *Biographies of scientific objects* (pp. 1–14). University of Chicago Press.

Daston, L. (2004). Attention and the values of nature in the enlightenment. In L. Daston & F. Vidal (Eds.), *The moral authority of nature* (pp. 100–126). Chicago University Press.

Daston, L. (2005). Fear and loathing of imagination in science. *Daedalus, 134*(4), 16–31.

Daston, L. (2007). Condorcet and the meaning of enlightenment". In *Proceedings of the British Academy, Volume 151, 2006 Lectures.*

Daston, L., & Gallison, P. (2007). *Objectivity*. Zone Books.

Daston, L., & Sibum, O. (2003). Introduction: Scientific personae and their histories. *Science in Context, 16*(1–2), 1–8.

Daudin, H. (1926). *Cuvier et Lamarck. Les classes zoologiques et l'idée de série animale, De Linné à Lamarck. Méthodes de la classiacation et idée de série en botanique et en zoologie.* F. Alcan.

Davy, G. (1919). Émile Durkheim. *Revue de Métaphysique et de Morale, 26*(2), 181–198.
Davy, G. (1960). Emile Durkheim. *Revue Française de Sociologie, 1*(1), 3–24.
Dawson, G., & Lightman, B. (Eds.). (2014). *Victorian scientific naturalism: Community, identity, continuity*. Chicago University Press.
Dawson, G., & Topham, J. R. (2004). Science in the nineteenth-century periodical. *Literature Compass, 1*(VI-056), 1–11.
Dawson, G., Lintott, C., & Shuttleworth, S. (2015). Constructing scientific communities: Citizen science in the nineteenth and twenty first centuries. *Journal of Victorian Culture, 20*(2), 246–254.
De Block, A. V. A. (2005). Freud as an evolutionary psychiatrist. the foundations of a Freudian philosophy. *Philosophy, Psychiatry & Psychology, 12*(4), 315–324.
De Bont, R. (2010). Organisms in their milieu: Alfred Giard, his pupils, and early ethology, 1870–1930. *Isis, 101*(1), 1–29.
De Bont, R. (2015). *Stations in the field: A history of place-based animal research, 1870–1930*. Chicago University Press.
de Lilienfeld, P. (1896). *La pathologie sociale*, preface R. Worms. V. Giard et E. Brière.
De Tarde, G. (1884). Le Darwinisme naturel et le Darwinisme social. *Revue Philosophique, 17*, 607–637.
de Tarde, G. (1890). *Les lois de L'imitation: étude sociologique*. Felix Alcan.
De Tarde, G. (1894). Biologie et sociologie – réponse à Dr. Bianchi. *Archives d'anthropologie criminelle*.
de Tarde, G. (1895). Criminalité et santé sociale. *Revue Philosophique, 39*, 148–162.
de Bont, R. (2010). Organisms in Their Milieu: Alfred Giard, His Pupils, and Early Ethology, 1870–1930. *Isis, 101*(1), 1–29.
de Lanessan, J. (1901). La Morale Scienti.que. *Revue Scientifique, 16*.
de Lanessan, J. (1914). *La crise de la République*. Felix Alcan.
de Tocqueville, A. (1988 [1833]). *Journeys to England and Ireland* (J. P. Mayer, K. P. Mayer, G. Lawrence, Eds.). Taylor & Francis.
de Tracy, D., & Claude, A. L. (1817–1818). *Élémens d'idéologie*, 3rd edn. Ist Part. Courcier.
Deacon, T. D. (2003). The hierarchic logic of emergence: Untangling the interdependence of evolution and self-organization. In B. H. Weber & D. J. Depew (Eds.), *Evolution and learning: The Baldwin effect reconsidered* (pp. 273–308). MIT Press.
Delage, Y., & Goldsmith, M. (1912 [1909]). *The theories of evolution* (A. Tridon, Tran.). B. W. Huebsch.
De Lapouge, V. (1887). Les sélections sociales. *Revue d'Anthropologie, 2*(1), 519.
Delay, J. (1950). *Les dissolutions de la mémoire*. Presses Universitaires de France.
Delay, J. (1951). Ribot et le Jacksonism. *Dialectica, 5*(3/4), 413–444.
Depew, D. J. (2003). Baldwin and his many effects. In B. H. Weber & D. J. Depew (Eds.), *Evolution and learning: The Baldwin effect reconsidered* (pp. 3–32). The MIT Press.
Desmond, A. J. (1989). *The politics of evolution: Morphology, medicine and reform in radical London*. The University of Chicago Press.
Desmond, A. J. ([1994] 1997]). *Huxley: From Devil's disciple to evolution's high priest*. Addison-Wesley.
Desmond, A. J. (2001). Redefining the X axis: 'Professionals,' 'Amateurs' and the making of mid-victorian biology – A progress report. *Journal of the History of Biology, 34*, 3–50.
Dewhurst, K. (1982). *Hughlings Jackson on psychiatry*. Sandford Publications.
D'Hombres, N. (1989). *Les Savants en Révolution: 1789–1799*. Cité des sciences et de l'industrie.
Dictionnaire encyclopédique des sciences médicales. (1865–1889). publié sous la direction de MM. les docteurs Raige-Delorme et A. Dechambre. Paris: P. Asselin, S' de Labé, V. Masson et fils.
Digeon, C. (1959). *La crise allemande de la pensée Française*. Presses Universitaires de France.
Dixon, L. B. (1988). *Diderot, philosopher of energy*. Oxford University Press.
Dixon, T. M. (2003). *From passions to emotions: The creation of a secular psychological category*. Cambridge University Press.

Dixon, T. M. (2008). *The invention of altruism: Making moral meanings in Victorian Britain*. Oxford University Press for the British Academy.
Doolittle, F., & Bapteste, E. (2007). Pattern pluralism and the tree of life hypothesis. *Proceedings of the National Academy of Sciences of the United States of America, 104*(7), 2043–2049.
Douglas, M. (1986). *How institutions think*. Syracuse University Press.
Dowbiggin, I. (1985). Degeneration and hereditarianism in French mental medicine 1840-90: Psychiatric theory as ideological adaptation. In W. F. Bynum, R. Porter, & M. Shepard (Eds.), *The anatomy of madness: Essays in the history of psychiatry* (Vol. 1, pp. 188–232). Routledge.
Dowbiggin, I. (1991). *Inheriting madness: Professionalization and psychiatric knowledge in nineteenth-century France*. University of California Press.
Drouin-Hans, A.-M. (2011). Classer les emotions, hiérarchiser les hommes. In D. Becquemont & D. Ottavi (Eds.), *Penser Spencer* (pp. 163–178). Presses Universitaires de Vincennes.
Droz-Mendelzweig, M. (2010). La plasticité cérébrale de Cajal à Kandel: Cheminement d'une notion constitutive du sujet cérébral. *Revue d'histoire des sciences, 63*(2), 661–698.
Duchesneau, F. (1982). *La physiologie des lumie'res: Empirisme, mode'les et theories*. Martinus Nijhoff.
Dugas, L. (1924). *Le philosophe Théodule Ribot*. Payot.
Dupont, J.-C. (2010). L'histoire de l'integration: de Spencer Herbert, Sherrington Charles Dcott et apres. In Jean Gayon, Armand de Ricqlès, with Matteo Mossio (Eds.), *Les fonctions: Des organismes aux artefacts* (pp. 211–26). Pressesniversitaires de France.
Dupuis, C. (1997). *Actualité de Lamarck taxinomiste* (pp. 1744–18. Jean-Baptiste Lamarck.
Durkheim, E. (1882). L'agrégation de philosophie. *Études durkheimiennes*, fall 1993.
Durkheim, E. 1884. *Cours de philosophie fait au lycée de Sens en 1883-1884: notes prises entre 1883-84 par le philosophie français André Lalande*. ARTFEL-FRANTEXT 2002.
Durkheim, E. (1885a). Fouillée, A.: La propriété sociale et la democratie. *Revue Philosophique, 19*, 446–453.
Durkheim, E. (1885b). Gumplowicz, Ludwig: Grundrisse der Soziologie. *Revue Philosophique, 20*, 627–634.
Durkheim, E. (1885c). Schaeffle, A. Bau und Leben des sozialen Körpers: erster band. *Revue Philosophique, 19*, 840–101.
Durkheim, E. (1886a). De Greef, Guillaume: Introduction a' la sociologie. *Revue Philosophique, 22*, 658–663.
Durkheim, E. (1886b). Les Études de science sociale. *Revue Philosophique, 22*, 61–80.
Durkheim, E. (1887a). La philosophie dans les universités allemandes. *Revue internationale de l'enseignement, 23*(313–38), 423–440.
Durkheim, E. (1887b). La science positive de la morale en Allemagne. *Revue Philosophique, 24*, 33–58, 113–42, 275–84
Durkheim, E. (1887c). Suicide et natalité: étude de statistique morale. *Revue Philosophique, 26*, 446–463.
Durkheim, E. (1888a). Cours de science sociale: leçon d'ouverture. *Revue internationale de l'enseignement, 15*, 23–48.
Durkheim, E. (1888b). Introduction de la sociologie de la famille. *Annales de la faculté des Lettres de Bordeaux, 10*, 257–281.
Durkheim, E. (1889). Tönnies, F. Gemeinschaft und Gesellschaft. *Revue Philosophique, 27*, 416–422.
Durkheim, E. (1895a). L'enseignement philosophique et l'agrégation de philosophie. *RP, 39*, 121–147.
Durkheim, E. (1895b). Crime et santé sociale. *Revue Philosophique* xxix, 518–523.
Durkheim, E. (1897). Labriola, Antonio: essais sur la conception matérialiste de l'histoire. *RP, 44*, 645–651.
Durkheim, E. (1902). Espinas, A. 'Être ou ne pas être ou du postulat de la sociologie. *L'Année Sociologique, 5*, 127–129.
Durkheim, E. (1913). Wundt. W. Elemente der Voelkerpychologie (Leipzig,1912). *L'Année Sociologique, 12*, 50–61.

Durkheim, E. ([1892] 1921). "La famille conjugale." Revue philosophique 90:1–14.
Durkheim, E. ([1893] 1930a). *De la division du travail social (étude sur l'organisation des sociétés superieures)*. Presses Universitaires de France.
Durkheim, E. ([1897] 1930b). *Le suicide: étude de sociologie*. Presses universitaires de France.
Durkheim, E. ([1894–1995] 1937). *Les re'gles de la méthode sociologique*. Presses Universitaires de France.
Durkheim, E. (1938). *L'évolution pédagogique en France*. Introduction by M. Halbwachs. Presses Universitaires de France.
Durkheim, E. ([1890–1900] 1950). *Leçons de sociologie: physique de mœurs et du droit*. Avant-propos by H.N. Kubali. Presses Universitaires de France.
Durkheim, E. ([1900a] 1953). La sociologie et son domaine scientiaque. *Rivista Italiana de Sociologia*. Translated into French. Cuvillier.
Durkheim, E. ([1912] 1960). *Les formes élementaires de la vie religieuse: Les syste'me totemique en Australie*. Presses Universitaires de France.
Durkheim, E. ([1925] 1963). *L'éducation morale*. F. Alcan.
Durkheim, E. (1969). *Journal sociologique*. Edited by J. Duvignaud. Presses Universitaires de France.
Durkheim, E. ([1888–1914] 1970). *La science sociale et l'action*, introduction and presentation by J.C. Filloux. Presses Universitaires de France.
Durkheim, E. ([1928] 1971). *Le Socialisme; sa définition, ses débuts, la doctrine saint-simonienne* (M. Mauss, Ed.). Presses Universitaires de France.
Durkheim, E. ([1898–1912] 1974). *Sociologie et philosophie*, introduction by C. Bouglé. Presses Universitaires de France.
Durkheim, E. ([1895b] 1975). L'état actuel des études sociologiques en France. V. Karady, *Textes* I. Editions de Minuit. Translation of *La Riforma Sociale, 2*(1895), 691–707.
Durkheim, E. (1974–1975). Textes, edited by Victor Karady, 3 volumes. Les éditions de minuit.
Durkheim, E. ([1885–1912] 1978). *Émile Durkheim on institutional analysis*. (Mark Traugott, Trans.). University of Chicago Press.
Durkheim, E. ([1896–1912] 1980). *Contributions to the L'Année sociologique* (Y. Nandan, Ed.). Free Press.
Durkheim, E. ([1895] 1982). *The rules of sociological method (and other writings on sociology)* (W. D. Halls, Trans.). Free Press.
Durkheim, E. (1983). *Durkheim and the law* (S. Lukes & A. Scull, Ed.). Martin Robertson.
Durkheim, E. ([1893] 1984). *The division of labor in society* (W. D. Halls, Trans.). Free Press.
Durkheim, E. (1900a). Rôle des universités dans l'education sociale du pays. In *Congrès international de l'education sociale* (pp. 128–138).
Durkheim, E. (1900b). La sociologie en France au xix siècle. *Revue Bleue, xii*, 609–613, 647–652.
Durkheim, E. ([1887] 1992). *Suicide and fertility a study of moral statistics* (H. L. Sutcliffe, Trans.). European Journal of Population, *8*(3), 175–97.
Durkheim, E. ([1912] 1995). *The elementary forms of religious life* (K. E. Fields, Trans.). Free Press.
Durkheim, E. (1998). *Lettres a Marcel Mauss* (Philippe Besnard et Marcel Fournier, Ed.). Presses Universitaires de France.
Durkheim, E. ([1903–1917] 2003). *L'évaluation en comité* (Stephane Baciocchi & Jennifer Mergy, Ed.). Berghahn Books.
Durkheim, E. ([1883–1884] 2004). *Durkheim's philosophy lectures – Notes from the Lycée de Sens course, 1883-1884* (Neil Gross, Ed. and Trans., Robert Alun Jones, Ed.). Cambridge University Press.
Durkheim, E. ([1914] 2011). L'avenir de la religion. *Les Cahiers de Psychologie Politique*, (18), E01–E01.
Durkheim, E., & Fauconnet, P. (1903). Sociologie et sciences sociales. *Revue philosophique, 45*, 465–497.
Durkheim, E., & Mauss, M. ([1903] 1967). *Primitive classification* (R. Needham, Trans.). University of Chicago Press.

Eccles, J. C. (1989). Hughlings Jackson's views on consciousness. In C. Kennard & M. Swash (Eds.), *Hierarchies in neurology: A reappraisal of a Jacksonian concept* (pp. 39–48). Springer.
Edelman, G. M. (2006). The embodiment of mind. *Daedalus, 135*(3), 23–32.
Eisen, S. (1968). Frederic Harrison and Herbert Spencer: Embattled unbelievers. *Victorian Studies, 12*(1), 33–56.
Ellis, J. D. (1990). *The physician-legislators of France: Medicine and politics in the early Third Republic 1870–1914*. Cambridge University Press.
Ellis, H. (2017). *Masculinity and science in Britain 1831–1918*. Palgrave Macmillan.
Elwick, J. (2003). Herbert Spencer and the disunity of the social organism. *History of Science, 41*, 35–72.
Elwick, J. (2007). *Styles of reasoning in the British life sciences: Shared assumptions, 1820–1858*. Pickering and Chatto.
Elwick, J. (2014). Containing multitudes: Herbert Spencer, organisms social, and orders of individuality. In M. Francis & M. W. Taylor (Eds.), *Herbert Spencer: Legacies*. Routledge.
Elwitt, S. (1975). *The making of the Third Republic: Class and politics in France 1868–1884*. Louisiana State University Press.
Elwitt, S. (1986). *The Third Republic defended – Bourgeois reform in France 1880–1914*. Louisiana State University Press.
Engelhardt, H. T. (1975). John Hughlings Jackson and the mind body relation. *Bulletin of the History of Medicine, 49*(2), 137–151.
Eros, J. (1955). The positivist generation of French republicanism. *The Sociological Review, 3*, 255–277.
Espinas, A. V. (1878). *Des sociétés animales: étude de psychologie comparée* (2iéme édition). Germer Bailliére et cie.
Espinas, A. (1882). Les études sociologiques en France. *Revue philosophique, 13*, 565–604, and *14*, 337–67, 509–28.
Espinas, A. (1884). L'Agrégation de philosophie. *Revue internationale de l'enseignement, 7*(1), 585–607.
Faber, D. P. (1997). Theodule Ribot and the reception of evolutionary ideas in France. *History of Psychiatry, 8*(29–32), 445–458.
Fabian, J. (1983). *Time and the other: How anthropology makes its object*. Columbia University Press.
Fabiani, J.-L. (1988). *Des philosophes de la République*. Editions de Minuit.
Fabiani, J.-L. (2003). Clore enfin l'e`re des généralitees. In S. Baciocchi & J. Mergy (Eds.), *L'Évaluation en comité* (pp. 151–189). Berghahn Books.
Fabiani, J.-l. (2006). À quoi sert la notion de discipline? In J. Boutier, C. Passeron, & J. Revel (Eds.), *Qu'est-ce qu'une discipline* (pp. 11–34). Ecole des hautes études en sciences sociales.
Fabiani, J.-L. (2020). Durkheim and the philosophy of his time. In H. Joas & A. Pettenko (Eds.), *The oxford handbook of emile Durkheim*. (pp.1–14). https://doi.org/10.1093/oxfordhb/9780190679354.001.0001
Farber, P. L. (1975). Buffon and Daubenton: Divergent traditions within histoire naturelle. *Isis, 66*(1), 63–74.
Farber, P. L. (1981). Research traditions in 18th century natural history. In W. Bernardi & A. La Vergata (Eds.), *Lazzaro Spallanzani e la biologia del Settecento* (pp. 397–403). L.S. Olschki.
Farquhar, B. E. (1934). Hughlings-Jackson and his influence on neurology. *Lancet, 224*(5800), 909–913.
Ferrier, D. (1912). John Hughlings Jackson. *Proceedings of the Royal Society of London, Medicine, 84*, 18–25.
Feuerhahn, W. (2011). Les 'sociétés animales': un défi à l'ordre savant. *Romantisme, 154*, 35–51.
Feuerhahn, W. (2016). De la sorbonne au collège de France, enjeux du titre des chaires de Ribot. *Revue Philosophique, 141*(4), 477–488.
Feuerhahn, W., & Fueuerhahn, P. (2010). La science à l'échelle internationale. *Revue Germanique International*, 5–15.

Feuerhahn, W., & Trochu, T. (2016). Autoportraits de Thédule Ribot en correspondant. *Revue Philosophique, 206*(4), 521–540.

Filloux, J.-C. (1970). Introduction. In E. Dukheim (Ed.), *La science sociale et l'action* (pp. 5–68). Presses Universitaires de France.

Filloux, J.-C. (1977). *Durkheim et le socialisme*. Droz.

Finn, Michael A. (2012). *The West riding lunatic asylum and the making of the modern brain sciences in the nineteenth century* (PhD dissertation). University of Leeds.

Finn, M. R. (2017). *Figures of the pre-Freudian unconscious from Flaubert to Proust*. Cambridge University Press.

Fischer, J. L. (1979). Yves Delages (1854-1920). *Revue de synthèse, 95–96*, 443–461.

Fissel, M. E. (1991). The disappearance of the Patient's narrative and the invention of hospital medicine. In R. French & A. Wear (Eds.), *British medicine in an age of reform* (pp. 92–109). Routledge.

Fitzi, G., & Marcucci, N. (2017). Interview with Hans-Joas on the reception of Émile Durkheim in Germany. Berlin: Humboldt University of Berlin, 6 October 2014. *Journal of Classical Sociology, 17*(4), 382–398.

Fleck, L. ([1935] 1979). *Genesis and development of a scientific fact*. The University of Chicago Press.

Forrester, J. (1996). If p, then what? Thinking in cases. *History of the Human Sciences, 9*(3), 1–25.

Forster, R., & Ranum, O. (1980). *Medicine and society in France*. Johns Hopkins University Press.

Fouillée, A. (1905). La morale biologique, sa portée, ses limites- la rêve des cours scientifique. *Revue Scientifique, 4*.

Fournier, M. (2013). *Émile Durkheim: A biography* (D. Macey, Trans.). Polity Press.

Fox, R. (2012). *The savant and the state: Science and cultural politics in nineteenth-century France*. John Hopkins University Press.

Francis, M. (2007). *Herbert Spencer and the invention of modern life*. Acumen.

Francis, M. (2014). The evolutionary turn in positivism: G.H. Lewes and Leslie Stephen. In W. J. Mander (Ed.), *Oxford handbook of British philosophy in the 19th century* (pp. 285–305). Oxford University Press.

Francis, M. (2015). The problem with Star dust: Spencer's psychology and William James. In M. Francis & M. W. Taylor (Eds.), *Herbert Spencer: Legacies* (pp. 154–183). Routledge.

Freud, S. (1965). *A psycho-analytic dialogue: The letters of Sigmund Freud and Karl Abraham, 1907–1926* (H. C. Abraham & E. L. Freud, Ed., B. Marsh & H. C. Abraham, Trans.). Basic Books.

Freud, S. (1975). *The standard edition of the complete psychological works of Sigmund Freud*. Hogarth Press.

Freud, S. (1985). *The complete letters of Sigmund Freud to Wilhelm Fliess, 1887–1904* (J. Mussaieff-Masson, Trans. and Ed.). Belknap Press.

Freud, S. ([1915] 1987). *A phylogenetic fantasy: Overview of the transference neuroses* (Axel and Peter T. Hoffer, Trans., Ilse Grubrich-Simitis, Ed.). Belknap.

Freud, S. (1993–2000). *The correspondence of Sigmund Freud and Sándor Ferenczi* (Eva Brabant, Ernst Falzeder, and Patrizia Giampieri-Deutsch, under the supervision of André Haynal, Ed., Peter T. Hoffer, Trans.). 3 vol. Belknap Press.

Frith, C. D., & Rees, G. (2017). A brief history of the scientific approach to the study of consciousness. In S. Schneider & M. Velmans (Eds.), *The Blackwell companion to consciousness* (pp. 3–44). Blackwell Publication.

Fuller, T. (2003). The integrative biology of phenotypic plasticity. *Biology and Philosophy, 18*, 381–389.

Fullinwider, S. P. (1983). Sigmund Freud, John Hughlings Jackson, and Speech. *Journal of the History of Ideas, 44*(1), 151–158.

Furet, F., Ozouf, M., & with B. Baczko. (1988). *Dictionnaire critique de la révolution française*. Flammarion.

Fyfe, A., McDougall-Waters, J., & Moxham, N. (2015). 350 years of scientific periodicals. *Notes and Records of the Royal Society of London, 69*(3), 227–239.

Gane, M. (1988). *On Durkheim's rules of sociological method*. Routledge.
Gane, M. (1992). Durkheim: The sacred language. In M. Gane (Ed.), *The radical sociology of Durkheim and Mauss* (pp. 61–84). Routledge.
Gane, M. (1994). A fresh look at Durkheim's sociological method. In W. S. F. Pickering & H. Martins (Eds.), *Debating Durkheim* (pp. 66–85). Routledge.
Gane, M. (2010). Durkheim's theory of violence. In S. Romi Mukherjee (Ed.), *Durkheim and violence* (pp. 41–50). Wiley-Blackwell.
Gasser, J. (1988). La notion de mémoire organique dans l'oeuvre de T. Ribot. *History and Philosophy of the Life Sciences, 10*(2), 293–313.
Gauchet, M. (1992). *L'inconscient cérébral*. Éditions du Seuil.
Gayon, J. (2010). Raisonnement fonctionnel et niveaux d'organization en biologie. In J. Gayon et Armand de Ricqles with Matteo Mossio (Ed.), *Les fonctions: Des organismes aux artefacts* (pp. 125–38). Presses Universitaires de France.
Geiger, R. (1975). The institutionalization of sociological paradigms: Three examples from early French sociology. *Journal of the History of Behavioural Sciences, 11*, 235–245.
Geiger, R. (1977). Democracy and the crowd (France and Italy 1890-1914). *Societas, 7*, 47–71.
Geison, G. L. (1969). The protoplasmic theory of life and the vitalist-mechanist debate. *Isis, 60*(3), 273–292.
Gentner, D. (1988). Metaphor as structure mapping: The relational shift. *Child Development, 59*(1), 47–59.
Gentner, D., & Markman, A. B. (1997). Structure mapping in analogy and similarity. *The American Psychologist, 52*(1), 45–56.
Gere, C. (2011). 'Nature's experiment': Epilepsy, localization of brain function and the emergence of the cerebral subject. In F. Ortega & F. Vidal (Eds.), *Neurocultures: Glimpse into an expanding Uni-verse* (pp. 235–247). P. Lang.
Gere, C. (2013). Curating aphasia: Pierre Paul Broca's Museological science. *Interdisciplinary Science Reviews, 38*(3), 200–209.
Gere, C. (2015). Plasticity, pathology, and pleasure in cold war America. In D. Bates & N. Bassiri (Eds.), *Plasticity and pathology* (pp. 35–64). Fordham University Press.
Geroulanos, S. (2014). The plastic self and the prescription of psychology; ethno-psychology, crowd psychology and Psychotechnics 1890-1920. *Republics of Letters, 3*, 2. http://arcade.stanford.edu/rofl/plastic-self-and-prescription-psychology.
Geroulanos, S., & Meyers, T. (2018). *The human body in the age of catastrophe: Brittleness, integration, science, and the great war*. Chicago University Press.
Gerring, J. (2017). *Case study research: Principles and practices* (2nd ed.). Cambridge University Press.
Giard, A. M. (1877). Preface, *T.H. Huxley: Éléments d'anatomie comparée des animaux invertébrés* (G. Darin, Trans.) (pp I–IV). V.Adrien Delahaye et cie.
Giard, A. M. (1888a). L'évolution des êtres organisés. Leçon inaugurale. *Bulletin scientifique de la France et Belgique, 20*, 1–26.
Giard, A. M. (1888b). Histoire du transformisme. *Revue scientifique., 43*, 689–699.
Giard, A. M. (1889). Les facteurs d'évolution. Leçon d'ouverture – 2ieme année. *Revue scientifique, 44*, 641–648.
Giard, A. M. (1890). Le Principe de Lamarck et l'heredité des modifications somatiques. Leçon d'ouverture. *Revue scientifique*, 45–46.
Giard, A. (1904). *Controverses transformistes*. C. Naud.
Gibbs, R. W. (2008). *The Cambridge handbook of metaphor and thought*. Cambridge University Press.
Giddens, A. (1978). *Émile Durkheim*. Harvester.
Giglioni, G. (2013). Jean-Baptiste Lamarck and the place of irritability in the history of life and death. In S. Normandin & C. T. Wolfe (Eds.), *Vitalism and the scientific image in post enlightenment life science, 1800–2010*. Springer.
Gilbert, M. (1989). *On social facts*. Routledge.

Gilbert, M. (2014). Joint commitment: How we make the social world.
Gilbert, S., & Sahorta, S. (2000). Embracing complexity: Organicism for the 21st century. *Developmental Dynamics, 219*, 1–9.
Gillispie, C. C. (1956). The formation of Lamarck's evolutionary theory. *Archives internationals d'histoire des sciences, 9*, 323–338.
Gillespie, C. C. (1959). The Encyclopédie and the Jacobin Philosophy of Science: A Study in Ideas and Consequences. In M. Clagett (Ed.), *Critical problems in the history of science* (pp. 255–290). University of Wisconsin Press.
Gillispie, C. C. (1980). *Science and polity in France at the end of the old regime*. Princeton University Press.
Ginsburg, R., & Pardes, I. (Eds.). (2006). *New perspectives on Freud's "Moses and Monotheism"*. Conditio Judaica.
Gissis, S. B. (2002). Late nineteenth century Lamarckism and French sociology. *Perspectives on Science, 10*(1), 69–122.
Gissis, S. B. (2005). Herbert Spencer on biological heredity and cultural inheritance. In Hans-Jörg Rheinberger and Staffan Müller-Wille (Eds.), *The 3rd workshop on "The Cultural History of Heredity,"* preprint volume (pp. 137–52).
Gissis, S. B. (2009). Interactions between social and biological thinking: The case of Lamarck. *Perspectives on Science, 17*(39), 237–306.
Gissis, S. B. (2010). Lamarck on feelings: From Worms to humans. In C. T. Wolfe & O. Gal (Eds.), *The body as object and instrument of knowledge* (pp. 211–239). Springer.
Gissis, S. B. (2011). Lamarckism and the constitution of sociology. In S. B. Gissis & E. Jablonka (Eds.), *Transformations of Lamarckism, from subtle fluids to molecular biology*. M.I.T. Press.
Gissis, S. B. (2017). Collectivity' from a historical perspective. In S. B. Gissis, E. Lamm, & A. Shavit (Eds.), *Landscapes of collectivity in the life sciences*.
Gissis, S. B. (2018). Spencer's evolutionary entanglement: From liminal individuals to implicit Collectivities. In S. Scott Lidgard & K. Lynn (Eds.), *Biological individuality, integrating scientific, philosophical and historical perspectives*. Chicago University Press.
Gissis, S. B. (2019). 'Biologie'- Lamarck's endeavor of a science of living entities. In C. Bognon-Küss & C. T. Wolfe (Eds.), *Philosophy of biology before biology* (pp. 170–187). Routledge.
Gissis, S. B. (2020). Transfer of Lamarckisms and emerging 'scientific psychologies': 19th-early 20th centuries Britain and France. *Studies in History and Philosophy of Biological and Biomedical Sciences, 83*(October), 1–13.
Gissis, S. B. (2023). The Neo Lamarckian tools deployed by the young Durkheim: 1882–1892. *Journal of the History of Biology* (April) 1–37. https://doi.org/10.1007/s10739-023-09708-w
Glass, B., Temkin, O., & Straus, W. L., Jr. (Eds.). (1959). *Forerunners of Darwin: 1745–1859*. Johns Hopkins Press.
Gliboff, S. (2008). In H.G. (Ed.), *Bronn, Ernst Haeckel, and the origins of German Darwinism: A study in translation and transformation*. MIT Press.
Godfrey-Smith, P. (1996). *Complexity and the function of mind in nature*. Cambridge University Press.
Godfrey-Smith, P. (2003). Between Baldwin Skepticism and Baldwin Boosterism. In B. H. Weber & D. J. Depew (Eds.), *Evolution and learning: The Baldwin effect reconsidered* (pp. 53–68). The MIT Press.
Godfrey-Smith, P. (2016). Individuality, subjectivity, and minimal cognition. *Biology and Philosophy, 31*(6), 775–796.
Godfrey-Smith, P. (2019). Evolving across the explanatory gap. *Philosophical Theory Practical Biology, 11*(1), 1–24.
Goetz, R. (1986). La 'Science de l'homme' des ideologues. In *Doctrines,sciences ou pratiques socials?* Colloque. Nancy: Presses universitaires de Nancy.
Goetz, R. (1993). *Destutt de Tracy*. Droz.
Goetz, C. G., Bonduelle, M., & Gelfand, T. (1995). *Charcot: Constructingn neurology*. Oxford University Press.

Gohau, G. (1979). Alfred Giard (1846–1908). *Revue de synthèse, 95–96*, 393–406.
Gohau, G. (1997a). L'hydrogéologie et l'histoire de la géologie. In G. Laurent (Ed.), *Jean Baptiste Lamarck 1744–1829* (pp. 137–147). Editions de CTHS.
Gohau, G. (1997b). Le cadre mineral et l'evolution Lamarckiennne. In J. Schiller (Ed.), *Colloque international Lamarck* (pp. 31–46). Blanchard.
Gohau, G. (2006). Lamarck philosophe? In P. Corsi, J. Gayon, G. Gohau, & S. Tirard (Eds.), *Lamarck philosophe de la nature* (pp. 9–36). Presses Universitaires de France.
Goldstein, J. (1987). *Console and classify: The French psychiatric profession in the nineteenth century*. Cambridge University Press.
Goldstein, J. (2002). Mutations of the self in old regime and post-revolutionary France. In L. Daston (Ed.), *Biographies of scientific objects*. Chicago University Press.
Goldstein, J. (2005). *The post-revolutionary self: Politics and psyche in France, 1750–1850*. Harvard University Press.
Goldstein, J. (2013). Neutralizing Freud: The *Lycée* philosophy class and the problem of the reception of psychoanalysis in France. *Critical Inquiry, 40*(1), 40–82.
Goodfield, J. (1960). *The growth of scientific physiology; physiological method and the mechanist-vitalist controversy*. Hutchinson.
Gould, S. J. (1977). *Ontogeny and phylogeny*. Belknap Press of Harvard University Press.
Gould, S. J. (1992). Ontogeny and phylogeny revisited and reunited. *BioEssays, 14*(4), 275–279.
Gould, S. J. (1999a). A division of worms: Jean-Baptiste Lamarck contribution to evolutionary theory, Part I. *Natural History* (February).
Gould, S. J. (1999b). Branching through a wormhole: Reclassifying the types of 'Worms, Part II. *Natural History* (March).
Gould, S. J. (2002). *The structure of evolutionary theory*. Belknap Press.
Gould, S. J. (2011). Freud's evolutionary fantasy. In S. J. Gould (Ed.), *I have landed: The end of a beginning in natural history* (pp. 147–160). Cambridge Mass.
Gowers, W. R. (1880). The Gulstonian lectures, on epilepsy. *British Medical Journal, 1*(3), 353.
Granshaw, L. (1989). 'Fame and fortune by means of bricks and mortar': The medical profession and specialist hospitals in Britain, 1800–1948. In L. Granshaw & R. Porter (Eds.), *The hospital in history*. Routledge.
Granshaw, L. (1992). The rise of the modern hospital in Britain. In A. Wear (Ed.), *Medicine in society: Historical essays*. Cambridge University Press.
Grassé, P. P. (1944). La biologie: texte inedit de Lamarck. *Revue scientifique, 82*, 267–276.
Gray, T. S. (1996). *The political philosophy of Herbert Spencer: Individualism and organicism*. Avebury.
Greene, J. C. (1959). *The death of Adam; Evolution and its impact on western thought*. Ames, Iowa State University Press.
Green, C. D. (2014). James Mark Baldwin, the Baldwin effect, organic selection, and the American 'immigrant crisis' at the turn of the twentieth century. In G. Barker, E. Desjardins, & T. Pearce (Eds.), *Entangled life: Organism and environment in the biological and social sciences*. Springer.
Greenblatt, S. H. (1970). Hughlings Jackson's first encounter with the work of Paul Broca: The physiological and philosophical background. *Bulletin of the History of Medicine, 44*(6), 555–570.
Greenblatt, S. H. (1999). John Hughlings Jackson and the conceptual foundations of the neurosciences. *Physis, 36*(2), 367–386.
Grewe, P., & Bien, C. G. (2017). 150th anniversary of the Bethel epilepsy Center in Germany: An important milestone in the evolution of epilepsy care. *Seizure, 53*(December), 110–113.
Griffiths, P. (2003). Beyond the Baldwin effect: James Mark Baldwin's 'social heredity,' epigenetic inheritance, and niche construction. In B. H. Weber & D. J. Depew (Eds.), *Evolution and learning: The Baldwin effect reconsidered* (pp. 193–218). The MIT Press.
Griffiths, P. (2007). The Baldwin effect and genetic assimilation. In P. Carruthers, S. Laurence, & S. Stich (Eds.), *The innate mind: Culture and cognition* (Vol. 2). Oxford University Press.

Grimault, C. (1998). *Evolutionisme et fixisme en France: Histoire d'un combat 1800–1882*. Editions CNRS.

Gross, M. (1979). The lessened locus of feelings: A transformation in French physiology in the early nineteenth century. *Journal of the History of Biology, 12*(2), 231–271.

Gross, N. (2004). Introduction. In R. A. Jones (Ed.), *Durkheim's philosophy lectures: Notes from the Lycée de Sens course, 1883–1884* (pp. 1–30). Cambridge University Press.

Grubrich-Simitis, I. (1987). An essay on Freud. In I. Grubrich-Simitis (Ed.), *A phylogenetic fantasy*. Belknap.

Guenther, K. (2015). *Localization and its discontents: A genealogy of psychoanalysis and the neuro disciplines*. Chicago University Press.

Guillin, V. (2004). Théodule Ribot's ambiguous positivism. *Journal of History of Behavioural Sciences, 40*(2), 165–181.

Guillin, V. (2016). Mill, Ribot et la science du caractère. *Revue Philosophique de la France et de L'étranger, 141*(4), 489–500.

Guillin, V. (2018). Comte and social science. In M. Bourdeau, M. Pickering, & W. Schmaus (Eds.), *Love, order, and progress: The science, philosophy, and politics of Auguste Comte* (pp. 128–160). University of Pittsburgh Press.

Guillo, D. (2000). L'empreinte de l'histoire naturelle sur la sociologie comtienne: l'héritage de l'anatomie comparée des premie'res décennies du xix sie'cle dans le cours de philosophie positive. *L'Année Sociologique, 50*(1), 195–216.

Guillo, D. (2002). La sociologie d'inspiration biologique au XIXe siècle: une science de l'organisation sociale. *Revue Française de Sociologie, 41*(2), 241–275.

Guillo, D. (2006). La place de la biologie dans les premiers textes de Durkheim: un paradigme oublié? *Revue Française de Sociologie, 47*(3), 507–535.

Gumplowicz, L. (1885). *Grundriss der soziologie*. Manz.

Gumplowicz, L. (1893). *La lutte des races, recherches sociologiques* (C. Bayet, Trans.). Guillaumin.

Hacking, I. (1983). Nineteenth century cracks in the concept of determinism. *Journal of History of Ideas, 44*, 455–475.

Hacking, I. (1990). *The taming of chance*. Cambridge University Press.

Hacking, I. (1994). How numerical sociology began by counting suicides: From medical pathology to social pathology. In I. Bernard Cohen (Ed.), *The natural sciences and the social sciences* (pp. 101–133). Kluwer Academic Publishers.

Hacking, I. (1998). *Rewriting the soul: Multiple personality and the science of memory*. Princeton University Press.

Hacking, I., Barham, P., Osborne, T., & Lynch, M. (1995). Rewriting the soul. *History of the Human Sciences, 8*(4), 107–130.

Hagner, M. (1992). The soul and the brain between anatomy and Naturphilosophie in the early nineteenth century. *Medical History, 36*, 1–33.

Haines, V. A. (1988). Is Spencer's theory an evolutionary theory? *American Journal of Sociology, 93*(5), 1200–1223.

Hale, P. (2014). *Political descent: Malthus mutualism and the politics of evolution in Victorian England*. Chicago University Press.

Hallyn, F. (Ed.). (2002). *Metaphor and analogy in the sciences*. Kluwer Academic Publishers.

Hamlett, J. (2015). *At home in the institution: Material life in asylums, lodging houses and schools in Victorian and Edwardian England*. Palgrave Macmillan.

Hammet, I. (1984). Durkheim and the study of religion. In S. Fenton, R. Reiner, & I. Hammet (Eds.), *Durkheim and modern sociology* (pp. 202–218). Cambridge University Press.

Hamy, E. T. (1893). Les derniers jours du Jardin du Roi: et la fondation du Muséum d'Histoire naturelle. In E. T. Hamy (Ed.), *Centenaire de la fondation du Muséum d'Histoire naturelle* (pp. 1–162). Imprimerie Nationale.

Hamy, E. T. (1907). Les Débuts de Lamarck, matériaux inédits pour servir à sa biographie. *Paris: Bibiliothèque d'histoire scienti*ªque., *11*(1).

Hanks, P., & Giora, R., Eds. (2012). *Metaphor and figurative language*. Vol. 1: *Theoretical issues*; Vol. 4: *Cognition and metaphor*; Vol. 6, part 28: *Applied metaphor studies*. Routledge.

Harré, R. (1996). Metaphor in the history of psychology. *The British Journal for the Philosophy of Science, 47*(1), 141–145.

Harré, R., & Martin, J. (1982). Metaphor in science. In D. Miall (Ed.), *Metaphors: Problems, and perspectives* (pp. 89–105). Humanities Press.

Harrington, A. (1985). Hysteria, hypnosis and the lure of the invisible: The rise of neo-mesmerism in fin-de-Siècle French psychiatry. In W. F. Bynum, R. Porter, & M. Shepard (Eds.), *The anatomy of madness: Essays in the history of psychiatry, vol. 3: The asylum and its psychiatry* (pp. 226–246). Routledge.

Harrington, A. (1987). *Medicine, mind, and the double brain: A study in 19th century thought*. Princeton University Press.

Harrington, A. (1991). Beyond phrenology: Localization theory in the era. In P. Corsi (Ed.), *The enchanted loom: Chapters in the history of neuroscience*. Oxford University Press.

Harrington, A. (2009). The brain and the Behavioral sciences. In Peter J. Bowler, & John V. Pickstone (Eds.), *The Cambridge history of science*, vol. 6: *The modern biological and earth sciences* (pp. 504–23). Cambridge University Press.

Harris, M. (1992). Social diseases? – Crime and medicine in the Victorian press. In W. F. Bynum, S. Lock, & R. Porter (Eds.), *Medical journals and medical knowledge: Historical essays*. Routledge.

Harris, J. C. (2005). A clinical lesson at the Salpêtrière. *Archives of General Psychiatry, 62*(5), 470–472.

Harrison, F. (1905). *The Herbert Spencer lecture*. Clarendon Press.

Harvey, J. (1984). L'Evolution transformée: The Social Context of Scientific Debates Originating in the Société d'Anthropologie de Paris du Seconde Empire a la Troisième Republique. In B. Rupp-Eisenreich (Ed.), *Histoire de L'anthropologie (xvi-xix siécles)* (pp. 387–410). Klincksieck.

Hatfield, G. (2014). The emergence of psychology. In W. J. Mander (Ed.), *The Oxford handbook of British philosophy in the nineteenth century*. Oxford University Press.

Hayward, J. E. S. (1959). Solidarity: The social history of an idea in nineteenth century France. *International Review of Social History, 2*, 261–284.

Hayward, J. E. S. (1960). Solidarist syndicalism: Durkheim and Duguit, part I. *Social Research Review, 8*, 185–202.

Hayward, J. E. S. (1961). The official philosophy of the French Third Republic: Léon Burgeois and Solidarism. *International Review of Social History, 6*, 20–32.

Head, H. (1915). Hughlings Jackson on aphasia and kindred affections of speech. *Brain, 38*(1-2), 27–42.

Head, H. (1926). *Aphasia and kindred disorders of speech* (Vol. 1). Cambridge University Press.

Heilbron, J. (1995). *The rise of social theory* (S. Gogol, Trans.). University of Minnesota Press.

Heilbron, J. (2015). *French sociology*. Cornell University Press.

Henri, V. (1891). Report of the international congress of physiological psychology at Paris. *Mind, 16*(61), 157–160.

Henri, V. (1893). Les laboratoires de psychologie expérimentale en Allemagne. *Revue philosophique, 36*, 608–622.

Henrich, J. (2016). *The secret of our success: How culture is driving human evolution, domesticating our species, and making us smarter*. Princeton University Press.

Henson, R. A. (1989). The Hughlings Jackson tradition at the London hospital. In C. Kennard & M. Swash (Eds.), *Hierarchies in neurology: A reappraisal of a Jacksonian concept* (pp. 31–38). Springer.

Hering, E. (1870). *Über das gedächtniss als eine allgemeine function der organisirten materie*. K.K. hofund staatsdruckerei.

Herr, L. (1932). *Choix d'écrits*. Éditions Rieder.

Hesse, M. B. (1966). *Models and analogies in science*. University of Notre Dame Press.

Hesse, C. (1991). *Publishing and cultural politics in revolutionary Paris 1789–1810*. University of California Press.
Hirsch, T. (2018). Émile Durkheim, l'Année sociologique et le 'Parti sociologique unifié': Formes et enjeux du travail collectif (1897–1913). *Mil Neuf Cent, 36*(1), 35–58.
Hodge, M. J. S., & Jonathan, S. (1971a). Species in Lamarck. In J. Schiller (Ed.), *Colloque international Lamarck* (pp. 31–46). Blanchard.
Hodge, M. J. S., & Jonathan, S. (1971b). Lamarck's science of living bodies. *The British Journal for the History of Science, 10*, 323–352.
Hodge, M. J. S., & Jonathan, S. (1995). Lamarck: un grand changement de cadre conceptuel. In C. Blanckaert, J. L. Fischer, & R. Rey (Eds.), *Nature, Histoire, Société: essais en homage à J. Roger* (pp. 229–239). Klincksieck.
Hoffer, P. T. (1992). The concept of phylogenetic inheritance in Freud and Jung. *Journal of the American Psychoanalytic Association, 40*(2), 517–530.
Hofstadter, R. (1955). *Social Darwinism in America*. Beacon Press.
Holmes, F. L. (1985). *Lavoisier and the chemistry of life: An exploration of scientific creativity*. University of Wisconsin Press.
Holmes, F. L. (2014). *Claude Bernard and Animal chemistry: The emergence of a scientist*. Harvard University Press.
Holowchak, A. M. (2012). *Freud: From individual psychology to group psychology*. Jason Aronson.
Hooykaas, R. (1959). *Natural law and divine miracle: A historical-critical study of the principle of uniformity in geology, biology, and theology*. Brill.
Hopwood, N. (2009). Embryology. In P. J. Bowler & J. V. Pickstone (Eds.), *The Cambridge history of science* (Vol. 6, pp. 285–315). Cambridge University Press.
Hum, P. (2014). *L'oubli 1880–1900* (PhD dissertation). Université Charles De Gaulle, Lille III.
Hunter, K. M. (2021). Remaking the case. *Literature and Medicine, 11*(1), 163–179.
Hunter, R. A., & Hurwitz, B. (1961). The case notes of the National Hospital for the paralyzed and epileptic, Queen Square, London, before 1900. *Journal of Neurology and Neurosurgery and Psychiatry, 24*, 187–194.
Hurren, E. T. (2012). *Dying for Victorian medicine: English anatomy and its trade in the dead poor, c.1834–1929*. Palgrave Macmillan.
Hurwitz, B. (2006). Form and representation in clinical case report. *Literature and Medicine, 25*(2), 216–240.
Hurwitz, B. (2017). Narrative constructs in clinical case reporting. *Studies in History and Philosophy of Science, 62*, 65–73.
Hutschinson, J. (1911a). Obituary Dr. Hughlings Jackson. *The British Medical Journal, 2650–2*(October 14), 950–954.
Hutschinson, J. (1911b). The late Dr. Hughlings Jackson: Recollections of a life long friendship. *The British Medical Journal, 9*, 1551–1554.
Hutton, F. W. (1899). *Darwinism and Lamarckism, Old and New; four lectures*. G. P. Putnam.
Huxley, T. H. (1875). On the hypothesis that animals are automata, and its history. *The Eclectic Magazine of Foreign Literature, 21*(1).
Huxley, T. H. (1882). *Science and culture and 0ther essays*. D.Appleton and Company.
Huxley, T. H. (1892). On animal individuality [1852]. In M. Foster & E. Ray Lankester (Eds.), *The scientific memoirs of Thomas Henri Huxley* (Vol. 1, pp. 146–151). Macmillan.
Ioannou, I., & Laskowski, K. L. (2023). A multi-scale review of the dynamics of collective behaviour: from rapid responses to ontogeny and evolution. *Philosophical Transactions of the Royal Society B, 378*(1874).
Ingold, T. (1986a). *The appropriation of nature: Essays on human ecology and social relations*. Manchester University Press.
Ingold, T. (1986b). *Evolution and social life*. Cambridge University Press.
Ingold, T., & Palsson, G. (Eds.). (2013). *Biosocial becomings: Integrating social and biological anthropology*. Cambridge University Press.
Jablonka, E., & Ginsburg, S. (2019). *The evolution of the sensitive soul: Learning and the origins of consciousness*. MIT Press.

Jablonka, E., & Lamb, M. (2014). *Evolution in four dimensions: Genetic, epigenetic, behavioral, and symbolic variation in the history of life*. The MIT Press.
Jackson, J. H. (1863a). An experimental inquiry into the effect of the application of ice to the back of the neck on the retinal circulation. *Medical Times Gazette, 2*, 90–91.
Jackson, J. H. (1863b). *Observations on defects of sight in brain disease, and ophthalmoscopic examination during sleep*. Harrison and Sons.
Jackson, J. H. (1864). On the study of diseases of the nervous system. *London Hospital Reports*, vol. 1 (Reprinted 1903, *Brain, 26*(3), 367–82).
Jackson, J. H. (1866). A Physician's note on ophthalmology. *The Royal London Ophthalmic Hospital Reports and Journal of Ophthalmic Medicine and Surgery, 5*, 279.
Jackson, J. H. (1867). Remarks on the disorderly movements of chorea and convulsion, and on localisation. *Medical Times and Gazette*, (December 14), 642–43, (December 21), 669–70.
Jackson, J. H. (1868). Notes on the physiology and pathology of the nervous system. *Medical Times and Gazette*, (2), 177–179.
Jackson, J. H. (1873a). Observations on the localisation of movements in the cerebral hemispheres, as revealed by cases of convulsion, chorea and 'aphasia.'. *West Riding Lunatic Asylum Medical Reports, 3*, 177–195.
Jackson, J. H. (1873b). On the anatomical, physiological, and pathological investigation of epilepsies. *West Riding Lunatic Asylum Medical Reports, 3*, 315–339.
Jackson, J. H. (1873c). Remarks on the double condition of loss of consciousness and mental automatism following certain epileptic seizures. *Medical Times and Gazette*, 63–64.
Jackson, J. H. (1867–1875). Diverse Papers in *The West Riding Lunatic Asylum Medical Reports.*
Jackson, J. H. (1888). Remarks on the diagnosis and treatment of diseases of the brain. *British Medical Journal, 1437*(2), 59–63.
Jackson, J. H. (1925). *Neurological fragments by J. Hughlings-Jackson* (James Taylor, Ed.). Oxford University Press.
Jackson, J. H. ([1931–1932] 1958). *Selected writings of John Hughlings-Jackson* (J. Taylor, G. Holmes, & F. M. R. Walshe, Eds.), 2 vols. Staples Press.
Jackson, J. H. (n.d.-a). *Untitled commentary on the factors of evolution*. Unpublished Papers ref. QSA/870. Queen Square Archives.
Jackson, J. H. (n.d.-b). *Untitled comments on Ferrier's and Soury's ideas on cortical centres.* Unpublished Papers, ref.QSA/867. Queen Square Archives.
Jacyna, L. S. (1981). The physiology of mind, the unity of nature and the moral order in Victorian thought. *British Journal for the History of Science, 14*(2), 109–132.
Jacyna, L. S. (1982). Somatic theories of mind and the interests of medicine in Britain, 1850-1879. *Medical History, 26*(3), 233–258.
Jacyna, L. S. (1984). Principles of general physiology: The comparative dimension to British neuroscience in the 1830s and 1840s. *Studies in History of Biology, 7*, 47–92.
Jacyna, L. S. (2000). *Lost words: Narratives of language and the brain, 1825–1926*. Princeton University Press.
Jacyna, L. S. (2008). *Medicine and modernism: A biography of Sir Henry Head.* Pickering & Chatto.
Jacyna, L. S. (2011). Process and progress: John Hughlings Jackson's philosophy of science. *Brain, 134*(10), 3121–3126.
James, W. (1889). The congress of physiological psychology at Paris- report. *Mind, 14*(56), 614–616.
James, W. (1890). *Principles of psychology*. Henry Holt and Co.
Janet, P. (1888). Une Chaire de psychologie expérimentale et comparée au Collège de France. *Revue des Deux Mondes, 86*(58), 518–549.
Janet, P. (1915). L'oeuvre Psychologique de Théodule Ribot. *Journal de Psychologie normale et Pathologique, 12*(4), 268–282.
Jastrow, J. (1902). Memory (defects of). In J. M. Baldwin (Ed.), *Dictionary of philosophy and psychology* (Vol. 2, pp. 64–66). MacMillan.

Joas, H. (2017). Interview by Gregor Fitzi and Nicola Marcucci with Hans Joas on the reception of Émile Durkheim in Germany. Berlin: Humboldt University of Berlin, 6 October 2014. *Journal of Classical Sociology, 17*(4), 382–398.

Jones, R. A. (1972). Durkheim's critique of Spencer's ecclesiastical institutions. *Sociological Inquiry, 44*(3), 205–221.

Jones, R. A. (1974). Durkheim's response to Spencer: An essay towards historicism in the historiography of sociology. *Sociological Quarterly, 15*(3), 341–358.

Jones, R. A. (1975). Durkheim in context: A reply to Perrin. *Sociological Quarterly, 16*(4), 551–559.

Jones, R. A. (1981). Robertson Smith, Durkheim and sacrifice: An historical context for 'the elementary forms of the religious life. *Journal of the History of the Behavioral Sciences, 17*, 184–205.

Jones, S. S. (2001). *Durkheim reconsidered.* Polity Press.

Jones, R. A. (2005). Practices and pre-suppositions: Some questions about Durkheim and Les Formes élémentaires de la vie religieuse. In J. Alexander & P. Smith (Eds.), *The Cambridge companion to Durkheim* (pp. 80–100). Cambridge University Press.

Jones, S. S. (2010). Durkheim, the question of violence and the Paris commune of 1871. In S. Romi Mukherjee (Ed.), *Durkheim and violence*, Chap. 4. Wiley-Blackwell.

Jordanova, L. (1981). La psychologie naturaliste et le 'probleme des niveaux' la notion du sentiment intérieur chez Lamarck. In J. Schiller (Ed.), *Lamarck et son temps, Lamarck et notre temps* (pp. 69–80). Vrin.

Jordanova, L. (1984). *Lamarck.* Oxford University Press.

Jordanova, L. (1989). Nature's powers: A reading of Lamarck's distinction between creation and production. In J. R. Moore (Ed.), *History, humanity and evolution* (pp. 71–98). Cambridge University Press.

Kaitaro, T. (2008). Can matter mark the hours? Eighteenth-century vitalist materialism and functional properties. *Science in Context, 21*(4), 581–592.

Karady, V. (1976). Durkheim, les sciences sociales et l'université: bilan d'un semi-échec. *Revue Française de Sociologie, 17*, 267–311.

Karady, V. (1979). Strategies de réussite et modes de faire valoir de la sociologie chez les durkheimiens. *Revue Française de Sociologie, 20*, 49–82.

Karady, V. (1980). Educational qualifications and university careers in science in nineteenth century France. In R. Fox & G. Weisz (Eds.), *The Organization of Science and Technology in France, 1808–1914* (pp. 95–124). Cambridge University Press.

Karady, V. (1983). The Durkheimians in academe. A reconsideration. In P. Besnard (Ed.), *The sociological domain* (pp. 71–89).

Keller, E. F. (1995). *Refiguring life: Metaphors of twentieth century biology.* Columbia University Press.

Keller, E. F. (2002). *Making sense of life: Explaining biological development with models, metaphors and machines.* Harvard University Press.

Keller, E. F. (2008). Organisms, machines, and thunderstorms: A history of self-organization, part one. *Historical Studies in the Natural Sciences, 38*(1), 45–75.

Keller, E. F. (2010). It is possible to reduce biological explanations to explanations in chemistry and/or physics. In F. J. Ayala & R. Arp (Eds.), *Contemporary debates in philosophy of biology* (pp. 19–31). Wiley-Blackwell.

Keller, E. F. (2016). Thinking about biology and culture. *The Sociological Review Monographs, 64*(1), 26–41.

Kennard, C., & Swash, M. (Eds.). (1989). *Hierarchies in neurology: A reappraisal of a Jacksonian concept.* Springer.

Kitcher, P. (1992). *Freud's dream: A complete interdisciplinary science of mind.* MIT Press.

Kittay, E. F. (1987). *Metaphor: Its cognitive force and linguistic structure.* Clarendon.

Klautke, E. (2013). The French reception of völkerpsychologie and the origins of the social sciences. *Modern Intellectual History, 10*(2), 293–316.

Kostić, D., Hilgetag, C. C., & Tittgemeyer, M., Eds. (2020). Unifying the essential concepts of biological networks: Biological insights and philosophical foundations. *Philosophical Transactions of the Royal Society B, 375*(1796).

Kronfeldner, M. E. (2009). If there is nothing beyond the organic...: Heredity and culture at the boundaries of anthropology in the work of Alfred L Kroeber. *Naturwissenschaften, Technik und Medizin, 17*(2), 107–133.

Kronfeldner, M. E. (2018). The right to ignore: An epistemic Defense of the nature/culture divide. In R. Joyce (Ed.), *The Routledge handbook of evolution and philosophy* (pp. 210–224). Routledge.

Kronfeldner, M. (2021). Digging the channels of inheritance: On how to distinguish between cultural and biological inheritance. *Philosophical Transactions, B.376/1828*. https://doi.org/10.1098/rstb.2020.0042

Kuklick, H. (2005). Interpreting aboriginal religion: From nineteenth century evolutionism to Durkheimian sociology. *Humanities Research, 12*(1), 13–25.

Kusch, M. (1999). *Psychological knowledge: A social history and philosophy*. Routledge.

La Décade philosophique, littéraire et politique. (1794–1804). + La Revue, ou Décade philosophique, littéraire et politique. Paris.

LaCapra, D. (1972). *Emile Durkheim: Sociologist and philosopher*. Cornell University Press.

Lacroix, B. (1981). *Durkheim et le politique*. Presses de la Fondation nationale des sciences politiques.

Lakoff, G., & Johnson, M. (1980). *Metaphors we live by*. The University of Chicago Press.

Lamarck, J.-B. (1778). *Flore françoise*. L'imprimerie royale.

Lamarck, J.-B. (1791). *Tableau encyclopédique et méthodique des trois règnes de la nature. Botanique*. Panckoucke.

Lamarck, J.-B. (1794). *Rechèrches sur les causes des principaux faits physiques*. Chez Maradan.

Lamarck, J.-B. (1796). *Réfutation de la théorie pneumatique, ou de la nouvelle doctrine des chimistes modernes*. L'auteur.

Lamarck, J.-B. (1797). *Memoires présentant les bases d'une nouvelle théorie, physique et Chimique,* Lus à la première classe de l'Institut national. Chez l'auteur.

Lamarck, J.-B. ([1799] 1800a). *Tableau encyclopédique et méthodique des trois regnes de la nature*. H. Agasse.

Lamarck, J.-B. (1800b). *Biologie ou Considérations sur la nature, les facultés, les développemens et l'origine des corps vivans* (P. P. Grassé, Trans.). *Revue Scientifique, 82*, 267–276.

Lamarck, J.-B. (1801a). *Discours d'ouverture prononcé le 12 Floreal a l'an 8*. http://www.lamarck.cnrs.fr/index.php?lang=fr.

Lamarck, J.-B. (1801b). *Système des animaux sans vertèbres; préceédée du discours d'ouverture du cours de zoologie, donn. dans le Muséum national d'histoire naturelle l'an 8 de la République*. Deterville.

Lamarck, J.-B. (1802). *Hydrogéologie, ou, Recherches sur l'influence qu'ont les eaux sur la surface du globe terrestre*. Chez l'auteur; Maillard.

Lamarck, J.-B. (1802–1806). *Mémoires sur les fossiles des environs de Paris*. Muséum national d'histoire naturelle.

Lamarck, J.-B. (1783–1808). *Encyclopédie méthodique botanique* (Preface, Introduction, Classes, Espèce, Caractère, Botanique). Panckoucke; Plomteux.

Lamarck, J.-B. (1809). *Philosophie Zoologique*. Dentu. http://www.lamarck.cnrs.fr/index.php?lang=fr.

Lamarck, J.-B. (1800–1810). Annuaire météorologique: Pour l'an VIII [-XVII] de l'ère de la République française. Chez l'Auteur: Maillard.

Lamarck, J.-B. (1812). *Extrait du cours de zoologie sur les animaux sans vertebres*. D'Hautel & Gabon. http://www.lamarck.cnrs.fr/index.php?lang=fr.

Lamarck, J.-B. 1817. Articles from the various volumes of the *nouveau Dictionnaire d'histoire naturelle*: Faculté, Fonctions organiques, habitude, homme, Idée, Irritabilité, Inteligence, imagination, instinct, judgement, Météorologie, nature. Deterville.

Lamarck, J.-B. (1820). *Système analytique des connaissances positives de l'homme*. Chez l'auteur et Belin.

Lamarck, J.-B. (1815–1822). *Histoire Naturelle des animaux sans vertèbres*. Deterville. http://www.lamarck.cnrs.fr/index.php?lang=fr

Lamarck, J.-B. (1907). *Discours d'ouverture: an VIII, an X, an XI et 1806*. Bulletin scientifique de la France et de la Belgique.

Lamarck, J.-B. (1933). *The Lamarck manuscripts at Harvard* (William Morton Wheeler & Thomas Barbour, Ed.). Harvard University Press.

Lamarck, J.-B. 1972. *Inédits de Lamarck*; d'après les manuscripts conservés à la Bibliothèque centrale du Muséum national d'histoire naturelle de Paris (Max Vachon, Georges Rousseau, & Yves Laissus, Eds.). Masson.

Lamarck, J.-B. ([1809] 1984). *Zoological philosophy* (Hugh Elliot, Trans.). University of Chicago Press.

Lamarck, J.-B. ([1802] 1986). *Recherches sur l'organisation des corps vivans: et particulièrement sur son origine, précédé du discours d'ouverture du cours de zoologie*. Fayard/Chez l'auteur, Maillard).

Lamarck, J.-B. (1990). *Les Introuvables* (D. Goujet, Ed.). Societe Francaise de Systematique.

Lamarck, J.-B., Bruguière, O., & et Pelletier, H. (1792). *Choix de mémoires sur divers objets d'histoire naturelle, formant les collections du journal d'histoire naturelle*. L'imprimerie du Cercle Social.

Landrieu, M. ([1809] 1909). *Lamarck, le fondateur du transformisme; sa vie, son œuvre*, a l'occasion du centenaire de la 'Philosophie zoologique.' la Société zoologique de France.

Lang, A. (1889). Lamarck und Darwin: ein Beitrag über Geschichte der Entwicklungslehre, I-IV. *Kosmos, 132–142*, 243–250, 408–417, 510–533.

Lapie, P. (1896). L'Anné sociologique 1895: la morale sociale. *Revue de Métaphysique et de Morale, 4*(3), 338–361.

Lassere, P. (1924). *Mes routes*. Plon-Nourrit et cie.

Latour, B. (1988). *The pasteurization of France (*A. Sheridan & J. Law, Trans.). Cambridge.

Laubichler, M. D. (2007). Does history recapitulate itself? Epistemological reflections on the origins of evolutionary developmental biology. In M. D. Laubichler & J. Maeinschein (Eds.), *From Embryoplogy to Evo Devo* (pp. 13–34). MIT Press.

Laubichler, M. D. (2009). Form and function in Evo-Devo: Historical and conceptual reflections. In M. D. Laubichler & J. Maienschein (Eds.), *Form and function in developmental evolution*. Cambridge University Press.

Laubichler, M. D. (2010). Evolutionary developmental biology offers a significant challenge to the neo-Darwinian paradigm. In F. J. Ayala & R. Arp (Eds.), *Contemporary debates in philosophy of biology* (pp. 199–212). Wiley-Blakcwell.

Laurent, G. (1987). *Paléontologie et évolution en Cuvier et Lamarck à Darwin*. Editions du Comité des travaux historiques et scientifiques.

LaVergata, A. (1966a). Giard. In P. Tort (Ed.), *Dictionnaire du darwinisme et de la évolution* (pp. 1911–1931). Presses Universitaires de France.

LaVergata, A. (1966b). Perrier. In P. Tort (Ed.), *Dictionnaire du darwinisme et de la évolution* (pp. 3403–3407). Presses Universitaires de France.

LaVergata, A. (1995). Herbert Spencer: Biology, sociology, and cosmic evolution. In S. Maasen, E. Mendelsohn, & P. Weingart (Eds.), *Biology as society, society as biology: Metaphors* (pp. 193–229). Kluwer Academic.

La Vergata, A. (1996). Giard. In P. Tort (Ed.), *Dictionnaire du Darwinisme et de la évolution* (pp. 1911–1931). Presses Universitaires de France.

Lawrence, C. (1985). Incommunicable knowledge: Science, technology and the clinical art in Britain 1850-1914. *Journal of Contemporary History, 20*, 503–420.

Laycock, T. (1845). On the function of the brain. *British and Foreign Medical Review, 19*, 298–311.

Laycock, T. (1860). *Mind and brain; or, the correlations of consciousness and organisation; with their applications to philosophy, zoology, physiology, mental pathology, and the practice of medicine*. Sutherland and Knox.

Le Bon, G. (1894). *Les lois psychologiques de l'évolution des peuples*. Felix Alcan.

Le Bon, G. (1895a). *Psychologie des foules*. Felix Alcan.

Le Bon, G. (1895b). *La Psychologie des Foules*. G. Masson.

Le Dantec, F. (1899). *Lamarckiens et Darwiniens, discussions de quelques theories sur la formation d'espe'ces*. F. Alcan.

Leclerc, G. (1972). *Anthropologie et colonialisme*. Fayard.

Leguay, P. (1910). *La Sorbonne*. B. Grasset.

Lehning, J. R. (2018). *To be a citizen: The political culture of the early French Third Republic*. Cornell University Press.

Lekka, V. (2015). *The neurological emergence of epilepsy: The National Hospital for the paralysed and epileptic (1870–1895)*. Springer.

Lenoir, R. (1919). La psychologie de Ribot et la pensée contemporaine. *Revue de Métaphysique et de Morale, 26*(6), 739–763.

Leon, X. (1917). Nécrologie de Théodule. Ribot. *Revue de Métaphysique et de Morale, 24*(1), 129–132.

Leonard, J. (1992). *Médecins, malades et société dans la France de xixieme siecle*. Sciences en situation.

Lepenies, W. (1988). *Between literature and science: The rise of sociology* (R. J. Hollingdale, Trans.). Cambridge University Press.

Lesch, J. E. (1984). *Science and medicine in France: The emergence of experimental physiology 1790–1855*. Harvard University Press.

Levins, R., & Lewontin, R. (2006). A program for biology. *Biological Theory, 1*(4), 333–335.

Lewes, G. H. (1855, November 3). Life and mind, 3rd review essay on Spencer's Principles of Psychology. *The Leader*.

Lewes, G. H. ([1879] 1891). *Problems of life and mind*, (first series: The foundations of a creed), vol. 2. Mifflin & Co.

Lewontin, R. (2000). *The triple helix: Gene, organism, environment*. Harvard University Press.

Lightman, B. (2007). *Victorian popularizers of science: Designing nature for new audiences*. Chicago University Press.

Lightman, B. (2011). Victorian periodicals, evolution and public controversy. *Spontaneous Generation, 5*(1), 5–11.

Lightman, B. (2016a). Spencer's American disciples: Fiske, Youmans, and the appropriation of the system. In B. Lightman (Ed.), *Global Spencerism: The communication and appropriation of a British evolutionist*. Brill.

Lightman, B. (2016b). Popularizers, from 1860s to the participation and the transformations of nineteenth century publishing: From the 1860s to the 1880s. *Notes and Records of the Royal Society of London, 70*(4), 343–359.

Lightman, B., & Reidy, M. R. (Eds.). (2014). *The age of scientific naturalism: Tyndall and his contemporaries*. Pickering and Chatto.

Von Lilienfeld, P. (1896). *La pathologie sociale, preface R. Worms*. V. Giard & E. Brière.

Limoges, C. (1980). The development of the Muséum d'Histoire naturelle of Paris 1800–1914. In R. Fox & G. Weizs (Eds.), *The organization of science and technology in France 1808–1914* (pp. 211–240). Cambridge University Press.

Limoges, C. (1994). Milne-Edwards, Darwin, Durkheim and the division of labour: A case study in reciprocal conceptual exchanges between the social and the natural sciences. In I. Bernard Cohen (Ed.), *The natural sciences and the social sciences* (pp. 317–343). Kluwer Academic.

Limoges, C. (2014). A second glance at evolutionary biology in France. In E. Mayr & W. B. Provine (Eds.), *The evolutionary synthesis: Perspectives on the unification of biology* (pp. 322–328). Harvard University Press.

Logue, W. (1983). *From philosophy to sociology: The evolution of French liberalism, 1870–1914*. Northern Illinois University Press.

Loison, L. (2009). Les conceptions embryologiques et phylogénétiques d'Alfred Giard (1846-1908) et Edmond Perrier (1844-1921), deux appropriations de la loi biogénétique fondamental. *Bulletin d'histoire et d'épistemologie des sciences de la vie, 16*(2), 165–183.

Loison, L. (2010). *Qu'est-ce-que le néolamarckisme? Les biologistes Français et la question de l'évolution des espe'ces*. Vuibert.

Loison, L. (2011a). French Roots of French Neo-Lamarckisms, 1879–1985. *Journal of the History of Biology, 44*(4), 713–744.

Loison, L. (2011b). The Notions of Plasticity and Heredity among French Neo-Lamarckians (1880–1940): From Complementarity to Incompatibility. In S. Gissis & E. Jablonka (Eds.), *Transformations of Lamarckism* (pp. 67–76). MIT Press.

Loison, L. (2018). Lamarckism and epigenetic inheritance: A clarification. *Biology and Philosophy, 33*, 29.

Loison, L. (2020). Disentangling genetic assimilation from the Baldwin effect. A philosophical perspective. *Paradigmi, 38*(3), 441–462.

López Piñero, J. M. (2010). The work of John Hughlings Jackson, edited and translated by GE Berrios. *History of Psychiatry, 21*, 85–99; *21*, 224–36.

Lorch Marjorie Perelman. (2004). The unknown source of John Hughlings Jackson's early interest in aphasia and epilepsy. *Cognitive Neurology, 17*(3), 124–132.

Lorch Marjorie Perelman. (2008). The merest logomachy: The 1868 Norwich discussion of aphasia by Hughlings Jackson and Broca. *Brain, 131*(6), 1658–1670.

Lubek, I. (1981). Histoire de psychologies sociales perdues: le cas de Gabriel Tarde. *Revue Française de Sociologie, 22*(3), 361–395.

Lukes, S. (1969). Durkheim's individualism and the intellectuals. *Political Studies, 17*(1), 14–30.

Lukes, S. (1973a). *Émile Durkheim: His life and work: A historical and critical study*. Allen Lane.

Lukes, S. (1973b). *Individualism*. Blackwell.

Lukes, S. (2018). Durkheim and the new sociology of morality. In H. Joas & A. Pettenkofer (Eds.), *The Oxford handbook of Émile Durkheim*. Oxford University Press. https://doi.org/10.1093/oxfordhb/9780190679354.013.30

Lustig, A. J. (2004). Ants and the nature of nature in Auguste Forel, Erich Wasmann, and William Morton wheeler. In L. Daston & F. Vidal (Eds.), *The moral authority of nature* (pp. 282–307). University of Chicago Press.

Maasen, S., & Weingart, P. (2000). *Metaphors and the dynamics of knowledge*. Routledge.

Maasen, S., Mendelsohn, E., & Weingart, P. (2002). *Biology as society, society as biology: Metaphors*. Kluwer Academic Publishers.

Macgraw, R. (2002). *France, 1800–1914: A social history*. Routledge.

MacLeod, R. M. (1970). The X-Club: A social network of science in late-Victorian England. *Notes and Records of the Royal Society of London, 24*(2), 305–322.

Malabou, C. (2018). Is psychic phylogenesis only a phantasy? New biological developments in trauma inheritance. In G. Sharvit, K. Feldman, et al. (Eds.), *Freud and monotheism* (pp. 177–198). Fordham University Press.

Marcel, J.-C., & Guillo, D. (2009). Durkheimian sociology, biology and the theory of social conflict. In S. Romi Mukherjee (Ed.), *Durkheim and violence* (pp. 83–100). Wiley-Blackwell.

Marinelli, L., & Mayer, A. (2006). Editors' introduction: Forgetting Freud? For a new historiography of psychoanalysis. *Science in Context, 19*(1), 1–13.

Marsh, R. M. (2010). Merton's sociology 215-216 course. *American Sociologist, 41*, 99–114.

Martin, J. P. (1975). Kinnier Wilson's notes of conversations with Hughlings Jackson. *Journal of Neurology, Neurosurgery, and Psychiatry, 38*, 313–316.

Martin, E. (1991). The egg and the sperm: How science has constructed a romance based on stereotypical male-female roles. *Signs: Journal of Women in Culture and Society, 16*(3), 485–501.

Martin, J., & Harré, R. (1982). Metaphor in science. In D. S. Miall (Ed.), *Metaphor* (pp. 89–105). Humanities Press.

Massis, H. (1931). *Evocations: Souvenirs 1905–1911*. Librairie Plon.

Mauss, M. (1908). L'Art et le mythe d'apres M. Wundt. *Revue Philosophique, 66*, 47–79.

Mauss, M. (1925). In memoriam, l'oeuvre inédite de Durkheim. In V. Karady (Ed.), *Marcel Mauss œuvres* (Vol. 3, pp. 473–488). Les éditions de minuit.

Mauss, M. (1939). Ribot et les sociologues. In *Centenaire de Théodule Ribot, Jubilé de la psychologie scientifique Française1839–1889-1939* (pp. 137–138). Agen.

Mauss, M., & Fauconnet, P. (1901). Sociologie. In *La Grande Encyclopedie* (Vol. 30). Société anonyme de la Grande Encyclopédie.

Mayer, E. (1972). Lamarck revisited. *Journal of the History of Biology, 5*, 55–94.

Meloni, M. (2016). The transcendence of the social: Durkheim, Weismann, and the purification of sociology. *Hypothesis and Theory, 1*, 11.

Meloni, M. (2019). *Impressionable biologies: From the archaeology of plasticity to the sociology of epigenetics*. Routledge.

Mendelsohn, A. (2017). Empiricism in the library: Medicine's case histories. In L. Daston (Ed.), *Science in the archives: Pasts, presents, futures* (pp. 85–110). Chicago University Press.

Mengal, P. (1994). Henri Pieron (1881-1964) et les Neo-lamarckiens français. In C. Debni, J. Gayon, & J. F. Picard (Eds.), *Les sciences biologiques et medicales en France, 1920–1950* (pp. 113–124). Editions CNRS.

Mercier, C. (1912). The late Dr. Hughlings Jackson. *British Medical Journal, 1*(12663), 85–86.

Mergy, J. (2009). The politics of the future-an unknown text by Émile Durkheim. *Durkheimian Studies, 15*, 7–14.

Merllié, D. (1989). Lévy-Bruhl et Durkheim: Notes biographiques en marge d'une correspondance. *Revue philosophique, 179*(4), 493–514.

Merllié, D. (1993). Les rapports entre Revue philosophique et la Revue de métaphysique et de morale-MM, Leon Xavier, Theodule Ribot, Lucien Levy Bruhl. *Revue de Métaphysique et de Morale, 1–2*, 59–108.

Merllié, D. (2012). Durkheim, Lévy-Bruhl et la 'pensée primitive': Quel différend? *Année Sociologique, 62*(2), 429–446.

Merllié, D. (2017). Correspondance d'Émile Durkheim avec Lucien Lévy-Bruhl. *Cahiers Vilfredo Pareto, 55*(2), 105–168.

Merz, J. T. ([1904–1912] 1976). *A history of European scientific thought in the nineteenth century*. P. Smith.

Mesoudi, A. (2016). Cultural evolution: A review of theory, findings and controversies. *Evolutionary Biology, 43*(4), 481–497.

Métraux, A. (1982). French crowd psychology in the nineteenth century. In R. William (Ed.), *The problematic science: Nineteenth century psychology: Psychology in nineteenth century thought*. Woodward and Mitchel G. Ash.

Metraux, A. (1982). French crowd psychology: Between theory and ideology (chapter 11). In W. R. Woodward & M. G. Ash (Eds.), *The problematic science: psychology in nineteenth-century thought*. Praeger.

Miall, D. (Ed.). (1982). *Metaphor*. The Harvester Press.

Micale, M. S. (2007). Two cultures revisited: The case of the fin de Siècle. In R. Bivins & J. V. Pickstone (Eds.), *Medicine, madness and social history*. Palgrave Macmillan.

Miller, W. W. (2012). *A Durkheimian quest: Solidarity and the sacred*. Berghahn Books.

Milne-Edwards, H. (1837). *Elémens de zoologie: Ou Leçons sur l'anatomie, la physiologie, la classification et les moeurs des animaux*. H. Dumont.

Milne-Edwards, H. (1851). *Introduction á la zoologie générale, ou considérations sur les tendences de la nature dans la constitution du règne animal, premiere partie*. Masson.

Milne-Edwards, A. (1882). *Élements de l'histoire naturelle des animaux- anatomie et physiologie animales: Programme du cours d'anatomie et de physiologie animales pour la classe de philosophie* (arre'té ministeriel de 2 août 1880). Masson.

Momigliano, A. (1970). The ancient city of Fustel de Coulanges. In *Essays in ancient and modern historiography* (pp. 325–344). Chicago University Press.

Montgomery, K. (1992). Remaking the case. *Literature and Medicine, 11*(1), 163–179.

Montgomery, K. (1993). *Doctors' stories: The narrative structure of medical knowledge*. Princeton University Press.

Morange, M. (2010). What history tells us: The French neo-lamarckians. *Journal of Bioscience, 35*(4), 515–517.

Moravia, S. (1972). Philosophie et medicine en France a la fin du xviiie siècle. *Studies on Voltaire, 472*(89), 1089–1151.

Moravia, S. (1980). Enlightenment and the sciences of man. *History of Science, 18*, 247–268.

Morgan, M. S. (2012a). One observation or many? Justification or discovery? *Philosophy of Science, 79*(5), 667–677.

Morgan, M. S. (2012b). Travelling facts. In P. Howlett & M. S. Morgan (Eds.), *How well do facts travel? The dissemination of reliable knowledge*. Cambridge University Press.

Morgan, M. S. (2014). Resituating knowledge: Generic strategies and case studies. *Philosophy of Science, 81*(5), 1012–1024.

Mosbah-Natanson, S. (2008). Internationalisme et tradition nationale: Le cas de la constitution de la sociologie Française autour de 1900. *Revue d'Histoire des Sciences Humaines, 1*(18), 35–62.

Mosbah-Natanson, S. (2011). La sociologie comme 'mode'? Usages éditoriaux de label 'sociologie' en France a` la fin du XIX siècle. *Revue Française de Sociologie, 52*(1), 103–132.

Moscovici, S. (1981). *L'Âge des foules: Un traité historique de psychologie des masses*. Fayard. Motabe 2020.

Mucchielli, L. (1994). Sociologie et psychologie en France, l'appel à un territoire Commun: Vers une psychologie collective (1890–1940). *Revue de synthese IVieme s., 3–4*, 445–483.

Mucchielli, L. (1995). Pourquoi réglementer la sociologie? Les interlocuteurs de Durkheim. In M. Borlandi & L. Mucchielli (Eds.), *La Sociologie et sa méthode* (pp. 15–48). L'Harmattan.

Mucchielli, L. (1998). *La découverte du social: Naissance de la sociologie en France*. Éditions la découverte.

Mucchielli, L. (2006). Aux origines de la psychologie universitaire en France (1870–1900): enjeux intellectuels, contexte politique, réseaux et stratégies d'alliance autour de la Revue philosophique de Théodule Ribot. *Annals of Science, 55*(3), 263–289.

Muelberger, A. (2017). The persuasive rhetoric of a manifesto (1870): Ribot's promise of an 'independent' psychological science. *Centaurus, 59*, 204–222.

Mukherjee, R. (2006). On violence as the negativity of the Durkheimian: Between anomie, sacrifice and effervescence. In S. Romi Mukherjee (Ed.), *Durkheim and violence* (pp. 5–39). Wiley-Blackwell.

Müller-Wille, S., & Rheinberger, H. J. (2012). *A cultural history of heredity*. Chicago University Press.

Neswald, E. (2014). Saving the world in the age of entropy: John Tyndall and the second law of thermodynamics. In B. Lightman & M. S. Reidy (Eds.), *The age of scientific naturalism: Tyndall and his contemporaries*. Pickering and Chatto.

News. (1878). *Mind, 3*(April 10), 295.

Nicholson, D., & Gawne, R. (2015). Neither logical empiricism nor vitalism, but organicism: What the philosophy of biology was. *History and Philosophy of the Life Sciences, 37*(4), 345–381.

Nicoglou, A. (2018). The concept of plasticity in the history of the nature–nurture debate in the early twentieth century. In J. Maurizio Meloni, D. F. Cromby, et al. (Eds.), *The Palgrave handbook of biology and society*. Palgrave Macmillan.

Nicolas, S. (1998). Aux origines de la psychologie universitaire en France (1870–1900). *Annals of Science, 55*, 263–289.

Nicolas, S. (1999). L'hérédité psychologique d'après Théodule Ribot (1873): La première thèse française de psychologie scientifique. *L'Année Psychologique, 99*(2), 295–348.

Nicolas, S. (2001). Introducing psychology as an academic discipline in France: Théodule Ribot and the Collège de France (1888–1901). *Journal of the History of the Behavioral Sciences, 37*(2), 143–164.

Nicolas, S. (2005). *Théodule Ribot .1839–1917*. L'Harmattan.

Nicolas, S., & Gounden, Y. (2017). Centenaire Ribot. deuxieme partie: la reception de l'œuvre de Théodule Ribot publiée chez l'éditeur Ballie're.1874-1883. *Bulletin de Psychologie, 549*(3), 179–195.

Nicolas, S., & Makowski, D. (2017). Centenaire Ribot (première partie). La réception de l'œuvre de Théodule Ribot publiée chez l'éditeur Ladrange (1870-1873). *Bulletin de Psychologie, 549*(3), 163–176.

Nicolas, S., Barnes, M. E., & Murray, D. J. (2014). A French description of German psychology laboratories in 1893 by Victor Henri, a collaborator of Binet. *Psychological Research, 79*(3), 361–370.

Nicolas, S., Sabourin, M., & Piolino, P. (2016). The seminal contributions of Théodule Ribot (1839-1916): The centenary of the passing of the founder of French psychology. *L'Année Psychologique, 116*(4), 519–546.

Nord, P. G. (2011). The Third Republic. In E. Berenson, V. Duclert, & C. Prochasson (Eds.), *The French Republic: History, values, debates* (pp. 44–55). Cornell University Press.

Normandin, S. (2006). Claude Bernard and an introduction to the study of experimental medicine: 'Physical vitalism,' dialectic, and epistemology. *Journal of the History of Medicine and Allied Sciences, 62*(4), 495–528.

Novicow, J. (1900). Les castes et la sociologie biologique. *Revue Philosophique, 50*, 361–373.

Nowell-Smith, H. (1995). Nineteenth-century narrative case histories: An inquiry into stylistics and history. *Canadian Bulletin of Medical History, 2*(1), 47–67.

Nungesser, F. (2020). The sociality of mind: Key arguments, inner tensions and divergent appropriations of Durkheim's sociology. In H. Joas & A. Pettenkofer (Eds.), *The Oxford handbook of Émile Durkheim*. Oxford University Press.

Nye, R. A. (1975). *The origins of crowd psychology: Gustave Le Bon and the crisis of mass democracy in the 3rd Republic*. Sage.

Nye, R. A. (1984). *Crime, madness, and politics in modern France*. Princeton University Press.

Nye, M.-J. (1993). *From chemical philosophy to theoretical chemistry*. University of California Press.

Nye, M.-J. (1996). *Before big science: The pursuit of modern chemistry and physics, 1800–1940*. Prentice Hall International.

Nye, R. A. (2007). Western masculinities in war and peace. *American Historical Review, 112*, 417–438.

Nyhart, L. K. (1995). *Biology takes form: Animal morphology and the German universities, 1800–1900*. University of Chicago Press.

Nyhart, L., & Lidgard, S. (2011). Individuals at the center of biology: Rudolf Leuckart's Polymorphismus Der Individuen and the ongoing narrative of parts and wholes. *Journal of the History of Biology, 44*(3), 373–443.

O'Neale, J. C. (1996). *The authority of experience: Sensational theory in the French enlightenment*. Pennsylvania State University Press.

Offer, J., Ed. (2000). *Herbert Spencer – Critical assessments*, 4 vol. Routledge.

Offer, J. (2010). *Herbert Spencer and social theory*. Palgrave Macmillan.

Offer, J. (2015). A new reading of Spencer on 'society', 'organicism' and 'spontaneous order'. *Journal of Classical Sociology, 15*(4), 337–360.

Ohayon, A. (2008). Psychologie clinique et psychologie expérimentale. Une union jamais consommée. *Sciences Humaines, 7*, 52–53.

Olen, P., & Turner, S. (2015). Durkheim, sellars, and the origins of collective intentionality. *British Journal for the History of Philosophy, 23*(5), 954–975.

Olmsted, J. M. D., & Olmsted, E. H. (1952). *Claude Bernard and the experimental method in medicine*. H. Schuman.

Omodeo, P. (1997). Le transformisme au xviiie siécle. In G. Laurent (Ed.), *Jean-Baptiste Lamarck 1744–182* (pp. 471–482). Editions de CTHS.

Oppenheim, J. (1991). *"Shattered nerves": Doctors, patients, and depression in Victorian England*. Oxford University Press.

Ortony, A. (Ed.). (1979). *Metaphor and thought*. Cambridge University Press.

Osborne, M. (1992). Applied natural history and utilitarian ideals: Jacobin science at the Muséum d'histoire naturellle 1789–1870. In B. T. Ragan & E. A. Williams (Eds.), *Recreating authority in revolutionary France* (pp. 125–143). Rutgers University Press.

Osborne, M. O. (1994). *Nature, the exotic, and the science of French colonialism*. Indiana University Press.

Osborne, M. A. (2019). *The emergence of tropical medicine in France*. University of Chicago Press.

Otis, L. (1994). *Organic memory: History and the body in the late nineteenth and early twentieth centuries*. University of Nebraska Press.

Outram, D. (1984). *Georges Cuvier: Vocation, science, and authority in post-revolutionary France*. Manchester University Press.

Outram, D. (1997). Le Muséum national d'Histoire naturelle après 1793: Institution scienti^aque ou champ de bataille pour les familles et les groupes d'ifluence? In C. Blanckaert, C. Cohen, C. P. Corsi, & J. L. Fischer (Eds.), *Le Muséum au premier sie'cle de son histoire* (pp. 25–30). Editions du Muséum d'histoire naturelle.

Ozouf, M. (1982). *L'école, l'eglise et la République 1871–1914*. Ed. Cana.

Ozouf, M. (1987). La révolution et l'idée de l'homme nouveau. In C. Lucas (Ed.), *The French revolution and the creation of modern political culture* (Vol. 2, pp. 213–232). Pergamon Press.

Ozouf, M. (1989). *L'homme régénéré: Essais Sur la Révolution française*. Gallimard.

Ozouf, J., & Ozouf, M. (1992). *La republique des instituteurs*. Julliard.

Packard, A. S. (1901). *Lamarck, the founder of evolution: His life and work*. Longmans, Green.

Paoletti, G. (2012). *Durkheim et la philosophie: représentation, realité et lien social*. Classiques Garnier.

Papineau, D. (2007). The Baldwin effect and genetic assimilation reply to Griffiths. In Peter Carruthers, Stephen Laurence, and Stephen Stich (Eds.), *The innate mind*, vol. 2: *Culture and cognition*. Oxford University Press.

Parodi, D. (1920). *La philosophie contemporaine en France: essai de classification des doctrines*. Felix Alcan.

Parot, F. (2012). France. In L. T. Benjamin & D. B. Baker (Eds.), *Oxford handbook of the history of psychology: Global perspective* (pp. 228–254). Oxford University Press.

Paskins, M., & Morgan, Mary S., Eds. (2020). *An anthology of narrative science, narrative science project ERC*. https://www.narrative-science.org/uploads/3/1/7/6/31762379/narrative_science_anthology_complete_final_draft_9_11_2019.pdf.

Pearce, T. (2010). From 'circumstances' to 'environment': Herbert Spencer and the origins of the idea of organism-environment interaction. *Studies in History and Philosophy of the Biological and the Biomedical Sciences, 41*, 241–252.

Pearce, T. (2020). *Pragmatism's evolution: Organism and environment in American philosophy*. University of Chicago Press.

Peel, J. D. Y. ([1971] 1992). *Herbert Spencer: The evolution of a sociologist*. Gregg Revivals.

Péguy, C. ([1906] 1959). *Oeuvres en prose 1898–1908*. Gallimard

Peirce, C. (1890). Review of Théodule Ribot's psychology of attention. *The Nation* (June 19).

Peirce, C. S. (2010 [1890]). *Writings of Charles S. Peirce: A chronological edition Volume 8, 1890–1892*. Indiana University Press.

Perrier, E. (1881). *Les colonies animales et la formation des organismes*. Masson.

Perrier, E. (1884). *La philosophie zoologique avant Darwin*. Felix Alcan.

Perrier, E. (1888). *Le transformisme*. librarie J.B. Bailliére et fils.

Perrier, E. (1893). Lamarck et le transformisme actuel. In *Centenaire de la fondation du Muséum d'Histoire naturelle, volume commémoratif* (pp. 469–528). Imprimerie Nationale.

Perrier, E. (1909). Jean de Lamarck. *Revue pedagogique, 55*(7).

Perrin, R. G. (1975). Durkheim's Misrepresentation of Spencer: a Reply to Jones' "Durkheim's Response to Spencer". *Sociological Quarterly, 16*(4), 544–550.

Perrin, R. G. (1993). *Herbert Spencer: A primary and secondary bibliography*. Garland Publications.

Persell, S. M. (1999). *Neo-Lamarckism and the evolution controversy in France, 1870–1920*. The Edwin Mellen Press.

Petit, A. (1993). Heurs et malheurs du positivisme: Philosophie des sciences et politique scientifique chez Auguste Comte et ses premiers disciples (1820–1900) (PhD dissertation). Université Paris 1.

Petit, A. (2018). Positivismes, écoles et mouvances. *Revue d'histoire des sciences humaines, 32*, 99–128.

Pick, D. (1989). *Faces of degeneration: A European disorder, c.1848-c.1918*. Cambridge University Press.

Pick, D. (2007). Maladies of the will: Freedom, fetters, and the fear of Freud. In R. Bivins & J. V. Pickstone (Eds.), *Medicine, madness and social history-essays in honour of Roy Porter* (pp. 197–209). Palgrave Macmillan.

Pickering, W. S. F. (2000). Representations as understood by Durkheim- an introductory sketch. In W. S. F. Pickering (Ed.), *Durkheim and representations* (pp. 11–23). Routledge.

Pickstone, J. V. (2007). Working knowledges before and after circa 1800. *Isis, 98*, 489–516.

Pickstone, J. V. (2007). Working knowledges before and After circa 1800: Practices and disciplines in the history of science, technology, and medicine. *Isis, 98*(3), 489–516.

Pieron, H. I. (1912). Le domaine psychologique. *L'Année Psychologique, 19*(1), 1–26.

Pieron, H. I. (1939). La Laboratoire de psychologie de la Sorbonne. In *Centenaire de Théodule Ribot, Jubilé de la psychologie scientifique Française1839–1889-1939* (pp. 185–194). Agen.

Plas, R. (2012). Psychology and psychical research in France around the end of the 19th century. *History of the Human Sciences, 25*(2), 91–107.

Plas, R. (2014). Un moment spencérien aux origines de la psychologie 'scientifique' française? L'Hérédité psychologique de Théodule Ribot. *Arts et Savoirs, 4*. http://journals.openedition.org/aes/280.

Pomata, G. (2010). Sharing cases: The *Observationes* in early modern medicine. *Early Science and Medicine, 15*(3), 193–236.

Pomata, G. (2013). The recipe and the case: Epistemic genres and the dynamics of cognitive practices. In K. Greyerz, S. Flubacher, & P. Senn (Eds.), *Connecting science and knowledge: Scenes of research* (pp. 131–154). V&R Unipress.

Pomata, G. (2014). The medical case narrative: Distant reading of an epistemic genre. *Literature and Medicine, 32*(1), 1–23.

Poovey, M. (1995). *Making a social body: British cultural formation 1830–1864*. Chicago University Press.

Popper, K. R., & Eccles, J. C. (1985). *The self and its brain*. Springer.

Porter, T. M. (1995). Statistical and social facts from Quetelet to Durkheim. *Sociological Perspectives, 38*(1), 15–26.

Porter, R. (2001). Victorian developments. In *Bodies politic: Disease, death and doctors in Britain, 1650–1900* (pp. 250–271). Cornell University Press.

Porter, T. (2014). The fate of scientific naturalism: From public sphere to professional exclusivity. In B. Lightman & G. Dawson (Eds.), *Victorian scientific naturalism: Community, identity, continuity*. Chicago University Press.

Postlethwaite, D. (1984). *Making it whole: A Victorian circle and the shape of their world*. Ohio State University Press.

Prichard, J. C. (1826 [1824]). *Researches into the physical history of mankind*. John and Arthur Arch.

Prost, A. (1968). *Histoire de l'enseignement en France 1800–1967*. A. Colin.

Purdon, M. J. (1975). Kinnier Wilson's notes of conversations with Hughlings Jackson. *Journal of Neurology, Neurosurgery, and Psychiatry, 38*, 313–316.

Quick, T. (2014). From phrenology to the laboratory: Physiological psychology and the institution of science in Britain (1830-1880). *History of the Human Science, 27*(5), 54–73.

Rabinbach, A. (2018). *The eclipse of the utopias of labor*. Fordham University Press.

Rabinow, P. (1989). *French modern: Norms and forms of the social environment*. MIT Press.

Radick, G. (2000). Language, brain function and human origins in the Victorian debates on evolution. *Studies in the History and Philosophy of the Biological and Biomedical Sciences, 31*(1), 55–75.
Radick, G. (2007). *The Simian Tongue: The long debate about animal language*. Chicago University Press.
Rainville, C., & Carter, D. (1998). La 'loi de ribot': Régression ou dissolution? *Bulletin de Psychologie, 51*(1), 45–66.
Raitiere, M. N. (2012). *The complicity of friends: How George Eliot, G. H. Lewes, and John Hughlings-Jackson Encoded Herbert Spencer's Secret*. Bucknell University Press.
Rawls, A. W. (1996). Durkheim's epistemology: The neglected argument. *American Journal of Sociology, 102*(2), 430–482.
Rawls, A. W. (1998). Durkheim's challenge to philosophy: Human reason explained as a product of enacted social practice. *The American Journal of Sociology, 104*(3), 887–901.
Rawls, A. W. (2001). Durkheim's treatment of practice: Concrete practice vs. representation as the foundation of reason. *Journal of Classical Sociology, 1*, 33–68.
Raymer, E. J. (2018). Human progress by human effort: Neo-Darwinism, social heredity, and the professionalization of the American social sciences, 1889–1925. *History and Philosophy of the Life Sciences, 40*(4), 1–23.
Reddy, W. M. (2001). *The navigation of feeling: A framework for the history of emotions*. Cambridge University Press.
Reil, H.-P. (2005). *Vitalizing nature in the enlightenment*. University of California Press.
Reiser, S. J. (1978). *Medicine and the reign of technology*. Cambridge University Press.
Renwick, C. (2009). The practice of Spencerian science: Patrick Geddes's biosocial program, 1876-1889. *Isis, 100*, 36–57.
Renwick, C. (2015a). Biology, social science and history: interdisciplinarity in three directions. *Palgrave Communications, 1*, 1–5.
Renwick, C. (2015b). Herbert Spencer, biology, and the social sciences in Britain. In M. Francis & M. W. Taylor (Eds.), *Herbert Spencer: Legacies*. Routledge.
Rey, R. (1994). Naissance de la biologie et redistribution des savoirs. *Revue de Synthese, 1*(2), 167–197.
Rey, R. (2000). *Naissance et développement du vitalisme en France de la deuxiéme moitié du 18e siécle a` la fin du Premier empire*. Voltaire Foundation.
Reynolds, A. (2007). Amoebae as exemplary cells: The protean nature of an elementary organism. *Journal of the History of Biology, 41*(2), 307–337.
Reynolds, A. S. (2018). *The third lens: Metaphor and the creation of modern cell biology*. University of Chicago Press.
Ribot, T. (1870). *La psychologie anglaise contemporaine – école expérimentale*. Ladrange.
Ribot, T. (1873a). *English psychology*. H.S. King& Co.
Ribot, T. (1873b). *L'Hereditè psychologique. étude psychologique sur ses phénoménes, ses lois, ses causes, ses consequences*. Ladrange.
Ribot, T. (1874). La psychologie physiologique en Allemagne- la measure de sensations. *Revue Scientifique, 7*, 553–563.
Ribot, T. (1875a). La Psychologie Allemande Contemporaine. *Revue Scientifique, 4*(31), 751–760.
Ribot, T. (1875b). Psychologie physiologique, Wundt. *Revue Scientifique, 5*(31), 544–549.
Ribot, T. (1875c). *Heredity: A psychological study of its phenomena, laws, causes, and consequences*. D. Appleton and Company.
Ribot, T. (1876a). "Introductory essay" to the 1st issue of the revue Pilosophique/originally anonymous. *Revue Philosophique de la France et de L'étranger, 1*(1), 1–4.
Ribot, T. (1876b). Review- physiologie: Hughlings-Jackson; clinical and physiological researches on the nervous system I. On the localisations of movements in the brain 1875. *Revue Philosophique, 1*, 214–216.
Ribot, T. (1877). Philosophy in France. *Mind, 2*, 366–386.

Ribot, T. (1881). *Les maladies de la mémoire*. Librarie Generale Ballie're et Cie.
Ribot, Théodule. (1882). *Diseases of memory: An essay in the positive psychology* (William Huntington Smith, Trans.). D. Appleton and company.
Ribot, T. (1883). *Les maladies de la volonté*. G. Ballière. English edition quoted 1894, Chicago: Open Court.
Ribot, T. (1884). *The diseases of the Will* (English translation). New York, J. Fitzgerald/(Humboldt library no. 52.
Ribot, T. ([1879] 1885a). *La Psychologie Allemande contemporaine.École expérimentale*. Paris: F. Alcan]Paris: librarie Germer baillie're et cie].
Ribot, T. (1885b). *Les maladies de la personnalité*. Felix Alcan.
Ribot, T. (1886). *German psychology of to-day: The empirical school* (J. M. Baldwin, Trans.). Charles Scribner Sons.
Ribot, T. (1887). *The diseases of personality* (J. Fitzgerald, Trans.). Humboldt.
Ribot, T. ([1887] 1888). La psychologie contemporaine: leçon d'ouverture de cours de psychologie experimentale et comparée du College de France. In *Revue de l'Hypnotisme expérimental et thérapeutique* (pp. 344–47).
Ribot, T. (1889). *Psychologie de l'attention*. Felix Alcan.
Ribot, T. (1891–1892). Enquête sur les idées générales. *Revue Philosophique, 376*.
Ribot, T. (1893). Heredité. In M. Berthelot, H. Derenburg, C. Defus, et al. (Eds.), *La Grande encyclopédie* (Vol. 19, pp. 1165–1169). H. Lamirault et C.
Ribot, T. (1894). *Diseases of the will* (Merwin-Marie Snell, Trans.). Open Court.
Ribot, T. (1896). *La psychologie des sentiments*. Felix Alcan.
Ribot, T. (1897). *L'évolution des idées générales*. Alcan.
Ribot, T. (1898). *Psychology of emotions*. W. Scott.
Ribot, T. (1900). La psychologie de 1889 a'1900; IIIe'me Congress International de psychologie. *Revue Scientifique, 44*(12), 177–178.
Ribot, T. (1900). *Essai sur l'imagination créatrice*. Felix Alcan.
Ribot, T. (1904). Sur la valeur des questionnaires en psychologie. *Journal de Psychologie Normale et Pathologique, 1*.
Ribot, T. (1905). *La logique des sentiments*. Felix Alcan.
Ribot, T. (1910). Chapter two. In H. Munsterberg (Ed.), *Subconscious phenomena* (pp. 33–39). Rebman.
Ribot, T. ([1909] 1928). Psychologie. In Henri Bouass et al. (Eds.), *De la Methode dans les sciences*, 1iere serie. Felix Alcan.
Ribot, T. (1957). Lettres a Espinas, edited by Raymon Lenoir. *Revue Philosophique, 147*, 1–14.
Ribot, T. (1962). Lettres a Espinas, edited by Raymon Lenoir. *Revue Philosophique, 152*, 337–40.
Ribot, T. (1964). Lettres a Espinas, edited by Raymon Lenoir. *Revue Philosophique, 154*, 79–84.
Ribot, T. (1970). Lettres a Espinas, edited by Raymon Lenoir. *Revue Philosophique, 160*, 339–348.
Ribot, T. (1975). Lettres a Espinas, edited by Raymon Lenoir. *Philosophie Générale, 165*(2), 157–172.
Ribot, T. (2005). Letters a Henri Pieron, edited by Jacqueline Carroy and Régine Plas. *Revue Philosophique, 195*(3), 335–356.
Richards, R. J. (1987). *Darwin and the emergence of evolutionary theories of mind and behavior*. University of Chicago Press.
Richards, E. (1989). The 'moral anatomy' of Robert Knox: The interplay between biological and social thought in Victorian scientific naturalism. *Journal of the History of Biology, 22*(3), 373–436.
Richards, R. J. (2008). *The tragic sense of life: Ernst Haeckel and the struggle over evolutionary thought*. University of Chicago Press.
Richerand, A. (1804). *Nouveau élémens de physiologie* (3rd ed.). Crapart, Caille & Ravier.
Riese, W. (1950). *Principles of neurology in the light of history and their use*. Nervous and Mental Diseases Monographs no. 80.

Riese, W. (1954). Hughlings Jackson's doctrine of consciousness: Sources, versions and elaborations. *Journal of Nervous and Mental Disease, 120*(5), 330–337.
Riese, W. (1956). The sources of Jacksonian neurology. *Journal of Nervous and Mental Disease, 124*(2), 125–134.
Riese, W. (1965). The sources of Hughlings Jackson's view on aphasia. *Brain, 88*(4), 811–822.
Riley, A. T. (2010). *Godless intellectuals? The intellectual pursuit of the sacred reinvented.* Berghahn Books.
Ringer, F. (1987). On segmentation in modern European educational systems: The case of French secondary education, 1865-1920. In D. K. Muller, F. Ringer, & B. Simon (Eds.), *The rise of modern educational system: Structural change and social reproduction 1870–1920.* Cambridge University Press.
Ringer, F. (1992). *Fields of knowledge: French academic culture in comparative perspective 1890–1920.* Cambridge University Press.
Rioux, J.-P. (1991). *Chronique d'une fin de siècle: France, 1889–1900.* Seuil.
Riskin, J. (2002). *Science in the age of sensibility: The sentimental empiricists of the French enlightenment.* University of Chicago Press.
Riskin, J. (2016). *The restless clock: A history of the centuries-long argument over what makes living things tick.* Chicago University Press.
Risse, G. B., & Warner, J. H. (1992). Reconstructing clinical activities: Patient records in medical history. *Social History of Medicine, 5*(2), 183–205.
Ritvo, L. B. (1990). *Darwin's influence on Freud: A tale of two sciences.* Yale University Press.
Rivett, G. (1986). *The development of the London hospital system, 1823–1982.* Oxford University Press (King Edward's Hospital Fund for London).
Roche, D. (1988). *Les républicains des lettres: Gens de culture et Lumières au xviii siécle.* Fayard.
Rocke, A. J. (2002). The theory of chemical structure and its application. In M. J. Nye & S. G. Brush (Eds.), *The Cambridge history of science, vol. 5: The modern physical and mathematical sciences* (pp. 255–271). Cambridge University Press.
Roger, J, Ed. (1979a). Les Néo-Lamarckiens Français. *Revue de Synthese, 95/96*, 279–469. (J. Roger, J.C. Sillard, A. Lagarde, Cl. Blanckaert, M. Viré, G. Gohau, A. Diara, K. Wellman, J.L. Fischer).
Roger, J. (1979b). Presentation. In *Les néo-lamarckiens français* (pp. 95–96), edited by Jacques Roger. Special issue of *Revue de synthèse* 3ie'me série.
Roger, J. [(1993] 1997). *The life sciences in eighteenth-century French thought* (Keith R. Benson, Ed., Robert Ellrich, Trans.). Stanford University Press.
Romand, D. (2010). Das Körper-Seele problem. *Revue de Synthèse, 131*(1), 35–51.
Romand, D. (2012). Fechner as a pioneering theorist of unconscious cognition. *Consciousness and Cognition, 21*(1), 562–572.
Roth, M. S. (2011). *Memory, trauma and history: Essays on living with the past.* Columbia University Press.
Royer, C. (1862). *De l'origine des espèces, ou Des lois du progrès chez les êtres organisés.* Guillaumin (: A. Moulin).
Royer, C. (1870). Remarques sur le transformisme. *Bulletins de la Société d'anthropologie de Paris*. II Série. *5*, 265–317.
Royce, J. (1904). *Herbert Spencer: An estimate and review.* Fox, Duffiled & Co.
Rudwick, M. J. S. (1985). *The meaning of fossils: Episodes in the history of palaeontology.* University of Chicago Press.
Rudwick, M. J. S. (2005). *Bursting the limits of time: The reconstruction of geohistory in the age of revolution.* University of Chicago Press.
Rudwick, M. J. S. (2008). *Worlds before Adam: The reconstruction of geohistory in the age of reform.* University of Chicago Press.
Rudwick, M. J. S. (2009). Biblical flood and geological deluge: The amicable dissociation of geology and genesis. *Geological Society Special Publication, 310*(1), 103–110.
Rudwick, M., & Taylor, K. L. (2019). Worlds before Adam: The reconstruction of geohistory in the age of reform. *Isis, 100*(4), 872–880.

Ruse, M. (2004). Adaptive landscapes and dynamic equilibrium: The Spencerian contribution to twentieth-century American evolutionary biology. In A. J. Lustig, R. J. Richards, & M. Ruse (Eds.), *Darwinian heresies* (pp. 131–150). Cambridge University Press.
Ruse, M. (2013). Wrestling with biological complexity: From Darwin to Dawkins. In C. H. Lineweaver, P. C. W. Davies, & M. Ruse (Eds.), *Complexity and the arrow of time* (pp. 279–307). Cambridge University Press.
Russo, R. P. (1981). La notion d'organisation chez Lamarck. In J. Schiller (Ed.), *Lamarck et son temps, Lamarck et notre temps* (pp. 111–142). Vrin.
Rutherford, A. (2015). Maintaining masculinity in mid twentieth century American psychology. *Osiris, 30*, 250–271.
Rylance, R. (2000). *Victorian psychology and British culture, 1850–1880*. Oxford University Press.
Rylance, R. (2006). The theatre and the granary: Observations on nineteenth-century medical narratives. *Literature and Medicine, 25*(2), 255–276.
Sapiro, G., Brun, E., & Fordant, C. (2019). The rise of social sciences and humanities in France. In C. Fleck, M. Duller, & V. Karady (Eds.), *Shaping human sciences disciplines: Institutional developments in Europe and beyond* (pp. 25–68). Palgrave Macmillan.
Sawyer, R. K. (2002). Durkheim's dilemma: Towards a sociology of emergence. *Sociological Theory, 20*(2), 227–247.
Schabas, M. (1997). Victorian economists and the science of mind. In B. Lightman (Ed.), *Victorian science in context* (pp. 72–93). Chicago University Press.
Schacter, D. L. (1996). *Searching for memory: The brain, the mind, and the past*. Basic Books.
Schacter, D. L. (2001). *Forgotten ideas, neglected pioneers: Richard Semon and the story of memory*. Psychology Press.
Scharbert, G. (2009). Freud and evolution. *History and Philosophy of the Life Sciences, 31*(2), 295–311.
Scheiner, S. M. (2014). The Baldwin effect: Neglected and misunderstood. *American Naturalist, 184*, ii–iii.
Schiller, J. (1969). Physiologie et classification dans l'oeuvre de Lamarck. *Histoire et biologie, 2*, 35–57.
Schiller, J. (1971). L'echelle des êtres et la serie chez Lamarck. In J. Schiller (Ed.), *Colloque international Lamarck* (pp. 87–103). Blanchard.
Schiller, J. (1978). *La notion d'organisation dans l'histoire de la biologie*. Maloine.
Schloegel, J. J., & Schmidgen, H. (2002). General physiology, experimental psychology and evolutionism. *Isis, 93*, 614–645.
Schmaus, W. (1996). Lévy Bruhl, Durkheim and the positivist roots of the sociology of knowledge. *Journal of the History of the Behavioral Sciences, 32*(4), 424–440.
Schmaus, W. (2000). *On Sens lectures: Downloaded*. http://www.relst.uiuc.edu/durkheim/reviews/Sens.Schamus.html.
Schmaus, W. (2004). *Rethinking Durkheim and his tradition*. Cambridge University Press.
Schmaus, W., Pickering, M., & Bourdeau, M. (2018). Introduction: The significance of Auguste Comte. In M. Bourdeau, M. Pickering, & W. Schmaus (Eds.), *Love, order, and progress: The science, philosophy, and politics of auguste Comte* (pp. 3–24). University of Pittsburgh Press.
Schütt, H. W. (2002). Chemical atoms and chemical classification. In M. J. Nye (Ed.), *The Cambridge history of science, vol. 5: The modern physical and mathematical sciences* (pp. 237–254). Cambridge University Press.
Schweber, S. S. (1980). Darwin and the political economists. *Journal of the History of Biology, 13*(2), 195–289.
Schweber, L. (2006). *Disciplining statistics: Demography and vital statistics in France and England, 1830/1885*. Duke University Press.
Scott, D. F. (1992). The discovery of anti-epileptic drugs. *Journal of the History of the Neurosciences, 1*(2), 11–118.
Searle, J. R. (1995). *The construction of social reality*. MIT Press.
Semon, R. (1921). *The Mneme*. George Allen and Unwin.
Semon, R. W. ([1908] 1922). *The Mneme* (Louis Simon, Trans.). G. Allen and Unwin.

Semon, R. W. (1923). *Mnemic psychology*. G. Allen & Unwin, Ltd.

Sewell, W. H. (1987). Le citoyen/la citoyenne activity passivity and the revolutionary concept of citizenship. In C. Lucas (Ed.), *The French revolution and the creation of modern culture* (Vol. 2, pp. 105–124). Pergamon Press.

Shamdasani, S. (2017). Questioning the unconscious. *Annals of the New York Academy of Sciences, 1406*(1), 86–89.

Sharvit, G., Feldman, K. S., Assman, J., Berenstein, R., Gotschel, W., & Hendel, R. (Eds.). (2018). *Freud and monotheism: Moses and the violent origins of religion*. Fordham University Press.

Shmuely, S. (2017). *The bureaucracy of empathy: Vivisection and the question of animal pain in Britain, 1876–1912* (PhD dissertation). STS, MIT.

Shmuely, S. D. (2018). *The bureaucracy of empathy: Vivisection and the question of animal pain in Britain*, 1876–1912. PhD. dissertation. MIT.

Shorter, E. (1995). The borderland between neurology and history. *Neurologic Clinics, 13*(2), 229–239.

Shorvon, S. (2014). The evolution of epilepsy theory and practice at the National Hospital for the relief and cure of epilepsy, Queen Square between 1860 and 1910. *Epilepsy and Behavior, 31*, 228–242.

Shorvon, S., & Compston, A. (2018). *Queen square: A history of the national hospital and its institute of neurology*. Cambridge University Press.

Simiand, F. (1898). Sociologique- L'année 1897. *Revue de Métaphysique et de Morale, 6*(5), 608–653.

Simpson, G. G. (1953). The Baldwin effect. *Evolution, 7*(2), 110–117.

Slavin, M. O., & Kriegman, D. (1990). Toward a new paradigm for psychoanalysis: An evolutionary biological perspective on the classical-relational dialectic. *Psychoanalytic Psychology, 7*(S), 5–31.

Sloan, P. (1990). Natural history 1680-1802. In R. C. Olby, G. N. Cantor, J. R. R. Christie, & M. J. S. Hodge (Eds.), *Companion to the history of modern science* (pp. 295–313). Routledge.

Sloan, P. (1995). The gaze of natural history. In C. Fox, R. Porter, & R. Wokler (Eds.), *Inventing human sciences: Eighteenth century domains* (pp. 112–151). University of California Press.

Small, A. (1894). Organic concept of society. *The Annals of the American Academy of Political and Social Science, 5*, 740–746.

Small, A. (1905). *General sociology: An exposition of the main development of sociological theory from Spencer to Ratzenhofer*. Chicago University Press.

Smith, R. (1880). Animal worship and animal tribes among the ancient Arabs and in the old testament. *Journal of Philology, 9*(17), 75–100.

Smith, R. (1973). The background of physiological psychology in natural philosophy. *History of Science, 11*, 75–123.

Smith, C. U. M. (1982a). Evolution and the problem of mind. Part I: Herbert Spencer. *Journal of the History of Biology, 15*(1), 55–88.

Smith, C. U. M. (1982b). Evolution and the problem of mind: Part II. John Hughlings-Jackson. *Journal of the History of Biology, 15*(2), 241–262.

Smith, R. (1992). *Inhibition: History and meaning in the sciences of mind and brain*. University of California Press.

Smith, C. U. M. (1993). The use and abuse of metaphors in the history of brain science. *Journal of the History of the Neurosciences, 2*(4), 283–301.

Smith, R. (1995). The language of human nature. In C. Fox, R. Porter, & R. Wokler (Eds.), *Inventing human sciences: Eighteenth century domains*. California University Press.

Smith, D. L. (1999). *Freud's philosophy of the unconscious*. Kluver Academic publishers.

Smith, R. (2005). Does reflexivity separate the human sciences from the natural sciences? *History of the Human Sciences, 18*(4), 1–25.

Smith, R. (2007). *Being human: Historical knowledge and the creation of human nature*. Columbia University Press.

Smith, R. (2010). British thought on the relations between the natural sciences and the humanities c. 1870-1910. In U. Feest (Ed.), *Historical perspectives on Erklären and Verstehen* (pp. 161–186). Springer.

Smith, C. U. M. (2012). Philosophy's loss, neurology's gain: The endeavor of John Hughlings-Jackson. *Perspectives in Biology and Medicine, 55*(1), 81–91.

Smith, R. (2016). History of psychology, what for? In K. S. Hroar & R. Smith (Eds.), *Centrality of history for theory construction in psychology* (pp. 3–28). Springer Nature.

Smith, R. (2020). Inhibition and top-down metaphor of organization. *Studies in History and Philosophy of the Biological and Biomedical Sciences, 80*(October), 1–9.

Smith, C., & Wise, N. (1989). *Energy and empire: A biographical study of Lord Kelvin*. Cambridge University Press.

Soederqvist, T. (2002). Neurobiographies: Writing lives in the history of neurology and the neurosciences. *Journal of the History of the Neurosciences, 11*(1), 38–48.

Solms, M., & Saling, M. (1986). On psychoanalysis and neuroscience: Freud's attitude to the localizationist tradition. *International Journal of Psycho-Analysis, 67*, 397–416.

Sommer, M. (2016). *History within: The science, culture, and politics of bones, organisms and molecules*. University of Chicago Press.

Soskice, J. M., & Harré, R. (1995). Chapter 11: Metaphor in science. In Z. Radman (Ed.), *From a metaphorical point of view* (pp. 289–307). De Gruyter.

Sotirios, P. (2018). *Reappearing characters in nineteen century French literature*. Palgrave Macmillan.

Soulié, S. (2009). *Les philosophes en République-l'aventure intellectuelle de la revue de métaphysique et de morale et de la société française de philosophie.1891–1914*. Presses universitaires de Rennes.

Soulié, S. (2012). Le débat philosophique autour des 'Formes élémentaires' (1912-1914). *L'Année Sociologique, 62*(2), 447–463.

Spalding, D. (1873). Herbert Spencer's psychology. *Nature, 7*(February 20), 298–300.

Spary, E. (1996). Political, natural and bodily economies. In N. Jardine, J. A. Secord, & E. C. Spary (Eds.), *Cultures of natural history* (pp. 178–196). Cambridge University Press.

Spary, E. (2000). *Utopia's garden: French natural history from old regime to revolution*. University of Chicago Press.

Spencer, H. (1851). *Social statics: Or, the conditions essential to human happiness specified and the first of them developed*. J. Chapman.

Spencer, H. (1852a). A theory of population, deduced from the general law of animal fertility. *The Westminster Review, 57*(April), 468–501.

Spencer, H. (1852b). The development hypothesis. *The Leader, 3*(March 20), 280–281. (Reprinted in *Essays Scientific, Political and Speculative, I*, 1–7).

Spencer, H. (1853). The universal postulate. *The Westminster Review, 60*(October 1), 513–550.

Spencer, H. (1855). *The principles of psychology*. Longman, Brown, Green and Longman.

Spencer, H. (1857a). Progress, its law and cause. *The Westminster Review, 67*, 445–85 (Reprinted in *Essays Scientific, Political and Speculative, 1*, 8–62).

Spencer, H. (1857b). The ultimate laws of physiology/transcendental physiology. *National Review, 5*(October), 332–55 (Reprinted in *Essays Scientific, Political and Speculative, 1*, 63–107).

Spencer, H. (1858). Recent astronomy and the nebular hypothesis/the nebular hypothesis. *The Westminster Review, 70*(July 27), 185–225 (Reprinted in *Essays Scientific, Political and Speculative, 1*, 108–91.

Spencer, H. (1859). The Laws of organic form. *British and Foreign Medico-Chirurgical Review, 45*(January), 189–202.

Spencer, H. (1860a). Bain on emotions and the will. *The British and Foreign Medico-Chirurgical Review, 49*(January), 42–52 (Reprinted in *Essays Scientific, Political and Speculative, 1*, 241–64).

Spencer, H. (1860b). The social organism. *The Westminster Review, 17*, 90–121 (Reprinted in *Essays Scientific, Political and Speculative, 1*, 265–307).

Spencer, H. (1862). *First principles*. Williams and Norgate.
Spencer, H. ([1864] 1867). *The principles of biology*. Williams and Norgate.
Spencer, H. ([1870/1872]. *The principles of psychology*, 2nd edn. Longman, Brown, Green and Longman.
Spencer, H. (1874 [1873]). *The study of sociology*. Appleton.
Spencer, H. (1886). The factors of organic evolution. *Nineteenth Century*, *19*(April–May), 570–89, 749–70 (Reprinted in *Essays Scientific, Political and Speculative*, *1*, 389–466).
Spencer, H. (1891). *Essays scientific, political and speculative* (Vol. I). Williams & Norgate.
Spencer, H. (1891 [1876]). The comparative psychology of man. *Essays Scientific, Political and Speculative*, 351–370. Williams and Norgate.
Spencer, H. ([1892] 1893a). *The principles of ethics*, vols. 2 D. Appelton &
Spencer, H. (1893b). The Inadequacy of 'Natural Selection,' I and II. *Contemporary Review, 63*(153–66), 439–456.
Spencer, H. (1894). Weismannism once more. *Contemporary Review, 63*, 592–608.
Spencer, H. ([1874] 1896). *Principles of sociology*. Williams and Norgate.
Spencer, H. (1898). *The principles of biology* (2nd ed.). Williams and Norgate.
Spencer, H. (1899). The filiation of ideas. In David Duncan (Ed.), *The life and letters of Herbert Spencer*, Appendix B (pp. 533–76).
Spencer, H. (1904). *An autobiography* (Vol. 2). D. Appleton & Co.
Spencer, H. (1908). *Life and letters of Herbert Spencer* (David Duncan, Ed.). Methuen.
Spencer, B., & Gillen, F. J. (1899). *The native tribes of Central Australia*. Macmillan.
Stacey, M. (1992). *Regulating British medicine: The general medical council*. Wiley.
Staley, T. W. (2009). The journal mind in its early years, 1876–1920: An introduction. *Journal of the History of Ideas, 70*(2), 259–263.
Star, S. L. (1989). *Regions of the mind: Brain research and the quest for scientific certainty*. Stanford University Press.
Star, S. L. (1992). The skin, the skull, and the self: Towards a sociology of the brain. In A. Harrington (Ed.), *So human a brain: Knowledge and values in the neurosciences* (pp. 204–228). BirkHäuser.
Star, S. L. (2010). This is not a boundary object: Reflections on the origin of a concept. *Science, Technology, and Human Values, 35*(5), 601–617.
Star, S. L., & Griesemer, J. R. (1989). Institutional ecology, 'translations' and boundary objects: Amateurs and professionals in Berkeley's Museum of Vertebrate Zoology, 1907-39. *Social Studies of Science, 19*(3), 387–420.
Starobinski, J. (1973). *1789, les emblèmes de la raison*. University Press of Virginia.
Staum, M. (1980). *Cabanis: Enlightenment and medical philosophy in the French revolution*. Princeton University Press.
Staum, M. (1996). *Minerva's message: Stabilizing the French revolution*. McGill-Queen's University Press..
Staum, M. (2007). Ribot, Binet, and the emergence from the anthropological shadow. *Journal of the History of the Behavioral Sciences, 43*(1), 1–18.
Staum, M. (2011). *Nature and nurture in French social sciences 1859–1914 and beyond*. McGill-Queens University Press.
Staum, M. J. (2003). *Labeling people: French scholars on society, race and empire, 1815-1848*. McGill-Queen's University Press.
Stedman Jones, G. (2002). The new social history. In C. Jones & D. Wahrman (Eds.), *The age of cultural revolutions: Britain and France 1750–1820* (pp. 94–136). University of California Press.
Steichen, O. (2012). La médecine factuelle et les rapports de cas. In A. Fagot-Largeault (Ed.), *L'émergence de la médecine scientifique* (pp. 225–246). Éditions Matériologiques.
Steinberg, D. A., & York, G. K. (1994). Hughlings Jackson, concomitance and mental evolution. *Journal of the History of the Neurosciences: Basic and Clinical Perspectives, 3*(3), 169–176.
Stengel, E. (1954). The significance of Freud's study of aphasia for psychoanalysis. *Psyche (Stuttgart), 8*(1), 17–24.

Stengel, E. (1963). Hughlings Jackson's influence in psychiatry. *British Journal of Psychiatry, 109*, 348–355.

Sterling, J. (2010). *Representing epilepsy: Myth and matter*. Liverpool University Press.

Sterelny, K. (2012). *The evolved apprentice: How evolution made humans unique*. The MIT Press.

Sterelny, K. (2021). *The Pleistocene social contract: Culture and cooperation in human evolution*. Oxford University Press.

Stocking, G. W. (1962). Lamarckianism in American social science: 1890-1915. *Journal of the History of Ideas, 23*(2), 239–256.

Stocking, G. W. (1968). *Race, culture and evolution*. Free Press.

Stocking, G. W., Jr. (1987). *Victorian anthropology*. Free Press.

Stock-Morton, P. (1971). Students versus the University in Pre-World War France. *French Historical Studies, 7*, 83–110.

Stock-Morton, P. (1988). *Moral education for a secular society: The development of morale Laïque in nineteenth century France*. SUNY.

Strathern, M. (2004). *Commons and borderlands: Working papers on interdisciplinarity, accountability and the flow of knowledge*. Sean Kingston Publications.

Strathern, M. (2007). Interdisciplinarity: Some models from the human sciences. *Interdisciplinary Science Reviews, 32*(2), 123–134.

Strenski, I. (1997). *Durkheim and the Jews of France*. University of Chicago Press.

Strenski, I. (1998). Durkheim's bourgeois theory of sacrifice. In N. J. Allen et al. (Eds.), *On Durkheim's elementary forms of religious life* (pp. 116–126). Taylor & Francis.

Strenski, I. (2002). Durkheim, disciplinarity, and the 'sciences' religieuses (chap. 5). In A. Anderson & J. Valente (Eds.), *Disciplinarity at the fin de siècle* (pp. 153–173). Princeton University Press.

Strenski, I. (2002). Durkheim, Disciplinarity, and the 'Sciences' Religieuses. In A. Anderson & J. Valente (Eds.), *Disciplinarity at the Fin de siècle* (pp. 153–173). Princeton University Press.

Strenski, I. (2006). *The new Durkheim*. Rutgers University Press.

Sulloway, F. J. (1992). *Freud, biologist of the mind: Beyond the psychoanalytic legend*. Harvard University Press.

Surkis, J. (2006). *Sexing the citizen: Morality and masculinity in France, 1870–1920*. Cornell University Press.

Swash, M. (1989). John Hughlings Jackson: A historical introduction. In C. Kennard & M. Swash (Eds.), *Hierarchies in neurology: A reappraisal of a Jacksonian concept* (pp. 3–10). Springer.

Swedberg, R. (2019). How do you make sociology out of data? Robert K. Merton's Course in Theorizing (Soc 213–214). *The American Sociologist, 50*, 85–120.

Taguieff, P. A. (1984). Georges Vacher de la Pouge 1854–1936. In D. Huisman (Ed.), *Dictionnaire des Philosophes*. Presses Univérsitaires de France.

Taine, H. (1869). *De l'intelligence*. Hachette.

Taine, H. (1873). Review of Ribot's *Heredité*. *Journal des débats politiques et littéraires* (November 23).

Taine, H. (1877). On the acquisition of language by children. *Mind, 2*, 252–259.

Tattersall, M. (2014). Thermal degeneration: Thermodynamics and the heat-death of the universe in Victorian science, philosophy, and culture. In M. Härmänmaa & C. Nissen (Eds.), *Decadence, degeneration and the end: Studies in the European fin de Siècle* (pp. 17–34). Palgrave Macmillan.

Tauber, A. (2010). *Freud, the reluctant philosopher*. Princeton University Press.

Taylor, J. (1915). The ophthalmological observations of Hughlings Jackson and their bearing on nervous and other diseases. *Brain, 38*, 391–417.

Taylor, J. (1925). A biography. In J. Taylor (Ed.), *Neurological fragments of John Hughlings Jackson* (pp. 1–26). Oxford University Press.

Taylor, M. W. (2007). *The philosophy of Herbert Spencer*. Continuum.

Taylor, M. W. (2014). Herbert Spencer and the Metaphysical Roots of Evolutionary Naturalism, in Thomson, W. 1862. 'on the Age of the Sun's Heat.'. *Macmillan's Magazine, 5*, 288–293.

Temkin, O. (1994). *The falling sickness: A history of epilepsy from the Greeks to the beginnings of modern neurology*. Johns Hopkins University Press.

Terrall, M. (2007). Speculation and experiment in enlightenment life sciences. In S. Mueller-Wille & H.-J. Rheinberger (Eds.), *Heredity produced: At the crossroads of biology, politics and culture 1500–1870* (pp. 253–276). MIT Press.

Thomson, W. (1862). On the Age of the Sun's Heat. *Macmillan's Magazine, 5*, 288–293.

Thomas, M. (2003). Rethinking the history of ethology: French animal behaviour studies in the Third Republic (1870–1940) (PhD dissertation). University of Manchester.

Thomes, N. ([1985] 1988). The great restraint controversy; a comparative perspective on Anglo-American psychiatry in the 19th century. In W. F. Bynum, R. Porter, & M. Shepard (Eds.), *The anatomy of madness: Essays in the history of psychiatry*, vol. 3 (pp. 190–226). Routledge.

Thompson, K. (1993). Durkheim, ideology and the sacred. In P. Hamilton (Ed.), *Emile Durkheim: Critical assessments*.

Thomson, A. (2001). Materialistic theories of mind and brain. In W. Lefèvre (Ed.), *Between Leibnitz, Newton and Kant: Philosophy and science in the eighteenth century* (pp. 149–173). Kluwer Academic publishing.

Thomas, M. (2017). Biological and social new orders: Charles Robin (1821–1885) and Comteanism in the Third Republic. *Historical Studies in the Natural Sciences, 47*(5), 653–677.

Tilly, C. (1986). *The contentious French*. Belknap Press.

Tilly, C., & Shorter, E. (1974). *Strikes in France 1830–1968*. Cambridge University Press.

Tirard, S. (2006). Générations spontanées. In P. Corsi, J. Gayon, G. Gohau, & S. Tirard (Eds.), *Lamarck philosophe de la nature*. Presses Universitaires de France.

Tirard, S. (2013). Claude Bernard et les trois formes de vie. In F. Duchesneau, J. J. Kupiec, & M. Morange (Eds.), *Claude Bernard: la méthode de la physiologie* (pp. 49–62). Éditions Rue d'Ulm.

Tiryakian, E. A. (1979). L'école Durkheimienne a la recherch de la société perdue: La sociologie naissante et son milieu culturel. *Cahiers Internationaux de Sociologie, 66*, 97–114.

Tiryakian, E. A. (1988). Durkheim, Mathiez, and the French Revolution: The political context of a sociological classic. *Archives européennes de sociologie. European Journal of Sociology, 29*(2), 373–396.

Tiryakian, E. A. (2009). *For Durkheim: Essays in historical and cultural sociology*. Ashgate.

Tollefsen, D. P. (2002). Challenging epistemic individualism. *ProtoSociology, 16*, 86–117.

Tollefsen, D. P. (2015). *Groups as agents*. Polity Press.

Tomasello, M. (2009). *Why we cooperate*. MIT Press.

Tomasello, M. (2010). *Origins of human communication*. MIT Press.

Tomasello, M. (2014). *A natural history of human thinking*. Harvard University Press.

Tomlinson, S. (2015). The method of nature: Herbert Spencer and the education of the adaptive mind. In M. A. Francis & M. W. Taylor (Eds.), *Herbert Spencer legacies* (pp. 16–39). Routledge.

Tönnies, F. (1896). Émile Durkheim 'de la division du travail social.'. *Archiv für systematische Philosophie, 2*, 497–499.

Tuomela, R. (2007). *The philosophy of sociality: The shared point of view*. Oxford University Press.

Tuomela, R. (2013). *Social ontology: Collective intentionality and group agents*. Oxford University Press.

Turbill, C. (2018). Making heredity matter: Samuel Butler's idea of unconscious memory. *Journal of the History of Biology, 51*, 7–29.

Turner, W. A. (1897). The care and Management of Epileptics in colonies. *Lancet, 149*, 1733–1735.

Turner, J. H. (1985). *Herbert Spencer: A renewed appreciation*. Sage.

Turner, S. P. (1986). *The search for a methodology of social science: Durkheim, Weber, and the nineteenth century problem of cause, probability and action*. D. Reidel.

Turner, S. P. (1993). *Emile Durkheim: Sociologist and moralist*. Routledge.

Turner, F. M. (1993). *Contesting cultural authority: Essays in Victorian intellectual life*. Cambridge University Press.

Tweney, R. D. (1987). Programmatic research in experimental psychology: E. B. Titchener's laboratory investigations, 1891–1927. In M. G. Ash & W. R. Woodward (Eds.), *Psychology in twentieth century thought and society* (pp. 35–57). Cambridge University Press.

Tyndall, J. (1892). "The Belfast Address" [1874]. *Fragments of science: A series of detached essays, addresses and reviews*. *2*, 135–201. Appleton.

van Ginneken, J. (1992). *Crowds, psychology, and politics 1871–1899*. Cambridge University Press.

Van Wyhe, J. (2004). *Phrenology and the origin of Victorian scientific naturalism*. Ashgate.

Vial, F. (2009). *The unconscious in philosophy and French and European literature: 19th and early 20th century*. Rodopi.

Vicenti, D. (2019). Experience and experimentation: Medicine, psychiatry and experimental psychology in Paul Janet. *Perspectives on Science, 27*(5), 704–738.

Vicenti, D. (2020). Théodule Ribot and the spiritualist tradition: The philosophical roots of scientific psychology. *British Journal for the History of Philosophy of Science, 28*(5), 1009–1030.

Vidal, F. (1994). La place de la psychologie dans l'ordre des sciences. *Revue de Synthèse, 115*(3–4), 327–353.

Vidal, F. (2006). *Les sciences de l'âme* (pp. xvi–xviii). Honoré-Champion.

Vidal, F., & Ortega, F. (2017). *Being brains: Making the cerebral subject*. Fordham University Press.

Virey, J. J. (1817–1822). Articles from the *Nouveau Dictionnaire d'histoire naturelle*: Homme, Nature, Rapports naturels, Regnes de la nature. Deterville.

Virtanen, R. (1960). *Claude Bernard and his place in the history of ideas*. University of Nebraska Press.

Vogt, P. W. (1976). The uses of studying primitives: A note on the Durkheimians 1890-1940. *History and Theory, 15*, 33–44.

Vogt, P. W. (1982). Identifying scholarly and intellectual communities: A note on French philosophy 1900-1939. *History and Theory, 21*(2), 267–278.

von Hartmann, E. (1869). *Philosophie des Unbewussten: Speculative Resultate nach induc-tivnaturwissenschaftlicher Methode*. Carl Duncker.

Voutsinas, D. (1988). La jubile de la psychologie scientifique française. *Bulletin de Psychologie, 49*(385), 404–469.

Vovelle, M. (1985). *La mentalité révolutionnaire: société et mentalités sous la Révolution française*. Editions sociales.

Wacquant, L. (2000). Durkheim and Bourdieu: The common plinth and its cracks. In B. Fowler (Ed.), *Reading Bourdieu on society and culture* (pp. 105–120). Blackwell.

Wagner, A. (1908). *Geschichte des Lamarckismus: als Einführung in die psycho-biologische Bewegung der Gegenwart*. Franckh.

Wahrman, D. (2004). *The making of the modern self: Identity and culture in 18th century England*. Yale University Press.

Wallesch, C. W. (1989). Hughlings Jackson and European neurology. In C. Kennard & M. Swash (Eds.), *Hierarchies in neurology: A reappraisal of a Jacksonian concept* (pp. 16–24). Springer.

Wallwork, E. (1972). *Durkheim: Morality and milieu*. Harvard University Press.

Walshe, F. M. R. (1953). Some problems of method in neurology. *Canadian Medical Association Journal, 68*(1), 21–29.

Walshe, F. M. R. (1961). Contributions of John Hughlings Jackson to neurology: A brief introduction to his teachings. *Archives of Neurology, 5*, 119–131.

Wartelle, J. C. (2004). La société d'Anthropologie de Paris de 1859 à 1920. *Revue d'histoire des sciences humaines, 1*(10), 125–171.

Wassmann, C. (2005). The science of emotion: Studying emotions in Germany, France and the US, 1860–1920, vol. 1 (PhD dissertation). University of Chicago.

Wassmann, C. (2017). Forgotten origins, occluded meanings: Translation of emotion terms. *Emotion Review, 9*(2), 163–171.

Weismann, A. (1891–1892). *Essays upon heredity and kindred biological problems*. In E. B. Polton, S. Schonland, & A. E. Shipley (Eds.), 2 vol. Oxford University Press.

Weismann, A. (1893). The all-sufficiency of natural selection – A reply to Herbert Spencer. *Contemporary Review, 64*, 309–338.
Weissman, C. (2011). The first evolutionary synthesis of august Weismann and the origins of neo-Darwinism (PhD dissertation). Tel-Aviv University.
Weisz, G. (1982). Associations et manifestations: les étudiants français a' la Belle Époque. *Le Mouvement Social, 120*, 31–45.
Weisz, G. (1983). *The emergence of modern universities in France 1863–1914*. Princeton University Press.
Weisz, G. (2006). *Divide and conquer*. Oxford University Press.
Welch, G. R., & Clegg, J. S. (2010). From protoplasmic theory to cellular systems biology: A 150-year reflection. *American Journal of Physiology: Cell Physiology, 298*(6), C1280–C1290.
Wellman, K. (1979). Felix Le Dantec et le néo lamarckisme français. *Revue de Synthese, 95–96*, 423–442.
Wenger, E. (1998). *Communities of practice: Learning, meaning and identity*. Cambridge University Press.
West-Eberhard, M. J. (2005). Developmental plasticity and the origin of species differences. *PNAS, 102*(1), 6543–6549.
Wiest, G. (2012). Neural and mental hierarchies. *Frontiers in Psychology, 3*, 1–8.
Wieviorka, M. (1992). Case studies, history or sociology? In C. Ragin & H. S. Becker (Eds.), *What is a case? Exploring the foundations of social inquiry* (pp. 159–172). Cambridge University Press.
Williams, E. A. (2002). *The physical and the moral: Anthropology, physiology and philosophical medicine in France 1750–1850*. Cambridge University Press.
Winter, S. (1999). *Freud and the institution of psychoanalytic knowledge*. Stanford University Press.
Winther, R. G. (2001). August Weismann on germ plasm variation. *Journal of the History of Biology, 34*, 517–555.
Woese, C. R. (2004). A new biology for a new century. *Microbiology and Molecular Biology Reviews, 68*(2), 173–186.
Wokler, R. (1995). The enlightenment science of politics. In C. Fox, R. Portert, & R. Wokler (Eds.), *Inventing human sciences: Eighteenth century domains* (pp. 323–346). University of California Press.
Wolfe, C. T. (2014). On the role of Newtonian analogies in eighteenth century life science: Vitalism and provisionally inexplicable explicative devices. In Z. Biener & E. Schliesser (Eds.), *Newton and empiricism* (pp. 223–261). Oxford University Press.
Wolfe, C. T. (2016a). *Materialism: A historico-philosophical introduction*. Springer.
Wolfe, C. T. (2016b). Materialism and the 'soft substance of the brain': Diderot and plasticity. *British Journal for the History of Philosophy, 24*(5), 963–982.
Wolfe, C. T. (2019). *La philosophie de la biologie avant la biologie: Une histoire du vitalisme*. Classiques Garnier.
Wolff, P., & Gentner, D. (2011). Structure-mapping in metaphor comprehension. *Cognitive Science, 35*(8), 1456–1488.
Worboys, M. (2011). Practice and the science of medicine in the nineteenth century. *Isis, 102*, 109–115.
Yerushalmi, H. (1991). *Freud's Moses: Judaism terminable and interminable*. Yale University Press.
York, G. K., & Steinberg, D. A. (1993). Hughlings Jackson rejection of the unconscious. *Journal of the History of the Neurosciences, 2*(1), 65–78.
York, G. K., & Steinberg, D. A. (1994). Hughlings Jackson rejection of the unconscious. *Journal of the History of the Neurosciences, 3*(3), 153–168.
York, G. K., & Steinberg, D. A. (2002). The philosophy of John Hughlings Jackson. *Journal of the Royal Society of Medicine, 95*, 314–318.
York, G. K., & Steinberg, D. A. (2006). *An introduction to the life and work of John Hughlings Jackson*. Welcome Trust Centre for the history of medicine at UCL.
York, G. K., & Steinberg, D. A. (2013). Jackson's neurological ideas. *Brain, 134*(10), 3106–3113.

Young, R. M. (1967). The development of Herbert Spencer's concept of evolution. *Actes du XIe. Congrès International d'Histoire des Sciences, 2*, 273–278.
Young, R. M. (1968). Association of Ideas. In P. P. Wiener (Ed.), *Dictionary of the history of ideas*. Charles Scribner's Sons.
Young, R. M. (1970). *Mind, brain and adaptation in the nineteenth century: Cerebral localization and its biological context from gall to Ferrier*. Oxford University Press.
Young, R. M. (1973). The role of psychology in the nineteenth-century evolutionary debate. In M. Henle, J. Jaynes, & J. J. Sullivan (Eds.), *Historical conceptions of psychology* (pp. 180–204). Springer.
Young, A. (2006). Remembering the evolutionary Freud. *Science in Context, 19*, 175–189.

Index

S. B. Gissis, *Lamarckism and the Emergence of 'Scientific' Social Sciences in Nineteenth-Century Britain and France*, History, Philosophy and Theory of the Life Sciences 36, https://doi.org/10.1007/978-3-031-52756-2

R

W

MIX
Papier aus verantwortungsvollen Quellen
Paper from responsible sources
FSC® C105338

If you have any concerns about our products,
you can contact us on
ProductSafety@springernature.com

In case Publisher is established outside the EU,
the EU authorized representative is:
Springer Nature Customer Service Center GmbH
Europaplatz 3, 69115 Heidelberg, Germany

Printed by Libri Plureos GmbH
in Hamburg, Germany